9
8.11.06

# INTROO
# COASTSSES
# AND GECRPHOLOGY

## Gerhard Masselink

*Department of Geography, Loughborough University, UK*

and

## Michael G. Hughes

*School of Geosciences, University of Sydney, Australia*

# Hodder Arnold

A MEMBER OF THE HODDER HEADLINE GROUP

First published in Great Britain in 2003 by
Hodder Arnold, a member of the Hodder Headline Group,
338 Euston Road, London NW1 3BH

http://www.hoddereducation.com

Distributed in the United States of America by
Oxford University Press Inc.
198 Madison Avenue, New York, NY10016

The advice and information in this book are believed to be true and
accurate at the date of going to press, but neither the authors nor the publisher
can accept any legal responsibility or liability for any errors or omissions.

*British Library Cataloguing in Publication Data*
A catalogue record for this book is available from the British Library

*Library of Congress Cataloging-in-Publication Data*
A catalog record for this book is available from the Library of Congress

ISBN 10: 0 340 76410 4 (hb)
ISBN 10: 0 340 76411 2 (pb)
ISBN 13: 978 0 340 76411 4


3 4 5 6 7 8 9 10


Typeset in 10 on 12 Palatino by Dorchester Typesetting Group Ltd
Printed and bound in India

# CONTENTS

*It was west of the Clock, to the right and to the left of the place where the road, near the Old King's statue, disembogued itself of monkeys and goat carriages and toy-vendors and sweet-sellers and handcars and bath-chairs and fruit-trucks and children's perambulators, that the Weymouth sands proper stretched out, so nobly, so generously, so hospitably, and so astonishingly far into the sea! [. . .] On the dry sand sat, in little groups, the older people, reading, sewing, sleeping, talking to one another, while on the wet sand the children, building their castles and digging their canals were far too absorbed and content to exchange more than spasmodic shouts to one another. The free play of so many radiant bare limbs against the sparkling foreground-water and the bluer water of the distance gave the whole scene a marvellous heathen glamour, that seemed to take it out of Time altogether, and lift it into some ideal region of everlasting holiday, where the burdn of human toil and the weight of human responsibility no more lay heavy upon the heart.*

John Cowper Powys
Weymouth Sands (1934)

*Something miraculous happens, thinks David, when you dive into the surf at Bondi after a bad summer's day . . . The electric cleansing of the surf is astonishing, the colf efferverscing over the head and trunk and limbs. And the internal results are a great wonder. At once the spirits lift. There is a grateful pleasure in the last hour of softer December daylight. The brain sharpens. The body is charged with agility and grubby lethargy swept away.*

Robert Drewe
The Bodysurfers (1983)

# PREFACE

This textbook has been written for intermediate-level undergraduate students. We have attempted a level of explanation in this text that is introductory, but will also extend the student's analytical skills and hopefully their interest in coastal morphodynamics. The boxes in each chapter are specifically intended to generate further interest by providing either a further analysis or treatment of a particular issue, an interesting application of a principle just discussed in the body of the text, or a virtual field trip.

In this text we have not sought to describe in detail the variability in coastal geomorphology along the world's coastlines, although some explanation for geographic variation is implicit in what we have presented. Our approach to coastal geomorphology emphasizes the understanding of processes. In this approach, the coastlines of the world provide 'natural laboratories' for investigating the physical, chemical, and biological processes that produce the rich diversity of coastal landforms. We have therefore drawn on research examples from around the world to explain concepts. Given our personal experiences, however, it is not surprising that there is a small geographic bias favouring the United Kingdom and Australia. While this may seem limiting to some, the coastlines of these two countries do include two hemispheres, several climatic zones and the entire range of coastal environments addressed in the second part of the book.

The first five chapters of this book address generic concepts, global issues and processes that are common to most coastal environments – they could provide a basis for a short introductory course in coastal processes on their own. Topics include the morphodynamic paradigm, Quaternary sea-level fluctuations, tides, waves and sediment transport processes. It is one of our intentions that students should use this section of the book as a 'reference handbook' when undertaking practical exercises during their coursework. The following five chapters address the morphodynamics of the five main types of coastal environments, namely fluvial-, tide- and wave-dominated environments, rocky coasts, and coral reefs and islands. Many of the concepts and examples discussed under specific coastal environments refer back to, and/or build on the material presented in the first five chapters. Coastal environments are arguably the most important and intensely used of all areas settled by humans. In the final chapter, therefore, we briefly address the issue of coastal management, particularly the management of coastal erosion.

A number of friends and colleagues have commented on aspects of this book: Dr. Kitty Bos, Dr. Rob Brander, Dr. Jo Bullard, Dr. Bruce Hegge, Dr. Paul Kench, Dr. Gui Lessa and Dr. Wayne Stephenson. Their comments have been very useful and have greatly improved the readability of this book. We are, of course, responsible for remaining errors, omissions, inappropriate embellishments and bad prose.

We would also like to thank the following colleagues for providing original figures for this book: Prof. Edward Anthony for Figure 8.43c; Dr. Jo Bullard for Figure 8.27a; Dr. Jim Chandler for Figures 9.4b and 9.4c; Dr. Giovanni Coco for Figure 1.11; Dr. Peter Cowell for Figure 8.24a; Dr. Simon Haslett for Figure 11.3; Dr. Peter Keene for Figure 9.5c; Dr. Paul Kench for Figure 10.9a; Dr. Aart Kroon for Figure 8.35b; Prof. Julian Orford for Figures 4.13b and 4.16; Dr. Andy Short for Figures 8.24b, 8.24e and 8.24f; Prof. Matt Tomczak for Figure 7.11; and Dr. Wayne Stephenson for Figure 9.13. Thanks also to Mark Szegner for cartographic assistance.

Finally, we would like to thank the following for permission to reproduce figures: Academic Press for Figure 2.11; American Geophysical Union for Figure 7.21b; Blackwell Publishers for Figures 1.8, 2.4, 2.9, 5.4, 5.5, 5.7, 5.19a, 5.22, 5.23b, 6.6, 6.8, 6.11b, 7.17 and 11.2; Cambridge University Press for Figures 1.2, 1.13b, 1.14, 2.2, 8.33a, 8.38a, 8.40, 8.41, 10.8 and 10.11; Coastal Education and Research Foundation for Figures 4.13b, 6.5, 7.8, 8.34, 9.12, 11.4 and 11.7; Edward Arnold for Figures 3.13 and 8.13; Elsevier Science for Figures 8.43a, 8.43b, 10.9b, 10.9c, 10.9d and 10.13; Geographical Association for Figure 9.3; Geological Association of Canada for Figure 8.37; Geologist's Association for Figure 7.7; Geological Society of America for Figure 6.10; George Allen & Unwin for Figures 5.6, 5.11 and 6.9b; Geoscience Australia for Figure 6.13; International Association of Sedimentologists for Figures 5.21 and 6.16, ; John Wiley & Sons for Figures 1.13a, 4.4, 4.19, 8.1, 8.4, 8.9, 8.12, 8.18, 8.21, 8.22, 8.23, 8.29, 9.1, 9.15, 10.4, 10.6, 10.12b, 10.12c and 10.12d; Jones and Bartlett Publishers for Figure 3.12; Macmillan Magazines for Figure 10.2; Natural Resources Canada for Figure 4.16; New South Wales Department of Land and Property Information for Figure 6.14; Oxford University Press for Figure 7.9; Parabolic Press for Figures 5.10a, 6.9a and 6.11a; Pearson Education for Figures 8.10 and 8.11; Simmons Aerofilms for Figure 9.4a; Springer-Verlag for Figure 5.19b; Society for Sedimentary Geology for Figures 7.1, 7.2 and 7.3; Society for Sedimentary Petrology for Figure 5.23a; Taylor & Francis Books for Figures 6.2, 11.3, 11.5 and 11.10; The University of Chicago Press for Figure 1.12; and U.S. Army Corps of Engineers for Figures 4.5 and 11.8; Thomas Telford for Figure 7.21a; and Werner Barthel for Figure 11.6a. In a few cases, it was not possible to trace the copyright holders of figures used in this text. We apologize therefore for any copyright infringements that may have occurred.

# LIST OF SYMBOLS

| | |
|---|---|
| $A$ | constant; dimensional parameter |
| $a$ | least squares regression coefficient; height of saltation (m); amplitude of tide (m) |
| $a_i$ | amplitude of partial tide (m) |
| $B$ | height of the subaerial beach (m) |
| $b$ | spacing of wave rays (m); least squares regression coefficient |
| $b_o$ | spacing of wave rays in deep water (m) |
| $C$ | constant; wave velocity or wave celerity (m s$^{-1}$); sediment concentration (kg m$^{-3}$) |
| $C_d$ | drag coefficient (–) |
| $C_g$ | wave group velocity (m s$^{-1}$) |
| $C_o$ | deep water wave velocity (m s$^{-1}$) |
| $C_s$ | shallow water wave velocity (m s$^{-1}$) |
| CSF | Corey shape factor (–) |
| $D$ | fetch duration (minutes or hours); grain size (m) |
| $D_r$ | reference grain diameter (0.25 mm) |
| $D_*$ | non-dimensional grain diameter |
| $D_{50}$ | median sediment size (m) |
| $d$ | wave orbital diameter (m) |
| $d_0$ | wave orbital diameter at the sea bed (m) |
| $E$ | vertical shoreline displacement caused by eustatic change (m); wave energy density (N m$^{-2}$) |
| $e_b$ | bedload efficiency factor (–) |
| $e_s$ | suspended load efficiency factor (–) |
| $F$ | fetch length (km); tidal form factor (–) |
| $F_c$ | centripetal force (N) |
| $F_g$ | gravitational attractive force (N) |
| $F_{lg}$ | local gravitational force (N) |
| $F_T$ | tractive force (N) |
| $F_t$ | tide-generating force (N) |
| $F_r$ | Froude number (–) |
| $F'_r$ | densimetric Froude number (–) |
| $f$ | constant; frequency (Hz) |
| $f_i$ | frequency of partial tide (hr$^{-1}$); percentage of grains in grain size class interval |
| $f_w$ | wave friction factor (–) |

| | |
|---|---|
| $G$ | universal gravitational constant (N m$^2$ kg$^{-2}$) |
| $G_i$ | phase of partial tide (radians) |
| $g$ | gravitational acceleration (m s$^{-2}$) |
| $H$ | wave height (m) |
| $H_b$ | breaker height (m) |
| $H_l$ | wave height of the largest waves (m) |
| $H_o$ | deep water wave height (m) |
| $H_{rms}$ | root mean square wave height (m) |
| $H_s$ | significant wave height (m) |
| $\bar{H}_s$ | mean annual significant wave height (m) |
| $H_{1/3}$ | significant wave height (m) |
| $H_{sx}$ | nearshore storm wave height that is exceeded only 12 hours each year (m) |
| $h$ | bed elevation (m); water depth (m) |
| $\bar{h}$ | local tidally-averaged water depth (m) |
| $h'$ | thickness of river plume (m) |
| $h_b$ | breaker depth (m) |
| $h_c$ | thickness of the Earth's crust (m); closure depth (m) |
| $h_i$ | thickness of the ice sheet (m); limiting depth for significant cross-shore sediment transport of sand by waves (m) |
| $h_*$ | depth of the shoreface (m) |
| $I$ | alongshore length of island (m); intermediate axis length of sediment grain (m) |
| $J$ | distance between island and mainland (m) |
| $K$ | constant |
| $K_r$ | wave refraction coefficient (–) |
| $K_s$ | wave shoaling coefficient (–) |
| $k$ | wave number ($2\pi/L$) |
| $k_s$ | Nikuradse roughness length (m) |
| $k'$ | skin friction roughness length (m) |
| $k''$ | bedform roughness length (m) |
| $L$ | wave length (m); tidal wavelength (m); longest axis length of sediment grain (m) |
| $L_o$ | deep water wave length (m) |
| $L_s$ | shallow water wave length (m) |
| $M_\phi$ | midpoint of grain-size class interval on phi scale (phi) |
| $MSR$ | mean spring tide range (m) |
| $m$ | dimensionless exponent; mass (kg) |
| $m_E$ | mass of the Earth (kg) |
| $m_M$ | mass of the Moon (kg) |
| $n$ | wave transformation parameter (–) |
| $P$ | elevation of the raised feature above MSL (m); pressure (kg m$^{-1}$ s$^{-2}$); wave energy flux (N m$^{-1}$ s$^{-1}$ = J s$^{-1}$) |
| $P_l$ | longshore component of the wave energy flux (N m$^{-1}$ s$^{-1}$ = J s$^{-1}$) |
| $Q_l$ | volumetric longshore sediment transport rate (m$^3$ day$^{-1}$) |
| $q$ | sediment transport rate (kg m$^{-1}$ s$^{-1}$) |

| | |
|---|---|
| $q_b$ | bedload sediment transport rate (kg m$^{-1}$ s$^{-1}$) |
| $q_s$ | suspended load sediment transport rate (kg m$^{-1}$ s$^{-1}$) |
| $R$ | distance between centres of mass of two bodies (m) |
| $R_e$ | Reynolds number (–) |
| | boundary Reynolds number (–) |
| $R_g$ | grain Reynolds number (–) |
| $R_{2\%}$ | wave runup height exceeded by 2% of the runup events (m) |
| $RTR$ | relative tide range (–) |
| $r$ | radius of the Earth (m); radius of a sediment grain (m) |
| $S$ | rise in sea level (m); shortest axis length of sediment grain (m); horizontal extent of the swash motion or swash length (m) |
| $S_c$ | compressive strength of the rock (N m$^{-3}$) |
| $S_{xx}$ | cross-shore component of radiation stress (N m$^{-2}$) |
| $T$ | wave period (s) |
| $T_p$ | peak spectral wave period (s) |
| $T_s$ | significant wave period (s) |
| $\overline{T}_s$ | mean annual significant wave period (s) |
| $T_{sx}$ | wave period associated with $H_{sx}$ (s) |
| $T_z$ | zero-crossing or mean wave period (s) |
| $T_{1/3}$ | significant wave period (s) |
| $t$ | time (s) |
| tanβ | beach gradient (–) |
| $U$ | amount of uplift (m); wind speed at 10 m height (m s$^{-1}$) |
| $U_A$ | wind stress factor (m s$^{-1}$) |
| $u$ | horizontal current velocity (m s$^{-1}$) |
| $u_g$ | velocity of sediment grain (m s$^{-1}$) |
| $u_m$ | maximum orbital wave velocity (m s$^{-1}$) |
| $u_0$ | maximum orbital wave velocity at the sea bed (m s$^{-1}$) |
| $u_z$ | flow velocity at elevation $z$ above the surface (m s$^{-1}$) |
| $u_{100}$ | velocity measured 100 cm above the bed (m s$^{-1}$) |
| $u_*$ | shear velocity (m s$^{-1}$) |
| $u_{*t}$ | threshold shear velocity (m s$^{-1}$) |
| $V$ | sediment volume (kg) |
| $\overline{v}_l$ | longshore current velocity at the mid-surf zone position (m s$^{-1}$) |
| $w_0$ | maximum deflection of the land surface under the ice sheet (m) |
| $w_s$ | sediment fall velocity (m s$^{-1}$) |
| $w_*$ | width of the shoreface (m) |
| $x$ | cross-shore direction (m); distance (m) |
| $\overline{x}$ | mean (first-moment) of grain size distribution (phi) |
| $y$ | longshore direction (m) |
| $\Delta y$ | shoreline retreat due to sea-level rise (m) |
| $Z_{berm}$ | height of berm (m) |
| $Z_{step}$ | height of beach step (m) |
| $z$ | height above the bed (m) |
| $z_0$ | hydraulic bed roughness length (m) |
| $\alpha$ | wave angle (°) |

$\alpha_b$      breaker angle ($^\circ$)

$\alpha_o$      deep-water wave angle ($^\circ$)

$\beta$      bed slope-angle ($^\circ$)

$\gamma$      breaker index or breaking criterion (–)

$<\gamma>$      relative wave height (–)

$\delta^{18}O$      oxygen isotope ratio (–)

$\delta y$      shoreline retreat (m)

$\varepsilon$      factor accounting for sediment density and porosity (–); surf scaling parameter (–)

$\eta$      water surface elevation (m); bedform height (m)

$\bar{\eta}$      mean sea level (m); departure of the water level from still water level (m)

$\eta_r$      water level departure from predicted tide (m)

$\bar{\eta}_s$      wave set-up at the shoreline (m)

$\theta$      angle ($^\circ$)

$\theta_c$      critical Shields parameter required for sediment motion (–)

$\kappa$      von-Karman constant (–)

$\lambda$      bedform spacing (m); spacing of beach cusp morphology (m)

$\mu$      molecular viscosity of fluid (N s m$^{-2}$)

$\xi$      Irribarren number (–); eddy viscosity (N s m$^{-2}$)

$\rho$      water density (kg m$^{-3}$)

$\rho_a$      density of air (kg m$^{-3}$)

$\rho_c$      density of the Earth's crust (kg m$^{-3}$); density of coastal waters (kg m$^{-3}$)

$\rho_i$      density of ice (kg m$^{-3}$)

$\rho_m$      density of the Earth's mantle (kg m$^{-3}$)

$\rho_r$      density of river water (kg m$^{-3}$)

$\rho_s$      sediment density (kg m$^{-3}$)

$\sigma$      wave radian or angular frequency ($2\pi/T$); standard deviation (2nd moment) or sorting of grain size distribution (phi)

$\tau$      fluid shear stress (N m$^{-2}$)

$\tau_0$      bed shear stress (N m$^{-2}$)

$\tau_w$      bed shear stress under waves (N m$^{-2}$)

$\phi$      friction angle of sediment ($^\circ$)

$\psi$      transport stage (–)

$\Omega$      dimensionless fall velocity (-)

# 1

# COASTAL SYSTEMS

## 1.1 INTRODUCTION

This book is about coastal processes and geomorphology. The term 'coastal' is, however, rather ambiguous and it is important to indicate at the outset what we mean by it. The spatial boundaries of 'our' coastal zone are defined in Figure 1.1 and conform to the conventions set out by Inman and Brush (1973). The boundaries correspond to the limits to which coastal processes have extended during the Quaternary geological period and include the coastal plain, the shoreface and the continental shelf. During the Quaternary, which lasted from *c.* 1.8 million years ago until present, sea level fluctuated over more than 100 m vertically due to expansion and contraction of ice sheets. The landward limit of the coastal system therefore includes the coastal depositional and marine erosion surfaces formed when the sea level was high (slightly above present-day sea level). The lowest sea levels placed coastal processes close to the edge of the continental shelf on several occasions. The seaward limit of the coastal system is therefore defined by the edge of the continental shelf, which typically occurs in water depths of 100–200 m.

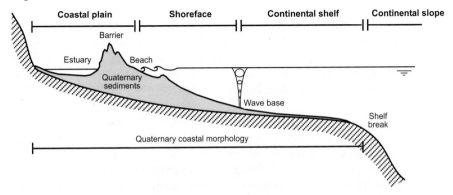

Figure 1.1 – Spatial boundaries of the coastal zone.

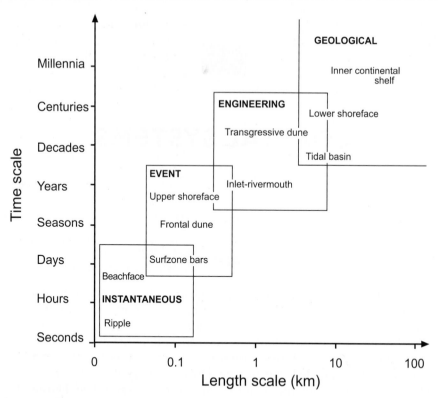

**Figure 1.2** – Definition of spatial and temporal scales involved in coastal evolution. Large-scale coastal landforms operate over long time scales, whereas small-scale coastal features respond over short time scales. [From Cowell and Thom, 1994.] [Copyright © 1994 Cambridge University Press, reproduced with permission.]

Having defined the spatial boundaries of the topic of this book, we now need to indicate the time scale we are interested in. Cowell and Thom (1994) group the time scales at which coastal processes operate into four classes (Figure 1.2):

- **Instantaneous time scales** – These involve the evolution of morphology during a single cycle of the forces that drive morphological change (waves, tides). The destruction of wave ripples under a group of high waves and the onshore migration of an intertidal bar over a single tidal cycle are examples of morphological change over instantaneous time scales.
- **Event time scales** – These are concerned with coastal evolution in response to processes operating across time spans ranging from that of an individual event, through to seasonal variation in driving forces. Examples of morphological change over event time scales are the scarping of coastal dunes in response to a major storm and the seasonal closure of an estuary by a sand bar.

- **Engineering time scales** – Coastal evolution resulting from many fluctuations in the driving forces takes place over engineering time scales. It is the time scale that coastal engineers are most concerned with and ranges from years to centuries. The migration of tidal inlets and the development of a foredune ridge are examples of coastal development over engineering time scales.
- **Geological time scales** – These time scales operate over decades to millennia. Whereas at the previous three time scales morphological change is mainly the result of fluctuations in the driving forces, on geological time scales, coastal evolution occurs more in response to mean trends in the driving forces (sea level, climate). Examples of coastal evolution at geological time scales are the infilling of a tidal basin or estuary, onshore migration of a barrier system and the switching of delta lobes.

This book will deal with coastal processes and geomorphology on all four time scales.

## 1.2 COASTAL CLASSIFICATION

A large variety of coastal landforms and morphologies exist in nature and for a long time coastal geomorphologists appeared to have been mainly concerned with describing and classifying these landforms and morphologies (*cf.*, Bird, 1984). **Coastal classification** schemes can be useful from a conceptual point of view and help to assess the different forcing factors and controls (*e.g.*, sea-level history, geology, climate, waves, tides) that lead to the great variety in coastal environments we encounter in nature. Classification also allows minor differences between coastal regions to be put aside in favour of their more significant similarities. Yet, classification does tend to describe rather than explain.

Most early classification schemes are based on the realization that coastal landforms are largely the product of sea-level variations. Such classifications distinguish between submerged and emerged coasts (Johnson, 1919). Typical **submerged coasts** are drowned river and glacial valleys, often referred to as rias and fjords, respectively. Coastal plains are characteristic of **emerged coasts**. Classifications based on sea-level variations are, however, of limited use because most coastlines are characterized by emerged, as well as submerged components. Shepard (1963) divided coasts into primary and secondary coasts. **Primary coasts** have a configuration resulting mainly from non-marine processes and include drowned river valleys and deltaic coasts. **Secondary coasts**, on the other hand, are coasts that have a configuration resulting mainly from marine processes or marine organisms. Examples of such coasts are barrier coasts, coral reefs and mangrove coasts.

The main shortcoming of these early classifications is that the emphasis on geological inheritance leaves only limited concern for the hydrodynamic processes. In contrast, Davies (1980) identified a number of coastal types solely based on wave height and tidal range. Because waves are generated by wind, the **worldwide distribution of wave environments** displays a strong latitudinal control, reflecting global climate zones (Figure 1.3). Coastal environments dominated by storm waves are located at the higher temperate and arctic latitudes, whereas swell-dominated environments are situated at the lower temperate and tropical latitudes. The influence of cyclones (hurricanes) is also observed in the tropics. The tidal range in the middle of the ocean is quite small (less than 1 m), but increases towards the coast and may attain values in excess of 10 m. The amplification of the tides depends on the gradient and width of the continental shelf, the location and shape of continents, and the presence of large embayments. Hence the **worldwide distribution of tidal range** is strongly controlled by the large-scale coastal configuration (Figure 1.4). Macro-tidal ranges exceeding 4 m are mostly observed in semi-enclosed seas and funnel-shaped entrances of estuaries. Micro-tidal ranges below 2 m occur predominantly along open ocean coasts and almost fully enclosed seas. Very small, but noticeable, tidal water fluctuations occur in large lakes (*e.g.*, the Great Lakes).

The global distributions of wave height and tidal range identified by Davies (1980) are often used to infer wave- and tide-dominance of coastal processes and morphology. However, as argued by Davis and Hayes (1984), the relative effects of waves and tides, rather than their absolute values, are more important in shaping the coast (Figure 1.5). Thus, environments dominated by tidal processes are not restricted to coasts experiencing macrotidal tide ranges, but may also be found along microtidal shores if the incident wave-energy level is low. Davis and Hayes (1984) further point out that there exists a rather delicate balance between wave and tide processes for

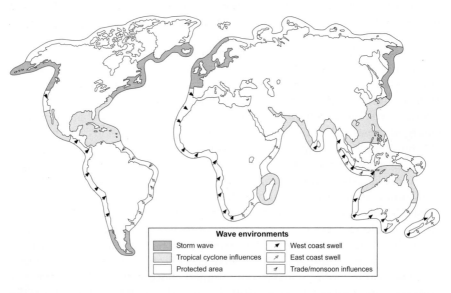

**Figure 1.3** – World distribution of wave environments. [Modified from Davies, 1980.]

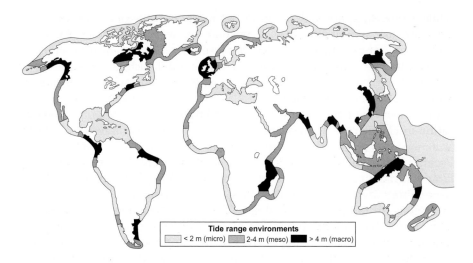

**Figure 1.4** – World distribution of mean spring tidal range. [Modified from Davies, 1980.]

low values of wave height and tide range. This is because the wave/tide regions in Figure 1.5 converge at the low end of the spectrum. Therefore, tide-dominated, wave-dominated or mixed-energy morphologies may develop with very little differences in wave and tide parameters.

It has long been recognized that the morphology of clastic coasts (*i.e.*, depositional coastal environments with mud, sand, and/or gravel material) responds to the relative dominance of river, wave and tidal factors. A ternary diagram can be constructed which expresses the relative importance of river outflow, waves and tidal currents. In such a diagram, deltas are positioned at the fluvial apex because a fluvial sediment source dominates,

**Figure 1.5** – Relationship between mean tidal range and wave height delineating different fields of wave/tide dominance. A particular coastal stretch may span several fields. [Modified from Davis and Hayes, 1984.]

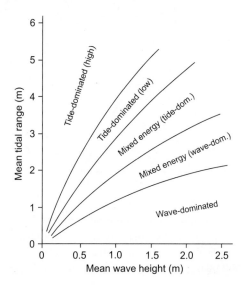

while prograding, non-deltaic coasts are located on the opposite wave-tide side, because sediment is moved onshore by waves and tides. Estuaries occupy an intermediate position, because they have a mixed sediment source and are affected by river, wave and tidal factors. This view has recently been extended by Dalrymple *et al.* (1992) to provide an evolutionary aspect to the river/wave/tide classification by including time as an additional component (Figure 1.6). According to this evolutionary **classification of clastic coastal environments**, time is expressed in terms of coastal accretion (or progradation) and coastal inundation (or transgression). Progradation generally occurs with a relatively constant sea level and a large sediment supply, whereas transgression takes place when sea level is rising. Progradation entails the infilling of the estuaries and their conversion into deltas, strand plains or tidal flats. This is shown in Figure 1.6 by movement to the back of the triangular prism. Changes associated with transgression, such as the flooding of river valleys and the creation of estuaries, are represented in Figure 1.6 by movement towards the front of the triangular prism.

Of all the coastal classifications discussed in this section, we consider the one illustrated in Figure 1.6 most in tune with the approach followed in this

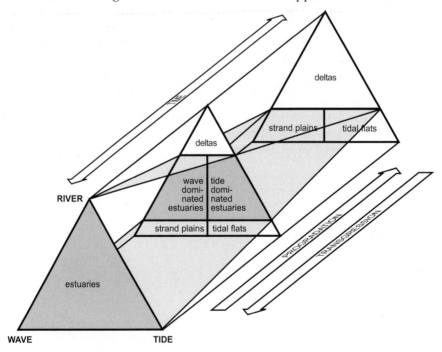

**Figure 1.6** – Evolutionary classification of coastal depositional environments. The long axis of the three-dimensional prism represents relative time with reference to changes in relative sea level and sediment supply (*i.e.*, transgression and progradation). The three edges of the prism correspond to conditions dominated by fluvial, wave and tidal processes. [Modified from Dalrymple *et al.*, 1992.]

book. In addition to considering the relative roles of the three main hydro-dynamic processes (waves, tides and river outflow), the conceptual model by Dalrymple *et al.* (1992) has the added advantage of acknowledging the dynamic nature of coastal environments.

## 1.3 MORPHODYNAMIC APPROACH

Coastal classifications indicate which environmental factors are important in shaping the coast. To understand more fully how these factors affect and control coastal morphology, a more in-depth approach is required. Wright and Thom (1977) were among the first to apply a systems approach to coastal morphology and evolution. They viewed the coastal environment as a dynamic geomorphic system with identifiable inputs and outputs of ener-gy and material, driven and controlled by environmental conditions (Figure 1.7). Wright and Thom (1977) were equally concerned with coastal processes and the associated morphological responses. They introduced the term 'coastal morphodynamics' for their approach and defined **morphodynam-ics** as 'the mutual adjustment of topography and fluid dynamics involving sediment transport'. The morphodynamic approach has subsequently become the paradigm for studying coastal evolution. We will now discuss the two main aspects of coastal morphodynamics, namely environmental conditions and the coastal system.

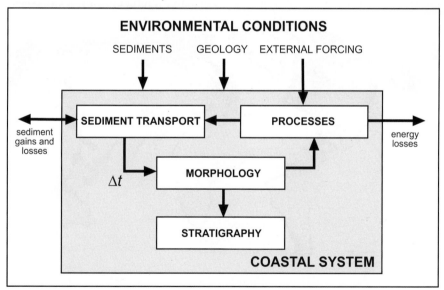

Figure 1.7 – Primary components involved in coastal morphodynamics. The feedback loop between morphology and processes is responsible for fundamental complexity in coastal evolution. Δt signifies the time dependence inherent in the morphodynamic evolution of coasts. [Modified from Cowell and Thom, 1994.]

## 1.3.1 Environmental conditions

**Environmental conditions** are the 'set of static and dynamic factors that drive and control coastal systems' (Wright and Thom, 1977). They are not affected by the coastal system itself and are referred to as the boundary conditions of the coastal system. Because environmental conditions drive and control coastal systems, spatial variations in these conditions are responsible for geographical variations in the coastal geomorphology (*cf.*, Davies, 1980). The three main types of environmental factors are geology, sediments and external forcing.

Geology comprises the initial state of the solid boundaries, including the regional or local geology and pre-existing morphologic state (shelf and shoreline configuration, lithology). On a global scale, the most important factor is the width and slope of the continental shelf, which is to a large extent controlled by global tectonics (Figure 1.8; Inman and Nordstrom, 1971). Wide and flat continental shelves (*e.g.*, east coast of the US) permit more rapid coastal progradation for any given rate of sediment supply than steep and narrow shelves (*e.g.*, southeast coast of Australia). In addition, wide shelves lead to greater reduction in wave height by frictional dissipation and are also responsible for the amplification of tides. Regionally, the configuration of the coastline can be important by controlling wave transformation processes. Reduced wave energy levels can be found in the lee of promontories and offshore islands and within coastal embayments. Lithology is also a significant factor, in particular along eroding rocky coasts

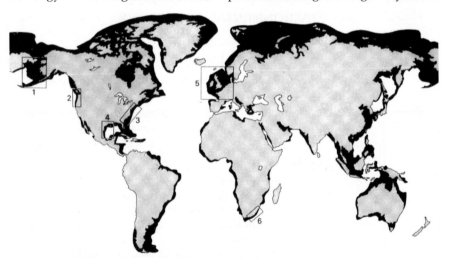

Figure 1.8 – World distribution of present-day continental shelves. Note the difference in shelf width between, e.g., the west and east coast of South America. Regions 1–6 represent some of the best studied shelf seas: 1 = Bering Sea; 2 = Oregon-Washington; 3 = Eastern U.S.A.; 4 = Gulf of Mexico; 5 = NW Europe; 6 = SE Africa. [From Johnson and Baldwin, 1986.] [Copyright © 1986 Blackwell Publishers, reproduced with permission.]

where it controls to a large extent the recession rate and cliff profile development.

**Sediments** are essential for coastal evolution. The two most important aspects are the nature and abundance of the unconsolidated material. Sediment availability depends on the location and volume of the sediment sources, and the processes of sediment transport between the sources and the coastal area. The sediments may have a marine, a fluvial/deltaic, a terrestrial, or a biological origin. Coastal sediments are composed of materials which vary physically and chemically from place to place, the most important variation being those of particle size and carbonate content. Gravel deposits are common to paraglacial areas where the coastal hinterland has been glaciated and/or periglaciated. Sandy sediments are characteristic of coastal zones and inner continental shelves in the lower middle latitudes. Muddy sediments are most common in humid temperate or tropical hot climatic zones, where they result in the infilling of estuaries and the accretion of deltas.

**External forcing** refers to those processes that provide the energy necessary to drive coastal processes and evolution. Important aspects are the frequency, magnitude and character of the external energy sources. The major sources of coastal energy are atmospheric (coastal winds and climate), terrestrial (river outflow) and marine (waves, tides, currents and other oceanographic phenomena). Of these, the sea is by far the most important source of coastal energy, although it should be pointed out that the marine energy regime is closely dependent on the atmospheric climate.

Temporal changes in environmental conditions drive coastal evolution over a range of time scales (Figure 1.2). Changes in the solid boundary conditions operate on the geological time scale. Such changes are mainly related to vertical land movements resulting in the emergence or submergence of coasts, and falling and rising relative sea levels, respectively. Changes in sediment type and abundance operate mainly on geological and engineering time scales. On the geological time scale, changing sea levels during the Quaternary significantly affected the sediment availability by redistributing sediment across the continental shelf. On the engineering time scale, one of the most significant contributors to changes in coastal sediment characteristics has been human tampering with coastal catchments and drainage basins, and their ability to deliver sediment to the coastal zone. The most significant temporal changes in external forcing occur on the instantaneous and event time scales. In particular, seasonal changes in weather and waves cause cyclic changes in coastal processes and morphology. Variations in the external forcing may also take place over longer time scales, for example due to climate change.

## 1.3.2 Coastal system

A **coastal system** comprises components that are linked by energy and material flows (Figure 1.7). The system is connected to the 'outside' world and is indeed controlled by the environmental conditions operating outside its boundaries. The coastal system itself consists of four main components (Cowell and Thom, 1994):

- **Processes** – This component includes all processes occurring in coastal environments that generate and affect sediment transport. The most important of these are hydrodynamic (waves, tides and currents) and aerodynamic (wind) processes. Along rocky coasts, weathering is an additional process that contributes significantly to sediment transport, either directly through solution of minerals, or indirectly by weakening the rock surface to facilitate entrainment by hydrodynamic processes. Biological, biophysical and biochemical processes are important on coral reefs, salt marshes and in mangrove environments.

- **Sediment transport** – The interaction between a moving fluid and a mobile bed induces bed shear stresses that may result in the entrainment and subsequent transport of sediment. The ensuing patterns of erosion and deposition can be assessed using the sediment balance. If the sediment balance is positive (*i.e.*, more sediment is entering a coastal region than exiting), deposition will occur, while a negative sediment balance (*i.e.*, more sediment is exiting a coastal region than entering) results in erosion.

- **Morphology** – The three-dimensional surface of a landform or assemblage of landforms (*e.g.*, coastal dunes, deltas, estuaries, beaches, coral reefs, shore platforms) is referred to as the morphology or topography. Changes in the morphology are brought about by erosion and deposition. They can be qualitatively assessed using sediment budgets or quantitatively using the sediment continuity equation (Box 1.1).

- **Stratigraphy** – As the landform develops over time, the integrated effect of morphologic change is recorded in the stratigraphy of the landform. Consider, for example, the infilling of an estuary over time with marine and fluvial sediments. These sediments will be deposited along the margins and on the bottom of the estuary and the resulting estuarine stratigraphic sequence then forms a partial record of evolution of the estuary. By studying the stratigraphy we can deduce, to some extent, the coastal evolution. It is important to realize that stratigraphic sequences are a record of the depositional history and that erosional events are only represented by gaps in the stratigraphic record.

The coupling mechanism between processes and morphology is provided by sediment transport and is relatively easy to comprehend. There is, however, also a link between morphology and processes to complete a feedback loop. As the morphology evolves, the conditions encountered by the hydrodynamic processes are progressively modified. For example, sand is transported on a beach in the onshore direction under calm weather conditions resulting in beach accretion. As the beach builds up, its seaward slope progressively steepens and this has a profound effect on the wave breaking processes and sediment transport. At some stage during beach steepening,

## Box 1.1 – Sediment Budgets and Sediment Balance

Morphological change is the direct result of sediment transport processes. **Sediment budgets** can be used to gain an understanding of the different sediment inputs (sources) and outputs (sinks) involved (Figure 1.9). Making up a sediment budget represents an accounting exercise whereby rather than using Euro, pounds or dollars, sediment volumes ($m^3$) are used. Key components of the sediment budget are the sediment fluxes, which are vectors representing the direction and amount of sediment transport by certain processes. Sediment fluxes are expressed as quantity of sediment moved per unit of time (e.g., $kg\ s^{-1}$, $m^3\ yr^{-1}$). Whether a sediment flux is considered input or output depends on the point of view. For example, the transport of sediment from beach to dunes entails a sediment loss for the beach, but a sediment gain for the dunes. If the sediment fluxes are known, sediment budgets can be used to quantitatively predict the morphological change over time.

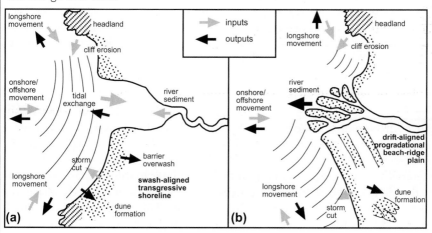

**Figure 1.9** – Sediment budgets on estuarine and deltaic coasts. (a) Riverine sediment may be deposited into an estuarine embayment, which may also receive sediment from the seaward direction through tidal processes. (b) On a deltaic coast, riverine sediment contributes to the coastal sediment budget, and is moved down the drift-aligned coast, giving rise to a beach ridge plain. [Modified from Carter and Woodroffe, 1994.]

As an example, consider an estuary with a surface area of $1\ km^2$ that receives an annual input of sediment from marine and fluvial sources of $10,000\ m^3\ year^{-1}$. If it is further assumed that this sediment is evenly spread over the estuary floor, then the depth of the estuary will decrease by

$$\frac{\text{sediment input}}{\text{surface area}} = \frac{10,000\ m^3\ year^{-1}}{1,000,000\ m^2} = 0.01\ m\ year^{-1}$$

If the average depth of the estuary is 10 m, then the estuary will be infilled in

$$\frac{\text{depth}}{\text{accretion rate}} = \frac{10\text{ m}}{0.01\text{ m year}^{-1}} = 1,000\text{ years}$$

This simple illustration assumes that the amount of sediment entering the estuary does not change while the estuary is infilling, and thus ignores one of the main principles of morphodynamic systems – that there is feedback between morphology and process. Nevertheless, such on-the-back-of-the-envelope calculations can be very insightful.

A more rigorous appreciation of morphological change is through the application of the continuity equation for sediment transport

$$\frac{\delta h}{\delta t} = \varepsilon \left( \frac{\delta q_x}{\delta x} + \frac{\delta q_y}{\delta y} \right) - \frac{\delta V}{\delta t}$$

where $\delta h/\delta t$ is the change in bed elevation over time, $\varepsilon$ accounts for sediment density and porosity, $\delta q_x/\delta x$ is the variation in sediment mass flux in the cross-shore direction, $\delta q_y/\delta y$ is the variation in sediment mass flux in the longshore direction and $\delta V/\delta t$ represents local sediment gains and losses, including engineering interventions such as beach nourishment and sand mining. The sediment continuity equation is a powerful means by which morphological change can be determined for different parts of a morphodynamic system. Computer models used to predict coastal morphological change invariably include the sediment continuity equation.

the hydrodynamic conditions may be sufficiently altered to stop further onshore transport of sediment. The feedback between morphology and processes can be negative or positive and is fundamental to coastal morphodynamics. It will be discussed in more detail later in this chapter. As a result of the close coupling between process and form in a morphodynamic system, cause and effect are not readily apparent. This gives rise to the 'chicken-and-egg' nature of coastal morphology and processes – it is often not clear whether the morphology is the result of the hydrodynamic processes, or vice versa. Importantly, in a developing morphodynamic system, process and form co-evolve. This makes it very difficult, if not impossible, to predict coastal development, especially over the longer time scales.

# 1.4 PROPERTIES OF COASTAL MORPHODYNAMIC SYSTEMS

Coastal morphodynamic systems possess a number of fundamental characteristics. It is important to grasp these, because they are essential for understanding coastal morphology, evolution and stratigraphy. For a more in-depth discussion of the properties of coastal morphodynamic systems the reader is referred to Cowell and Thom (1994).

## 1.4.1 Negative feedback and equilibrium

**Negative feedback** is a damping mechanism that acts to oppose changes in morphology. For example, beach erosion during storms generally results in the development of an offshore bar, but wave breaking on the bar signifi- cantly reduces the amount of wave energy reaching the shoreline, thereby limiting further beach erosion. Negative feedback is a stabilizing process and is the mechanism by which **equilibrium** is established. If a coastal sys- tem is in equilibrium, wind-, wave- and tidal energy still enters the system. However, the morphology is able to dissipate or reflect the incoming energy without the occurrence of net sediment transport and morphological change.

An example of morphological change leading to equilibrium conditions is shown in Figure 1.10 and illustrates the change in planform shape of an embayed coastline. The refracted waves approach the initially straight coastline such that they make an angle with the shoreline and drive relative- ly strong longshore currents and sediment transport (Figure 1.10a). The resulting change in morphology consists of accretion at both ends of the embayment and erosion in the centre of the embayment. The change in beach planform entails negative feedback, because it reduces the wave angle and longshore transport. Over time, the beach planform develops such that the angle between the coastline and the incident waves becomes zero. Under these conditions, longshore currents and sediment transport become insignificant and the beach is in equilibrium (Figure 1.10b). Other well-known examples of equilibrium conditions in coastal morphology have been identified for the shoreface profile (Dean, 1991), tidal inlets (O'Brien, 1969) and tidal basins (Eysink, 1990).

## 1.4.2 Positive feedback and self-forcing

The equilibrium concept suggests that coastal landforms are dominated by self-stabilizing behaviour due to negative feedback between form and process. However, many coastal systems do not always behave that pre- dictably and often the feedback loop between morphology and processes is

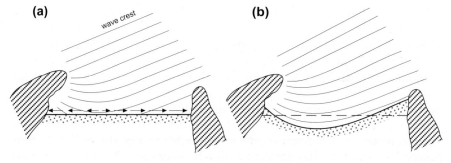

**Figure 1.10** – (a) Dis-equilibrium beach planform. (b) Equilibrium beach planform. The arrows in (a) represent longshore sediment transport.

ıtive, rather than negative. **Positive feedback** pushes a system away
m equilibrium by modifying the morphology such that it is even less
ompatible with the processes to which it is exposed. A morphodynamic
system driven by positive feedback seems to have a 'mind of its own' and
exhibits **self-forcing** behaviour.

An example of positive feedback is the infilling of deep estuaries by
marine sediments due to asymmetry in the tidal flow. In a deep estuary,
flood currents are stronger than ebb currents, resulting in a net influx of sediment and infilling of the estuary. As the estuary is being infilled, the tidal
asymmetry increases as friction and shoaling effects are enhanced by the
reduced water depths. In turn, the increase in tidal asymmetry speeds up
the rate of estuarine infilling. This constitutes positive feedback between the
estuarine morphology and the tidal processes, resulting in rapid infilling of
the estuary. Eventually, tidal flats start developing in the estuary and this
marks a reversal in feedback. As the tidal flats become more extensive, the
flood asymmetry of the tide progressively decreases so that the estuarine
morphology approaches steady state as sediment imports and exports
equilibrate.

## 1.4.3 Positive–negative feedback and self-organization

Morphodynamic systems often display a sequence of positive feedback,
which drives the morphodynamic change towards a new state, followed by
negative feedback, which stabilizes the sedimentation regime resulting in
equilibrium. This is referred to as **self-organization** and the result of this
process is a rather orderly (*i.e.*, organized) arrangement of sediments, sedimentary facies and morphodynamic sub-systems.

An example of self-organization on the event time scale is the formation
of beach cusp morphology (Box 1.2). Beach cusps are rhythmic shoreline
features formed by swash action and are characterized by steep-gradient,
seaward-pointing cusp horns and gentle-gradient, seaward-facing cusp
embayments (Figure 1.12). The self-organization theory of beach cusp
formation considers beach cusps to be the result of feedback between
morphology and swash flow (Werner and Fink, 1993). Positive feedback
enhances random morphologic irregularities in the following way. Small
topographic depressions on the beachface are amplified by attracting and
accelerating water flow, thereby promoting erosion. At the same time, small
positive relief features are enhanced by repelling and decelerating water
flow, thereby promoting accretion. The sequence of positive feedback is followed by negative feedback processes, which inhibit erosion and accretion
on well-developed cusps, and maintain equilibrium. The important feature
is that the morphological regularity arises from the internal dynamics of the
system. In our beach cusp example, the dimension of the rhythmic morphology is scaled by the horizontal swash excursion – the more extensive
the wave action on the beach, the larger the cusps.

# Box 1.2 – Self-Organization

When clastic sediments (mud, sand and gravel) are subjected to hydrodynamic processes (wind, waves and currents), they tend to become organized into distinct sedimentary facies and morphodynamic sub-systems. The order arises from within the morphodynamic system and hence the term 'self-organization' is employed. It is very difficult to demonstrate convincingly that a landform is indeed a self-organizing feature. The problem here is to demonstrate the emergence of order as a result of the internal dynamics, rather than resulting from some outside forcing.

One way to investigate self-organizing behaviour is computer modelling. An

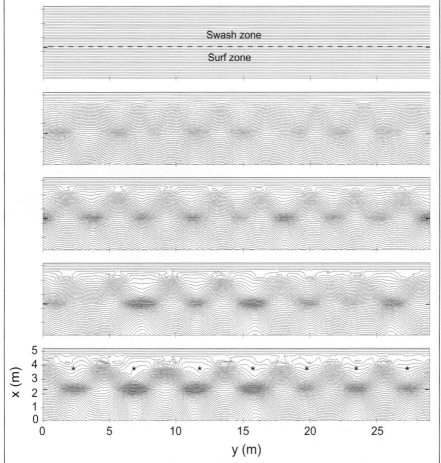

**Figure 1.11** – Numerical simulation of beach cusp formation. The five panels show contour plots of the morphology at the start of the simulation and after 50, 200, 400 and 800 simulated wave cycles. The swash excursion is 1.8 m and the resulting cusp spacing is c. 4 m. The small bold ticks in the panels indicate MSL and the asterisks in the bottom panel indicate the location of the cusp horns. [Modified from Coco et al., 2000.]

example of such a model is that by Coco *et al.* (2000) who investigated the formation and maintenance of beach cusps. In their computer model, an initial planar beach with small random morphological perturbations is subjected to swash action. Some of these perturbations become enhanced (small holes become big holes; small bumps become big bumps), while others are suppressed. Ultimately, the feedback processes between morphology and swash hydrodynamics result in the formation of beach cusp morphology (Figure 1.11). The important point is that the spacing of the cusps is not hard-wired into the model, but emerges after running the model. Apparently, the cusp spacing does depend on the details of the computer algorithm (for example the equation linking swash flow velocity to sediment transport rate), but is not an input variable of the model. The computer-generated cusps are self-organizing features and it seems possible that natural cusps are as well.

Self-organization is a relatively new concept in geomorphology, but is increasingly employed to explain and describe a wide range of geomorphic systems (Werner, 1999). The concept has been applied to analyse, describe and explain a large number coastal features, including gravel barriers, estuaries, beach cusps, coastal dunes and nearshore bars.

**Figure 1.12** – Gravel beach cusps in Alum Bay, Isle of Wight, England. The spacing of the cusps is *c.* 8 m. [From Kuenen, 1948.] [Copyright © 1948 The University of Chicago Press, reproduced with permission.]

## 1.4.4 Relaxation time

Morphological adjustment involves a redistribution of sediment, which requires a finite amount of time. The time required for the morphologic adjustment to occur is referred to as the **relaxation time** and is a measure of the morphological inertia within the system. The relaxation time of landforms depends on three factors:

- **Energy level** – The higher the energy level, the larger the sediment transport rates and the shorter the relaxation time. Therefore, coastal morphology responds faster to increasing energy levels (storm erosion) than decreasing energy levels (post-storm recovery).
- **Sediment mobility** – Relaxation times increase with decreasing sediment mobility. The material comprising cliffs and shore platforms has a large resistance to change and hence these landforms are characterized by the longest relaxation times. Sand is more easily moved than gravel and therefore sandy coastal landforms are generally characterized by shorter relaxation times than those consisting of gravel material.

- **Spatial scale of the landform** – The relaxation time strongly depends on the volume of sediment involved in the morphologic adjustment. The spatial scale of a landform is related to the temporal scale of its forcing (Figure 1.2). Therefore, large coastal landforms, such as coastal barriers, have longer relaxation times than small morphological features, such as beach cusps.

Generally, the relaxation time exceeds the time between changes in environmental conditions. It is therefore unlikely that a steady-state equilibrium is ever reached, in particular for large coastal landforms.

An example illustrating relaxation time over a relatively short time scale (decades) is shown in Figure 1.13a. Moruya beach in south-east Australia

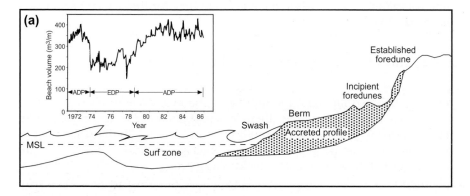

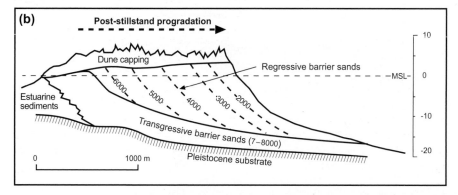

Figure 1.13 – Examples from south-east Australia illustrating the presence of relaxation times. (a) Relaxation time on the decadal time scale is apparent in the time series of beach volume change on Moruya beach, New South Wales, Australia. Two distinct periods can be identified and are designated accretion-dominated period (ADP) and erosion-dominated period (EDP). [From Thom and Hall, 1991.] [Copyright © 1991 John Wiley & Sons, reproduced with permission.] (b) Relaxation time on the millenium time scale is indicated in the chronostratigraphy of a prograded barrier system. The isochrons (dashed lines) are in years BP. [From Cowell and Thom, 1994.] [Copyright © 1994 Cambridge University Press, reproduced with permission.]

suffered major erosion during severe storms in 1974 and 1975, as clearly indicated by a dramatic decrease in subaerial beach volume from 400 to 200 m³ per metre beach width (Thom and Hall, 1991). The recovery of the beach to pre-storm conditions took about 6 years. After these years of beach accretion, the beach remained relatively stable. Figure 1.13b shows the relaxation time associated with coastal sand barriers adjusting to relatively stable sea-level conditions (Cowell and Thom, 1994). Sea level attained its present level approximately 6,000 years ago after having risen by about 130 m over the preceding 10,000 years. Following the stabilization of the sea level, the coastline kept prograding despite the fact that sea level was relatively stable. Figure 1.13 clearly demonstrates that the relaxation time increases with the size of the morphological feature. For Moruya beach, the relaxation time was about 5 years, while the relaxation time for the coastal barrier was more than 1,000 years.

## 1.4.5 Cumulative evolution and inheritance

Coastal evolution is a cumulative process, because the morphological outputs are included amongst the inputs for the next cycle of change (Figure 1.7). Coastal landforms and their morphodynamic processes therefore reflect both the imprint of present-day processes and the imprint of past processes. To some degree, coastal morphology and morphodynamics are thus inherited from the past. Rocky coasts are particularly affected by inheritance.

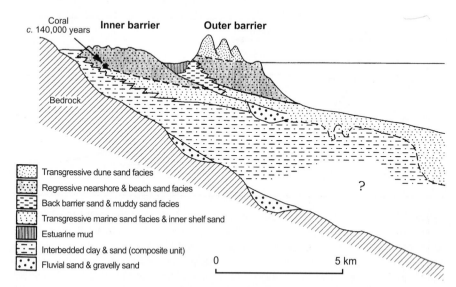

Figure 1.14 – Generalized stratigraphic section of Newcastle Bight embayment, 100 km north of Sydney on the southeast coast of Australia. The inner and outer barrier are of Last-Interglacial and Holocene age, respectively. [From Cowell and Thom, 1994.] [Copyright © 1994 Cambridge University Press, reproduced with permission.]

An example of **cumulative evolution** and **inheritance** is given in Figure 1.14 by the Quaternary history of the southeast Australian coast. The two main events responsible for the coastal evolution shown were the Last-Interglacial transgression and the Holocene transgression. Both episodes caused deposition of transgressive (rising sea level) and regressive (coastal progradation) stratigraphic sequences. The stratigraphy shows that the Last-Interglacial barrier (the inner barrier) developed on a substrate that was different from that on which the Holocene barrier (outer barrier) was deposited. Cowell and Thom (1994) argue that the Holocene barrier developed under more energetic wave conditions than the Last-Interglacial barrier due to a steeper substrate and a more exposed setting, despite the fact that the offshore wave conditions may have been similar.

## 1.5 SUMMARY

- The spatial boundaries of the coastal zone correspond to the limits to which coastal processes have extended during the Quaternary geological period and include the coastal plain, the shoreface and the continental shelf.
- Coastal morphodynamics are defined as the mutual adjustment of topography and fluid dynamics involving sediment transport. The coastal morphodynamic system consists of four linked components: process, sediment transport, morphology and stratigraphy. An essential ingredient to coastal morphodynamics is the feedback between morphology and processes.
- Environmental conditions drive and control the processes within a morphodynamic system, but are not affected by the system. The three major types of environmental conditions are geology, sediment supply/characteristics and external energy sources.
- There is a close coupling between temporal and spatial scales in coastal morphodynamics. Therefore, the modification of large-scale morphological features occurs over long time periods, whereas small-scale features change over short time periods.
- Coastal morphodynamic systems possess a number of fundamental characteristics that make prediction of coastal evolution difficult. These characteristics include negative feedback, equilibrium, positive feedback, self-forcing, self-organization, relaxation time, cumulative evolution and inheritance.

## 1.6 FURTHER READING

Carter, R.W.G. and Woodroffe C.D., 1994. Coastal evolution: An introduction. In: R.W.G. Carter and C.D. Woodroffe (editors), *Coastal Evolution*, Cambridge University Press, Cambridge, 1–31. [An authoritative introduction into coastal evolution.]

Cowell, P.J. and Thom, B.G., 1994. Morphodynamics of coastal evolution. In: R.W.G. Carter and C.D. Woodroffe (editors), *Coastal Evolution*, Cambridge University Press, Cambridge, 33–86. [An excellent treatment of the principles of coastal morphodynamics. Section 1.4 of this chapter is largely based on this work.]

Davies, J.L., 1980, *Geographical Variation in Coastal Development* (2nd Edition), Longman, New York. [This text is somewhat outdated, but still provides a very good overview of the global distribution of coastal environments and environmental conditions.]

Wright, L.D. and Thom, B.G., 1977. Coastal depositional landforms: A morphodynamic approach. *Progress in Physical Geography*, **1**, 412–459. [Benchmark paper outlining the morphodynamic approach to coastal landforms. Section 1.3 of this chapter is largely based on this work.]

# 2

# SEA LEVEL

## 2.1 INTRODUCTION

The coastline is formed by the intersection of the sea level with the land. If sea level rises, the coastline is shifted landward, whereas if sea level falls, the coastline is shifted seaward. Rising sea levels are referred to as **transgressive conditions** and generally result in the drowning and/or onshore migration of coastal landforms. Falling sea levels are referred to as **regressive conditions** and commonly result in the emergence of coastal landforms and coastal progradation. Over short time scales (seconds–months), sea level rises and falls with waves, tides, changes in atmospheric pressure and wind. However, when these fluctuations are averaged out, a stable value can be obtained. This constant value is referred to as the mean sea level (MSL).

At the outset, we need to be aware of the distinction between two types of sea-level changes:

- **Relative sea-level change** refers to changes in the sea level relative to that of the land and operates on a regional/local level. It can be brought about by a change in the sea level and/or a change in the level of the land.
- **Eustatic sea-level change** (sometimes referred to as absolute sea-level change) refers to a world-wide or global change in sea level and is unrelated to local/regional effects. The most common cause of eustatic sea-level variation is change in the ocean water volume.

When MSL is measured at a single location over an extended period, only a relative sea-level curve can be derived. The eustatic change in sea level at this location can only be determined from the relative sea-level curve if the vertical movement of the land level is known. If sea-level measurements indicate a constant relative sea level, this does not imply that both the land level and eustatic sea level are constant – it is more likely that the change in eustatic sea level is exactly balanced by the movement of the land.

## 2.2 INDICATORS OF FORMER SEA LEVELS

MSL is known to have varied through time and ample evidence of past sea-level variations can be found along most coastlines. Pirazzoli (1996) provides an excellent review of the different types of **sea-level indicators**.

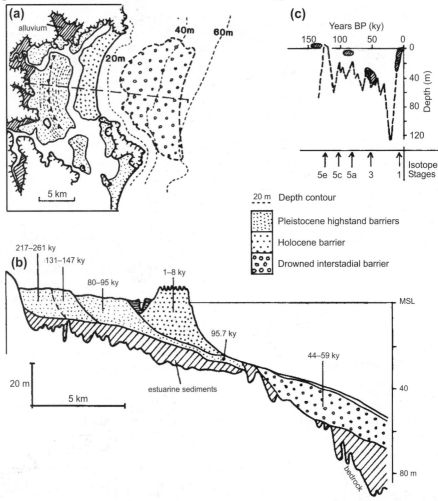

Figure 2.1 – Stratigraphic evidence of sea-level change in the Tuncurry embayment, central New South Wales, Australia. (a) Distribution of barrier systems onshore and on the inner shelf. (b) Shore-normal cross-section (dashed line in (a)) through the various barriers showing their dimensions and ages ('ky' refers to thousand years before present). (c) Group ages of the various barrier systems (shaded) and the estimated positions of sea level when they formed. The numbers indicate the oxygen Isotope Stages and the sea-level curve (dashed line) is that of Chappell and Shackleton (1986). [From Roy et al., 1994.] [Copyright © 1994 Cambridge University Press, reproduced with permission.]

Indicators of sea levels higher than present (sea-level highstands) are easier to find than those for lower sea levels (sea-level lowstands). This is because evidence of sea-level lowstands are generally located below present sea level and are hence not readily accessible. In addition, indicators of lower sea levels are often obliterated during subsequent submergence by higher sea levels.

Some of the most accurate sea-level indicators are marine organisms that produce shells or hard surfaces and live at known tidal levels (*e.g.*, lichens, barnacles, coral, worms, oysters). These organisms form communities that are commonly arranged in horizontal bands. Information about sea-level change can be obtained by comparing the elevation of present-day bands (with living organisms) with those of similar fossil organisms which now stand at a different level.

Former sea levels in depositional coastal environments are recorded in the coastal geomorphology and stratigraphy. An example of this is presented in Figure 2.1 which demonstrates the presence of a series of coastal barriers of different ages. Barriers are the main depositional element along wave-dominated coasts and, in our example, the present-day shoreline is formed by a relatively young barrier system of Holocene age (1,000–8,000 years BP; 'BP' refers to before present). Landward of this barrier is an older Pleistocene barrier complex comprising three distinct units with their ages increasing in the landward direction (80,000–95,000, 131,000–147,000 and 217,000–261,000 years BP). Therefore, four periods of barrier formation can be recognized in the stratigraphic record and during each of these, sea level was close to the present MSL. Evidence of a fifth barrier system can be found on the inner shelf. This barrier was build between 44,000 and 59,000 years BP, a time when sea level was *c.* 40 m below present MSL.

Rocky coasts generally lack depositional features and indicators of sea-level change in these settings consist of eroded shore platforms, cliffs, notches and benches. As an example, Figure 2.2 shows a fossil notch, found in close proximity to the contemporary shoreline, but at a significantly higher elevation (*c.* 2 m above MSL) compared with active notches (*c.* 0.5 m above MSL). This indicates that the fossil notch developed under higher relative sea levels than present. Along some rocky shores, marine terraces (shore platforms) can be found high above present-day sea level, providing compelling evidence of higher relative sea levels in the past.

**Figure 2.2** – Fossil notch located *c.* 2 m above MSL on Rottnest Island, Western Australia, providing evidence of sea-level highstand. [Photo G. Masselink.]

# 2.3 CAUSES OF SEA-LEVEL CHANGE

There are many causes of sea-level change. Some of these contribute to a eustatic sea-level change, but all contribute to relative changes in the sea level. On the basis of their spatial extent, we can distinguish between global, regional and local causes of sea-level change.

## 2.3.1 Global causes: Changes in ocean water volume and thermal expansion

One of the principal causes of sea-level fluctuations is a change in the quantity of water in the oceanic basins. Such fluctuations are world-wide and are termed eustatic. An increase in the amount of water in the ocean results in a rise in sea level, whereas a decrease in the amount of oceanic water causes a lowering of the sea level. Water is present on the Earth in various forms and locations, but the total volume can be considered constant and is expressed by the **global water balance**

$$K = A + O + L + R + S + B + M + U + I \qquad (2.1)$$

where $K$ is the total water volume (a constant) and the other variables are indicated in Table 2.1. The vast majority of the Earth's water is contained in oceans and seas, but groundwater and frozen water (ice) also make up a significant proportion of the global water volume. These water volumes can

**Table 2.1** Estimates of storage volumes and equivalent water depth for components of the world water balance.

| Parameter | Present volume, km³ | Equivalent water depth |
|---|---|---|
| A  Atmospheric water | 13,000 | 36 mm |
| O  Ocean and seas | 1,370,000,000 | 3.8 km |
| L  Lakes and reservoirs | 125,000 | 35 cm |
| R  Rivers and channels | 1,700 | 5 mm |
| S  Swamps | 3,600 | 10 mm |
| B  Biological water | 700 | 2 mm |
| M  Moisture in soils and the unsaturated zone | 65,000 | 18 cm |
| U  Groundwater | 4,000,000 to 60,000,000 | 11 to 166 m |
| I  Frozen water | 32,500,000 | 90 m |
|     Antarctica | 29,300,000 | 81 m |
|     Greenland | 3,000,000 | 8 m |
|     Others | 200,000 | 1 m |

Note: there is a large uncertainty regarding the volume of water stored as groundwater (from Pirazzoli, 1996)

be expressed as an equivalent water depth. For example, if all the atmospheric water (water vapour) rained into the oceans, then sea level would rise by 36 mm. From Table 2.1 it is apparent that only groundwater and frozen water (and to a lesser extent water in lakes and reservoirs) can be important contributors to changes in the ocean water volume. In particular, the melting or growth of continental ice sheets is of paramount importance for eustatic sea-level changes. It is noted that melting of floating ice sheets (mainly in the Arctic) has no effect on the eustatic sea level since the weight of the ice is already supported by the water.

The melting/growth of ice sheets is the main mechanism for sea-level change and is referred to as **glacio-eustasy**. Later in this chapter, we will see that glacio-eustatic sea-level change is strongly linked to climate change. During cold climatic periods, sea water is progressively lost from the oceans via precipitation as snow on the continents. During warm climatic periods, the ice melts resulting in a rise in sea level.

Even if the quantity of sea water remains constant, sea level may change due to variations in the sea water temperature. Sea water density increases with decreasing water temperature up to its freezing point at $-1.75°C$. In other words, if a volume of water is heated (cooled), it will occupy a larger (smaller) volume. Therefore, a decrease in sea water temperature causes a fall in sea level, whereas an increase in sea water temperature induces a sea-level rise. The latter process is known as **thermal expansion** and we will see later in this chapter that it is contributing significantly to present sea-level rise. To get a feel for the importance of thermal expansion, an increase of $1°C$ over a water depth of 4,000 m produces a rise in sea level of 0.6 m.

## 2.3.2 Regional causes: Isostatic changes

The Earth's crust floats on a denser underlying layer (asthenosphere), similar to ice floating on water. This two-layer system is in **isostatic balance** when the total weight of the crust is exactly balanced by its buoyancy. Addition of a load to the crust (*e.g.*, water, ice or sediments) will upset the isostatic balance. To compensate for the increased weight of the crust, some of the asthenospheric material will flow away and the land level will fall.

**Glacio-isostasy** refers to isostatic adjustments of the Earth's crust due to the loading and unloading of ice sheets (Figure 2.3). The weight associated with thick ice sheets causes a depression of the underlying land surface to a level approximately equal to ¼ of the maximum thickness of the ice (Box 2.1). At the margins of the depressed land surface, known as the **forebulge**, the land is pressed slightly upward due to the flow of asthenospheric material towards this region. When the ice melts, the land surface will resort back to its former position, thus the area formerly covered by the ice will come up and the forebulge will go down. This process is known as **isostatic rebound**. Shorelines located in areas formerly covered by ice sheets (*e.g.*, Canada, Scandinavia, Scotland) will therefore experience a postglacial fall in relative sea level, whereas coastlines in the forebulge areas will undergo a relative sea-level rise.

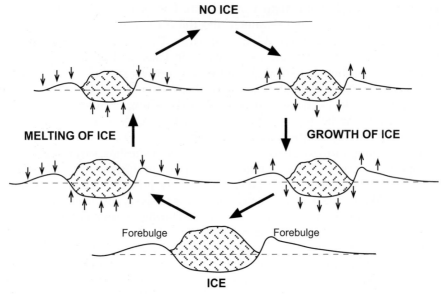

**Figure 2.3** – Vertical land movements associated with glacio-isostasy for growing and melting of ice sheets. Note the presence of forebulge areas near the margins of the ice sheet.

## BOX 2.1 – POSTGLACIAL REBOUND

If we ignore the response times involved with isostatic adjustment, it is relatively easy to determine the amount of postglacial rebound that can be expected following unloading. The following example is derived from Allen (1997). Figure 2.4 shows schematically the depression of the Earth's crust by an ice sheet. The **isostatic balance** above the depth of compensation can be performed for a column far from the ice sheet and a column through the centre of the ice sheet, and reads as

$$\rho_i h_i + \rho_c h_c = \rho_c h_c + \rho_m w_0$$

where $\rho_i$, $\rho_c$ and $\rho_m$ are the densities of ice, crust and mantle, respectively, and $h_i$, $h_c$ and $w_0$ are the thickness of the ice sheet, crust and the maximum deflection of the land surface under the ice sheet, respectively. If the thickness of the crust is considered constant, the isostatic balance can be simplified to

$$\rho_i h_i = \rho_m w_0$$

Assuming further a density of ice of 800 kg m$^{-3}$ and a density of the mantle material of 3,300 kg m$^{-3}$ we find

$$w_0 = \frac{\rho_i}{\rho_m} h_i = 0.24 h_i$$

Thus, depression of the Earth's crust is about a quarter of the thickness of the ice sheet. The Laurentian ice sheet over the Hudson Bay area of Canada had an average thickness of 5 km during the height of the glaciation, implying that the amount of depression, and consequent rebound during deglaciation, may have exceeded I km.

**Figure 2.4** – Schematic showing isostatic balance. [From Allen, 1997.] [Copyright © 1977 Blackwell Publishers, reproduced with permission.]

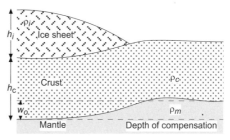

There are considerable time lags involved with glacio-isostatic movements due to the rigidity of the Earth's crust. These time lags are so long that, even at present, the Earth's crust is still isostatically adjusting to the melting of ice sheets that occurred more than 10,000 years ago. For example, **post-glacial uplift** in the Baltic region due to isostatic rebound has amounted to 300 m over the past 10,000 years, and the region is still being uplifted at a maximum rate of 10 mm $yr^{-1}$ (Figure 2.5). In Britain, isostatic adjustments are also still ongoing. Scotland and north England, which were covered by ice during the last glacial period, are being uplifted at rates of up to 2 mm $yr^{-1}$, whilst south England, which is located in the forebulge area, is subsiding by up to 2 mm $yr^{-1}$.

During deglaciation, melt water from ice sheets produces a considerable load on the sea floor of the continental shelves. This results in a subsidence of the sea floor due to **hydro-isostasy**. Along coasts with wide continental shelfs, the water load across the shelf is not uniform because the deeper part of the shelf near the shelf edge experiences a higher pressure than the shallow part near the coastline. As a result of this non-uniform pressure, a hinge point is formed somewhere in the middle of the shelf whereby the subsidence of the deeper part of the shelf results in an uplifting of the shallow part (Chappell *et al.*, 1982). As a result, the shoreline may be uplifted and a relative fall in sea level is experienced. Most of the Southern Hemisphere coastlines, where the effect of glacio-isostasy is not important or non-existent, have experienced such a fall in relative sea level during the second half of the Holocene (Section 2.5.1).

The geophysical effects of glacio- and hydro-isostasy have been investigated using numerical models and Lambeck (1993) has distinguished four different zones:

- **Near-field sites** are regions within the limits of the former ice sheets.
- **Ice-margin sites** are regions near the former ice margins.
- **Intermediate-field sites** are regions just outside former ice margins.
- **Far-field sites** are regions well away from the influence of the former ice sheets.

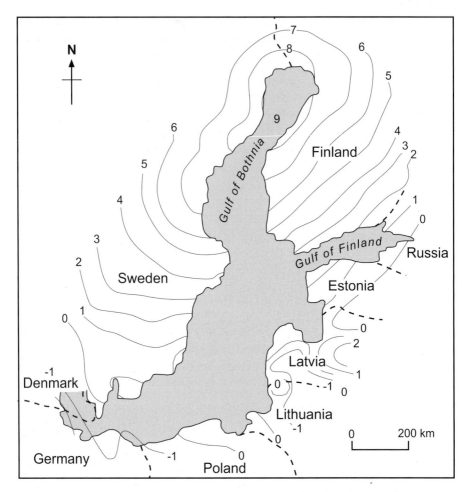

**Figure 2.5** – Isobases of present-day rates of uplift of land surfaces in Sweden and Finland due to postglacial rebound in mm year$^{-1}$. [Modified from Eronen, 1983.]

A remarkable outcome of global isostatic models is that the vertical land movements caused by deglaciation are not limited to formerly glaciated areas and nearby regions, but are assumed to extend more or less all around the globe.

## 2.3.3 Local causes: Tectonics and subsidence

Locally, **tectonic activity** can result in changes in the level of the land. Trends of vertical displacement of tectonic origin often appear to be continuous and gradual over the long term, but frequently consist of spasmodic movements associated with earthquake activity. Another local cause of sea-level change is compaction of sediment resulting in land **subsidence**.

Unconsolidated sediments, such as those found in deltaic environments, often contain up to 50% water by weight. In peat layers, **compaction** may reduce water content to as little as 10%. When the sediment is compacted, the land level drops. Also associated with deltaic environments is the process of **sediment-isostasy** whereby the Earth's crust is depressed due to the weight of the deltaic sediment (similar to glacio-isostasy). The combination of compaction and sediment-isostasy makes deltaic environments very prone to relative sea-level rise. The Mississippi delta, for example, has exhibited a drop in land level over the last 10,000 years of *c*. 165 m, equating to a rate of relative sea-level rise of 16.5 mm yr$^{-1}$ (Fairbridge, 1983). Similarly, the Holocene relative sea-level rise for the Yangtze delta due to land subsidence is estimated at 1.6–4.4 mm yr$^{-1}$ (Stanley and Chen, 1993). Land subsidence can also result from anthropogenic extraction of ground-water, oil or gas, and can be important in river-mouth or lagoonal areas. Land sinking due to water extraction has been reported from many coastal regions, for example, 4.6 m in Tokyo and 2.7 m in the Po delta (Pirazzoli, 1996).

## 2.4 PLEISTOCENE SEA LEVELS

The **Pleistocene**, which started at 1.8 million years BP, is commonly known as the **Ice Age** because it was a period characterized by relatively low temperatures compared to previous geological periods. During the Pleistocene, 17 alternations of cold climatic phases known as **glacials** and warm climatic phases referred to as **interglacials** have been identified (Shackleton and Opdyke, 1976). The glacials lasted around 100,000 years whereas the interglacials had a significantly shorter duration, approximately 10,000 years. The last glacial started about 70,000 years BP and finished 11,500 years BP. During the interglacials ice sheets melted and sea level rose, while during the glacials ice sheets grew and sea level fell. Interspersed within the glacials and interglacials were cold and warm intervals of shorter duration (*c*. 1,000 years), known as **stadials** and **interstadials**, respectively. These shorter warming/cooling cycles also caused eustatic sea-level fluctuations. Along coastlines that have undergone a steady, tectonically-induced uplift over the Pleistocene, evidence of interglacial sea levels can be observed in the form of raised shore platforms or coral reefs (Box 2.2). More commonly, however, surficial traces of the earliest lower sea levels associated with glaciations have been obliterated by subsequent rises of sea level. Hence, detailed sedimentological and stratigraphical evidence only exists for the youngest cycles. The next three sections will discuss in more detail the sea-level and climate variations that occurred during the Pleistocene.

## Box 2.2 – Derivation of the Pleistocene Eustatic Sea-Level Curve

There are two principal difficulties associated with determining the eustatic sea-level variation during the Pleistocene. First, it is very difficult to separate the eustatic sea-level variations from the effects of isostatic land movements. Second, there have been a large number of glacial/interglacial cycles and the eustatic sea-level variations associated with each of these have occurred within a similar range. Therefore, if the land level has remained constant during the Pleistocene, stratigraphic evidence of the older glacials/interglacials may have been obliterated by subsequent glaciations.

To overcome these two problems, most of the work on Pleistocene eustatic sea level has focused on locations distant from glaciated areas, and with a relatively simple history of tectonic uplift during the Pleistocene (Chappell and Shackleton, 1986; Fairbanks, 1989). Under conditions of steady uplift, stratigraphic evidence of older interglacials, such as beaches, shore platforms and coral reefs, are raised above subsequent interglacial sea levels and will thus be preserved to a large degree. The result is a 'flight' or 'staircase' of interglacial morphologies, with the age of the features increasing with elevation.

To derive the eustatic sea-level change from such a flight of interglacial beaches/platforms/reefs one needs to know the rate of uplift in the region and the dates associated with these features. Eustatic sea level is then derived with a simple formula

$$E = P - U$$

where $E$ is the amount of vertical shoreline displacement caused by eustatic change, $P$ is the present elevation of the raised feature above MSL and $U$ the amount of uplift that has occurred. For example, if a raised shore platform is found at 120 m above MSL, the age of the platform is 100,000 years and the rate of uplift is 1 mm yr$^{-1}$, then the platform has been uplifted by 100 m and the eustatic change in sea level is 20 m. In other words, when the shore platform developed, the eustatic sea level was 20 m higher than present.

### 2.4.1 Sea level in the Pleistocene: Evidence from deep-sea cores

Evidence for changes in the sea-water temperature and eustatic sea level during the Pleistocene can be derived from **deep-sea cores** by analysing the **oxygen isotope ratio** $\delta^{18}O$ (Box 2.3). This ratio can be used as a proxy for the volume of water stored in ice sheets and therefore sea level. Figure 2.6 shows the $\delta^{18}O$ record for the last 2.6 My deduced from benthic foraminifera present in a deep-ocean sediment core. Increasing values of $\delta^{18}O$ represent cooling periods and expansion of ice sheets (*i.e.*, sea-level fall), and decreasing values indicate warming periods and melting of the ice sheets (*i.e.*, sea-level rise). A large number of cooling/warming cycles can be identified in

## BOX 2.3 – OXYGEN-ISOTOPE RATIO

Analsyis of the oxygen isotope ratio makes use of the relative proportion of the two oxygen isotopes $^{18}O$ and $^{16}O$ in the skeletal material of calcareous microfossils (foraminifera) (Shackleton, 1987). The ratio $^{18}O/^{16}O$ in these micro-organisms is controlled by temperature-dependent fractionation of the isotopes from sea water into the skeleton of the micro-organism and the isotopic composition of the sea water. The latter is dependent on the amount of water stored in ice sheets. The oxygen isotope $^{16}O$ is lighter than $^{18}O$ and evaporation preferentially removes $^{16}O$. Growing ice sheets imply a net transfer of water from the ocean to the ice sheets, resulting in an enrichment of sea water with the heavier isotope $^{18}O$ (isotopically-heavy). Melting of the ice sheets on the other hand causes an enrichment of sea water with the lighter isotope $^{16}O$ (isotopically-light).

Oxygen-isotopic compositions are measured relative to a global standard known as the Standard Mean Ocean Water (SMOW). The oxygen isotopic composition of a sample can then be expressed as per mille differences relative to SMOW:

$$\delta^{18}O = 10^3 \left[ \frac{(^{18}O/^{16}O)_{sample} - (^{18}O/^{16}O)_{SMOW}}{(^{18}O/^{16}O)_{SMOW}} \right]$$

Positive values of $\delta^{18}O$ indicate enrichment of the sample in the heavier isotope, negative values indicate depletion. Changes in the shallow water temperature reflect to a large extent changes in the air temperature. Therefore, $\delta^{18}O$ of shallow-water foraminifera can serve as a proxy for surface water or air temperature. In the deep ocean, however, temperatures are likely to have fluctuated little during glacial–interglacial cycles, so $\delta^{18}O$ of deep-ocean (benthic) foraminifera is likely to represent changes in $\delta^{18}O$ of sea water due to changes in ice volume. An increase in $\delta^{18}O$ indicates an increase in global ice volume and hence a fall in sea level, whereas a decrease in $\delta^{18}O$ represents a decrease in global ice volume and hence a rise in sea level.

the record. In the literature, these cycles have been given **Isotope Stage** numbers to facilitate comparison between different oxygen isotope records. The odd numbers represent peaks in the $\delta^{18}O$ records (sea-level highstands and warm periods), whereas the even numbers indicate troughs in the record (sea-level lowstands and cold periods). The most recent cycles (Isotope Stages 1–18) are characterized by the largest ranges in $\delta^{18}O$ (greater than 1) and longest periods (c. 100,000 years). Presently, we appear to be at the start of another warming cycle.

As a rule of thumb, a difference of 0.1 in $\delta^{18}O$ is equivalent to 10 m of sea-level change. From Figure 2.6 we can thus infer that sea level fluctuated during the glacial-interglacial periods over a range of more than 100 m. This technique is strongly dependent on the assumption that deep-sea temperatures have remained constant. It is more desirable, therefore, to derive a palaeo sea-level curve directly from stratigraphic evidence, rather than the $\delta^{18}O$ record. Chappell and Shackleton (1986) determined a **eustatic sea-level**

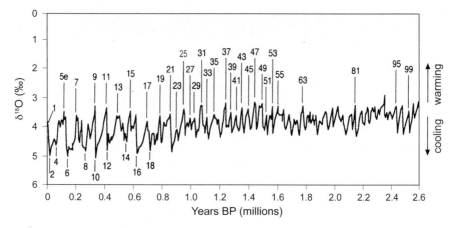

Figure 2.6 – Oxygen-isotope ($\delta^{18}O$) record for the last 2.6 My deduced from benthic foraminifera of deep-sea core ODP 677. The labels represent oxygen Isotope Stages that are used universally to facilitate comparison between $\delta^{18}O$ records. For example, Stage 5e is associated with the last interglacial, whereas Stage 2 is associated with the last glacial. [Modified from Shackleton, et al., 1990.]

**curve** over the last 250,000 years from a series of raised coral terraces on the Huon Peninsula, New Guinea (Figure 2.7). Their sea-level curve indicates that sea level at the height of the last interglacial was very similar to present-day sea level. Over most of the glacial cooling period, sea level

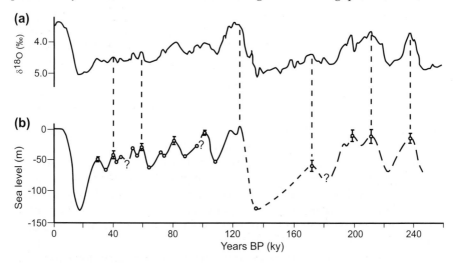

Figure 2.7 – (a) $\delta^{18}O$ record from east equatorial core V19-30. (b) Eustatic sea-level curve over the last 250,000 years for the Huon Peninsula, New Guinea, derived from a series of raised coral reefs. The time-axis of the sea-level curve was recalculated using the $\delta^{18}O$ record. Note the very good correspondence between the two records, indicating that $\delta^{18}O$ serves as a good proxy-indicator of sea level. [Modified from Chappell and Shackleton, 1986.]

showed a progressive fall, but was characterized by large fluctuations at least of the order of 20 m. Near the end of the last glacial, *c.* 18,000 years BP, sea level was 130 m lower than at present.

## 2.4.2 Climate in the Pleistocene: Evidence from ice cores

**Ice cores**, obtained by drilling through the layered accumulations of snow and ice deposited in polar or high-mountain ice sheets, permit a detailed reconstruction of past climate which can be compared with sea-level curves derived from deep-sea cores or with stratigraphic information. The ratios of hydrogen and oxygen isotopes in the ice provide an index of former temperatures (Box 2.3), while Greenhouse gas concentrations (mainly $CO_2$) can be determined from sealed air bubbles. The most widely published results of these ice core studies are from the Vostok drilling site in East Antarctica, and the GRIP (Greenland Ice-core Project) and GISP2 (Greenland Ice-Sheet Project) sites in southern Greenland.

The **Vostok ice core** provides the most extensive record and includes the past four glacial–interglacial cycles (420,000 years). The complete Vostok time series of $CO_2$ and temperature is shown in Figure 2.8. Each of the four glacial-interglacial cycles is characterized by a similar succession of changes. Rapid warming occurs at the onset of the interglacial and a slow, but intermittent cooling takes place during the glacial. The amplitude of the temperature change is *c.* 12°C and the coolest part of each glacial occurs just before the onset of the next interglacial. The temperature is strongly coupled to the $CO_2$ concentration. High temperatures during the interglacials are associated with large $CO_2$ concentrations (270–280 ppm) and low temperatures during the glacials coincide with low $CO_2$ concentrations (190–200 ppm). The close match between the temperature and $CO_2$ curve strongly supports a

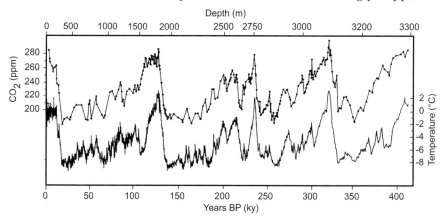

**Figure 2.8** – Variations in $CO_2$ (top line) and temperature (bottom line) during the past 420,000 years in the Vostok ice core from Antarctica. The core extends across four complete glacial-interglacial cycles. A strong positive correlation between $CO_2$ concentration and temperature is evident. [Modified from Petit *et al.*, 1999.]

link between Greenhouse gases and past climate change. It is further noted that the variations in temperature and $CO_2$ over the last two glacial–interglacial periods shown in Figure 2.8 match well with the sea-level record shown in Figure 2.7.

## 2.4.3 Causes of glacial/interglacial climate fluctuations

What has caused the temperature and sea level to fluctuate so widely and periodically over the Pleistocene period, in particular during the last 700,000 years? It is now widely accepted that the main trigger for climatic fluctuations was small variations in the Earth's orbit around the Sun causing changes in the radiative heat energy received from the Sun (Berger, 1992). There are three different mechanisms responsible for astronomical variations in solar radiation received at the top of the Earth's atmosphere (Figure 2.9):

- **Eccentricity** – The orbit of the Earth around the Sun is not a circle, but an ellipse. The orbit can be parameterized by the so-called eccentricity, which is based on the ratio of the major and minor of the ellipse (a value of 0 indicates that the Earth's orbit is a perfect circle). Temporal variations in the eccentricity have a periodicity of 100,000 years.
- **Obliquity** – The Earth's axis of rotation is not at right angles to the plane of the orbit, but is inclined to it. This accounts for the seasonal alterations of summer and winter. In the Northern Hemisphere, summer occurs when the axis of rotation is inclined towards the Sun, while winter occurs when the axis of rotation leans away from the Sun. At present the inclination, known as the obliquity, is nearly 23.5°, but the angle varies between 22.1° and 24.5°. The periodicity of this variation is about 41,000 years.
- **Precession** – The orientation of the (inclined) Earth's axis of rotation relative to the plane of the orbit is not constant, but changes with a period of about 21,000 years. This periodicity is referred to as the cycle of precession. At the present time, the Earth is closest to the Sun (perihelion) in the Northern Hemisphere winter. However, due to the cycle of precession, perihelion will coincide with the Northern Hemisphere summer in about 11,000 years.

Of these **orbital cycles** only changes in eccentricity cause the Earth as a whole to receive different amounts of solar radiation. The other two cycles (obliquity and precession) cause a redistribution of solar radiation between seasons in the different hemispheres. For example, a relatively large obliquity results in increased amounts of solar radiation during the Northern and Southern Hemisphere summers, and reduced amounts of radiation during the winters, but does not change the total amount of solar radiation received by the Earth.

Many attempts have been made to calculate the variation in solar radiation due to the orbital cycles as a function of latitude and season. The most notable of these have been those by the Serbian mathematician Milutin

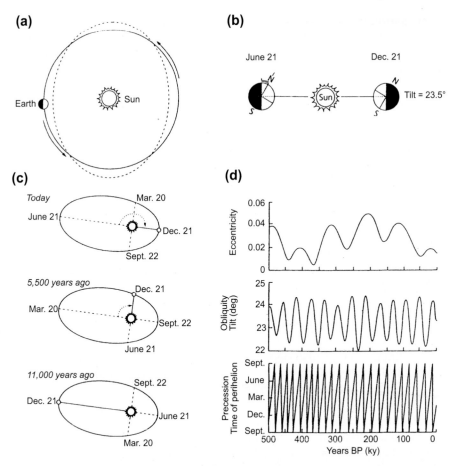

**Figure 2.9** – Three mechanisms responsible for variations in solar radiation received at the top of the Earth's atmosphere: (a) eccentricity, (b) obliquity, (c) precession and (d) the associated frequencies. [From Allen, 1997.] [Copyright © 1977 Blackwell Publishers, reproduced with permission.]

Milankovich (1879–1958) who produced detailed curves of solar radiation at various latitudes for the last 600,000 years. Milankovich found a conspicuous alternation of long periods of cool summers, which he considered would fail to melt all the winter snowfall in regions of appropriate latitude or elevation, and long periods of warm summers, which would hinder the accumulation of snow. Comparison of the calculated solar radiation curves with palaeo-temperature records has shown a remarkably close fit. In particular, the 100,000 years eccentricity cycle appears to be correlated with the glacial-interglacial variations of the last 700,000 years (Figure 2.6). Analysis of the Vostok ice-core record by Petit *et al.* (1999) demonstrated a very strong climatic forcing by orbital cycles for periods at 100,000 years, 41,000 years and to a lesser degree 21,000 years, providing further support for an orbital forcing of the glacial cycles.

The **astronomical theory** based on orbital cycles has certainly been valuable, but does not provide the sole explanation for the alternation of glacials and interglacials. There are three major observations that indicate that other factors are involved:

- Orbital variations in solar energy have always been present, but before the Pleistocene they did not bring about glacial and interglacial phases.
- Orbital variations in solar radiation cause disparate responses in both hemispheres. The amount of solar radiation received by the two hemispheres due to the orbital cycles shows an almost anti-phase relationship – when the Northern Hemisphere receives above-average amounts of solar radiation, the Southern Hemisphere receives below-average amounts. If variation in solar radiation is the only factor involved with driving climate change, Northern and Southern Hemisphere glaciations should alternate, not occur concurrently.
- Changes in the Earth's receipt of solar radiation due to orbital variations are relatively modest and only sufficient to cause direct cooling or warming of around 2°C. In reality, past temperatures fluctuated by much more than this amount.

The three observations listed above support the argument that variations in solar radiation are not directly and linearly linked to global variations in climate. Climate change also depends on the way in which heat is transferred across the Earth's surface via oceanic and atmospheric circulation. In order for the orbital variation in solar radiation to develop into full-blown glacials and interglacials, the signal needs to be amplified by the Earth's internal ocean/atmosphere system through positive feedback processes. It can therefore be concluded that the orbital cycles provide the external pacemaker for climate change which is modulated by the internal dynamics of the global atmosphere/ocean system (Hays *et al.*, 1976).

## 2.5 HOLOCENE TRANSGRESSION

The last glaciation reached a peak at around 18,000 years BP after which the earth warmed and the ice sheets started to melt. Deglaciation was interrupted by a period known as the 'Younger Dryas', which occurred from 11,000–13,000 years BP and was characterized by a cold climate and glacial advance. This period is generally considered part of the Pleistocene, hence the **Holocene** starts at 11,500 years BP (Roberts, 1998). The melting of the ice caps during the Holocene and the associated sea-level rise is known as the **Holocene transgression** (or Flandrian transgression). The melting histories of the various ice sheets were not synchronous, but by about 6,000 years BP almost all the ice involved with the glaciation had melted.

A number of eustatic sea-level curves associated with the Holocene transgression are shown in Figure 2.10. The general form of these curves indicates a relatively rapid rise in sea level of about 0.5 cm yr$^{-1}$ at the end of the

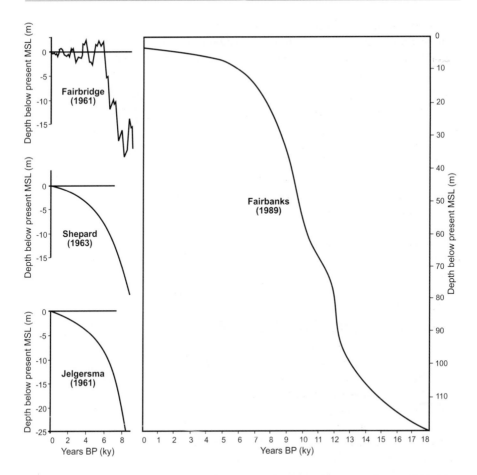

**Figure 2.10** – Late Pleistocene and Holocene sea-level curves according to Fairbridge (1961), Jelgersma (1961), Shepard (1963) and Fairbanks (1989).

Pleistocene (11,500–18,000 years BP), speeding up to about 2 cm yr$^{-1}$ at the start of the Holocene (7,000–11,500 years BP), then slowing over the mid-Holocene (5,000–7,000 years BP) and reaching present sea level at about 5,000 years BP. There is little controversy about this general form, however, there has been fierce debate regarding the fine detail of the curve. For example, according to Fairbridge (1961), the sea-level curve is characterized by significant fluctuations, whereas Jelgersma (1961) favours a gradual and smooth rise in sea level.

The lack of consensus with regards to the fine-tuning of the Holocene sea-level curve is largely ascribed to the difficulties associated with separating eustatic from isostatic effects. Determining the relative contributions of these two effects to sea-level change is very difficult, if not impossible, and the definition of eustasy as a world-wide, simultaneous, uniform change in sea level has recently been questioned. According to Tooley (1996), the definition is obsolete and of historical interest only, while Mörner (1987) has argued that eustatic curves should only be defined for regions, not for the whole Earth. Pethick (1984) adopts a pragmatic approach and points out that coastal geomorphologists should not be too concerned with differentiating between eustatic and isostatic sea-level changes, because what really matters is the relative sea level. Whether a sea-level rise is induced by an isostatic fall in land level or a eustatic rise in sea level, the effect in terms of coastal morphology is similar.

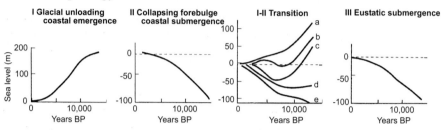

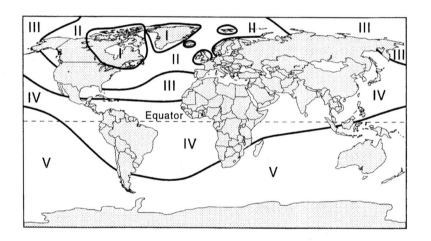

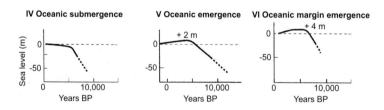

Figure 2.11 – Global variation in relative sea-level rise derived using an isostatic model under the assumption that no eustatic change has occurred since 5,000 years BP and that the Earth's mantle has a constant viscosity. Zone VI includes the landward margins of continental shelves in zones III, IV and V. Emerged beaches are predicted for Zones I, III, V and VI. [From Komar, 1998; modified from Clarke et al., 1978.] [Copyright © 1978 Academic Press, reproduced with permission.]

Along many coastlines, glacio- and hydro-isostatic effects have had a significant impact on the relative sea-level changes that occurred during the Holocene. Clarke *et al.* (1978) used a geophysical model to determine land movements due to glacio- and hydro-isostasy, and combined these with a eustatic sea-level curve to derive relative sea-level changes during the Holocene (Figure 2.11). The results indicate the existence of distinct groups of sea-level curves that are closely linked to the four different regions identified by Lambeck (1993; Section 2.3.2):

- At near-field sites (Zone I in Figure 2.11) the dominant contribution to sea-level change comes from ice-load effects, and late-glacial and post-glacial relative sea-level have been falling because of the rising land.
- The intermediate-field sites (Zone II in Figure 2.11) correspond to the forebulge around the former ice-margin which tends to subside in late-glacial and postglacial times to compensate for the uplift in nearby formerly glaciated areas. In these sites the relative sea level continues to rise even when deglaciation has ceased, though at gradually decreasing rates.
- At ice-margin sites (Transition Zone I–II in Figure 2.11) the relative sea-level curves vary greatly depending on the location, and range from a progressive fall (curve I–IIa in Figure 2.11) to a progressive rise (curve I–IIe in Figure 2.11).
- In far-field sites (Zones III–VI in Figure 2.11) glacio-eustatic changes in sea level are considerably greater than glacio- and hydro-isostatic effects. Relative sea-level rise predominates during the deglaciation period, often followed by a slight relative sea-level fall of hydro-isostatic origin during the late Holocene.

## 2.6 PRESENT SEA LEVEL

The geomorphology and sedimentology of the present-day coastline are primarily the result of the Holocene, and to a lesser extent Pleistocene, sea-level history. Of most interest to coastal communities, however, are the present and future changes in sea level. Over the last 100 years an extensive global network of tide gauges has developed, allowing an accurate assessment of global sea-level changes. At present, more than 1400 tide gauges are operative worldwide, although there is a strong bias towards the Northern Hemisphere. Monthly and yearly means from most tide-recording stations are stored in the data base of the Permanent Service for Mean Sea Level (PSMSL) at Birkenhead, UK, and are publicly available from their web site.

Tide gauges record changes in relative sea level and the central problem in identifying trends in eustatic sea level from tide gauge data has been in removing the isostatic trend from the data. This is clearly indicated in Figure 2.12, which shows local trends in sea level derived from a large number of long-term (more than 40 years) tidal records. Most stations exhibit a rise in sea level with rates of 1–3 mm yr$^{-1}$, but the tide record collected in

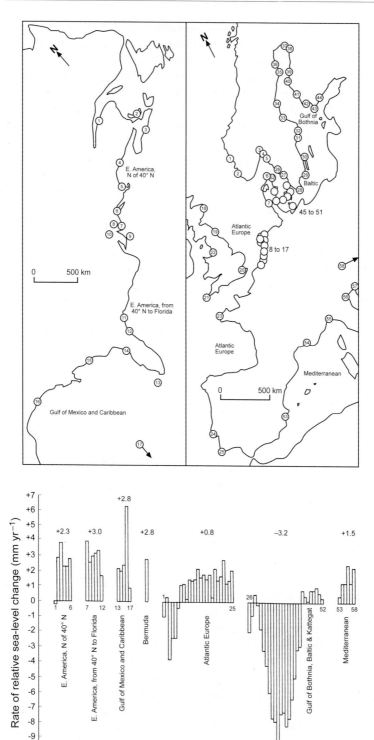

**Figure 2.12** – Locations of tide gauge stations with records starting earlier than 1925 on both sides of the North Atlantic and the associated trends of relative sea-level change. The numbers above the sea-level change graph are average rates of change in each region (+ = relative sea-level rise; – = relative sea-level fall). The numbers near the zero line in the sea-level change graph correspond to the tide gauge numbers in the location graph. [Modified from Pirazzoli, 1989.]

the Baltic region indicates a fall in sea level at a rate of up to 9 mm yr⁻¹. Clearly the Baltic region is still experiencing isostatic uplift due to deglaciation.

Due to uncertainties in determining the isostatic effects, it is difficult to obtain a reliable figure for eustatic sea-level rise. Nevertheless, efforts can be made to account for the isostatic land-level changes by using geological data directly from sites adjacent to tide gauges and subtracting trends in land level from the relative sea-level change. Figure 2.13 shows the results of two methods of estimating the eustatic component of sea-level rise. The first method takes the mean of a set of 130 station trends (corrected for changes in land level), resulting in a rate of sea-level rise over the last 100 years of 1.2 mm yr⁻¹. In the second method, the corrected data are averaged annually into a composite global mean sea-level curve and the slope of the curve is estimated to be 1.0 mm yr⁻¹.

The Intergovernmental Panel on Climate Change (IPCC) concludes that the average rate of eustatic sea-level rise during the 20th century is between 1 and 2 mm yr⁻¹. Comparison of the present rate of sea-level rise with the geological rate over the last two millennia (0.1–0.2 mm yr⁻¹) implies a relatively recent acceleration in the rate of sea-level rise. The onset of this acceleration appears to have occurred during the 19th century, and there is no clear evidence of any acceleration of sea-level rise over the 20th century alone (Church *et al.*, 2001).

There are a number of factors that could have contributed to the eustatic rise in sea level over the last century. It has become clear that the rise in sea level is partly due to the concurrent increase in global temperature of 0.3–0.6°C over the last 100 years. The climate- and human-related factors contributing to the observed sea-level rise of 1–2 mm yr⁻¹ and their ranges of uncertainty are shown in Figure 2.14. The main factors include thermal

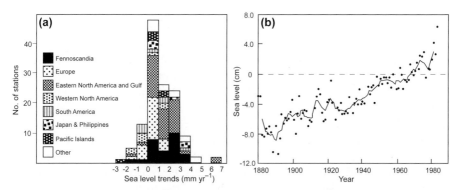

**Figure 2.13** – (a) Histogram of the number of tide gauge stations versus sea-level trends. [Modified from Gornitz, 1993.] (b) Composite eustatic mean sea-level curve over the last century (solid dots) with 5-year running mean (solid line). The period 1951–70 has been used as the reference period (*i.e.*, mean sea level from 1951–70 has been set to zero). [Modified from Gornitz, 1995.]

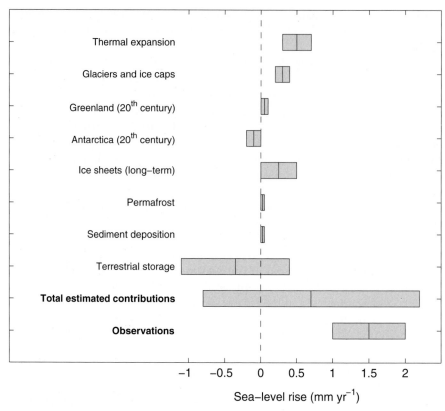

**Figure 2.14** – Ranges of uncertainty for the average rate of sea-level rise from 1910 to 1990 and the estimated contributions from different processes. [Modified from Church et al., 2001.]

expansion of the ocean (contributing 0.3–0.7 mm yr$^{-1}$), melting of glaciers and small ice caps (0.2–0.4 mm yr$^{-1}$) and long-term adjustment of ice sheets since the Last Glacial Maximum (0–0.5 mm yr$^{-1}$). The 20th century contributions of the ice sheets of Greenland and Antarctica have been estimated using ice sheet models. These models suggest that the Greenland contribution has been relatively modest (0–0.1 mm yr$^{-1}$), whereas Antarctica may have contributed negatively to the sea-level rise (–0.2–0 mm yr$^{-1}$) through an increase in its total ice mass due to increased precipitation. Changes in terrestrial storage of water may also have affected sea level, but the uncertainties associated with the estimates are very large (–1.1–0.4 mm yr$^{-1}$). An exact accounting of past sea levels is difficult due to the large uncertainties associated with the contributing factors.

# 2.7 FUTURE SEA LEVEL

The coastal zone is one of the most densely populated regions of the Earth and the ability to predict future change in sea level is of obvious importance to the global community. Not surprisingly, therefore, much research effort and funding has been, and still is, directed towards making such predictions. It is well accepted that simple extrapolation of existing trends in sea-level rise (1–2 mm yr$^{-1}$) into the future is inadequate. Such an approach ignores lag effects in the atmosphere and ocean to increased Greenhouse gases and neither considers future Greenhouse gas emission rates. To obtain reliable predictions of future sea-level rise, sophisticated computer models have been developed that consider the interactions between ocean and atmosphere. The advantage of such coupled **atmosphere-ocean general circulation models** (AOGCMs) is that they can be run for different scenarios thereby enabling an investigation of the effects of emission regulations. Unfortunately, despite the high level of sophistication of the current AOGCMs, their predictions include several major sources of uncertainty (Warrick, 1993), and the only statement that can be made with near-certainty is qualitative: global warming causes sea-level rise.

The importance of obtaining good estimates of **future sea-level rise** has been one of the main aims of the IPCC. The approach has been to run sophisticated AOGCMs for a range of emission scenarios. Table 2.2 shows the projected sea-level rise over the period 1990–2100 for one of these scenarios and the individual contributions of the different sea-level rise components. The IPCC has consistently placed strong emphasis on providing a range of uncertainties associated with the predictions and therefore minimum and maximum projections are given. The results indicate that sea level is predicted to be 0.11–0.77 m higher than today by 2100 (the global warming over the same period is *c*. 2°C). Most of the projected sea-level rise is due to thermal expansion of the oceans, followed by increased melting of

**Table 2.2** Sea-level rise due to climate change derived from running different AOGCM models following the IS92a scenario, including the direct effect of sulphate aerosols (from Church *et al.*, 2001)

| | Minimum, mm yr$^{-1}$ | Central value, mm yr$^{-1}$ | Maximum, mm yr$^{-1}$ |
|---|---|---|---|
| Thermal expansion | 0.11 | 0.27 | 0.43 |
| Glaciers/small ice caps | 0.01 | 0.12 | 0.23 |
| Greenland (20th century effects) | −0.02 | 0.04 | 0.09 |
| Antarctica (20th century effects) | −0.17 | −0.08 | 0.02 |
| •Sum | 0.11 | 0.44 | 0.77 |

•The sum includes contributions of permafrost, sedimentation and adjustments of ice sheets to past climate change. The role of terrestrial storage is not considered.

glaciers and small ice caps. On this time scale, the contributions made by major ice sheets in Greenland and Antarctica are relatively minor, but they may increase in importance over longer time scales.

## 2.8 SUMMARY

- Relative sea-level changes are changes in the sea level relative to that of the land and operate on a regional/local level. Such sea-level changes are mainly related to isostatic adjustments in the Earth's surface, in particular those associated with the increased (decreased) load by expanding (melting) ice caps.
- Eustatic sea-level changes are world-wide and are unrelated to local/regional effects. The principal cause of eustatic sea-level fluctuations is a change in the quantity of oceanic water.
- During the Pleistocene a large number of climatic cycles occurred, which in turn induced dramatic fluctuations in the eustatic sea level. During the warm periods (interglacials), sea levels were relatively high and comparable with present-day sea level. During the cold periods (glacials), vast quantities of sea water were stored in ice sheets and sea levels were more than 100 m lower than present.
- After the last glaciation of the Pleistocene, the Earth gradually became warmer. Ice caps started melting and caused a eustatic rise in sea level. The sea-level rise was initially rapid (2 cm $yr^{-1}$ from 7,000–11,000 years BP), but slowed down from 5,000–7,000 years BP, reaching present sea level at about 5,000 years BP.
- Eustatic sea level is presently rising by 1–2 mm $yr^{-1}$. This rise is generally attributed to global warming. The sea-level rise is expected to accelerate over the next century, and the Intergovernmental Panel on Climate Change predicts that by 2100 sea level will be 0.11–0.77 m higher than today.

## 2.9 FURTHER READING

Church, J.A., Gregory, J.M., Huybrechts, P., Kuhn, M., Lambeck, K., Nhuan, M.T., Qin, D. and Woodworth, P.L., 2001. Chapter 11: Changes in sea level. In: Intergovernmental Panel on Climate Change, *Climate Change 2001: the Scientific Basis*, Contribution of Working Group I to the Third Assessment Report of the Intergovernmental Panel on Climate Change, Cambridge University Press, Cambridge, 639–693. [Most up-to-date account of the relation between sea-level and climate change.]

Pirazzoli, P.A., 1996. *Sea-level changes: The last 20,000 years.* Wiley, Chichester. [Comprehensive review of Pleistocene and Holocene sea-level changes.]

# 3

# TIDES

## 3.1 INTRODUCTION

The tidal rise and fall of the ocean surface due to the gravitational attraction between the Earth, Moon, and Sun is barely noticeable in deep oceanic waters. On shallow continental shelves, along coastlines, and within estuaries, however, tidal processes can be the dominant 'shaper' of morphology. Even in environments where wind–wave or fluvial processes are dominant, tidal processes often play a key subordinate role. A geomorphologist's interest in tidal processes is generally limited to the movement of sediment by tidal currents in these shallow, coastal environments. In order to appreciate fully the scope of tidal interaction with morphology, from small-scale bedforms to large-scale morphological systems such as estuaries and deltas, it is important to have an appreciation of the fundamentals of tide-generation.

Our understanding of tides is based largely on the works of two very influential mathematicians: Isaac Newton (1643–1727) and Pierre-Simon Laplace (1749–1827). In his *Principia Mathematica*, Newton derived the fundamental astronomical forces that produce forced waves on an Earth covered by a uniform, infinitely deep ocean. Subsequently, in his *Mécanique céleste,* Laplace derived the fundamental hydrodynamic equations that govern the behaviour of these forced long waves on a more realistic rotating Earth with oceans of finite depth. A further contribution by William Thomson, also known as Lord Kelvin (1824–1907), demonstrated that Laplace's equations could describe tides in natural ocean basins surrounded by continental margins. Our discussion of tides will broadly follow this historical development of the topic. We will begin by describing the tide-generating force originally derived by Newton. The Equilibrium Theory grew from this base and is used to identify the major periodicities in the tide. Recent observations of tides on a global-scale then lead into the discussion of the Dynamic Theory, which emerged from the work of Laplace and Lord Kelvin. Smaller-scale interactions between tides and morphology, particularly relating to coastal bays and estuary channels, will be discussed in Chapter 7.

## 3.2 THE ASTRONOMICAL TIDE-GENERATING FORCE

It is the **gravitational attraction** between the Earth and other massive bodies in the solar system that produces what is known as the astronomical tide. This gravitational attraction can be thought of as a force $F_g$, which is written mathematically as

$$F_g = G\frac{m_1 m_2}{R^2} \tag{3.1}$$

where $G$ is the universal gravitational constant ($6.6 \times 10^{-11}$ N m$^2$ kg$^{-2}$), $m_1$ and $m_2$ are the respective masses of the two bodies that we are interested in, and $R$ is the distance between the centers of mass of each body. In essence, Equation 3.1 states that the larger the two bodies are (in terms of mass), the larger the gravitational attraction between them will be. It also states that the smaller the distance between the two bodies, the larger the gravitational attraction between them will be. In principle all planets and moons in the solar system can influence the behaviour of tides on Earth. In most cases, however, their effect is negligible, due either to their relatively small size or large distance from Earth. The astronomical tide-generating force acting upon the Earth is overwhelmingly dominated by the gravitational attraction of the Earth's Moon and the Sun, because of the Moon's close proximity and the Sun's large mass (Table 3.1).

Because of its greater importance, let us first consider the Moon's role in generating tides within the Earth's oceans. It is a common misconception that the Moon simply revolves around the Earth. In fact, the pair combine together as a system that rotates anticlockwise around their common centre-of-mass, known as the **barycentre**. Due to the massive size of the Earth in comparison to the Moon (Table 3.1), the location of the barycentre is actually inside the Earth, approximately 4,700 km out from its centre (Figure 3.1). The time taken for one rotation of the Earth–Moon system around its barycentre is 27.32 days – one **sidereal month**. The orbital path that the Moon traces during one rotation has a radius equal to the distance between its own centre and the barycentre, on average some 379,700 km. It can be seen in Figure 3.1 that this path is many times larger than the circumference of the Earth, and that is why it appears (incorrectly) to us that the Moon is revolving around the Earth's centre-of-mass. During one rotation of the

**Table 3.1** Properties of the Earth, Moon and Sun relevant to the derivation of the tide-generating force

|  | Mass, tonnes | Radius, km | Average distance from Earth, km |
|---|---|---|---|
| Earth | $5.97 \times 10^{21}$ | 6,378 | 0 |
| Moon | $7.35 \times 10^{19}$ | 1,738 | 384,400 |
| Sun | $1.99 \times 10^{27}$ | 696,000 | 149,600,000 |

**Figure 3.1** – The Earth–Moon system rotates on its common centre-of-mass, known as the barycentre. The barycentre (solid dot) is located approximately 4,700 km out from the Earth's own centre of mass (open dot). View is looking down upon the Earth's north pole, hence the sense of rotation is anticlockwise.

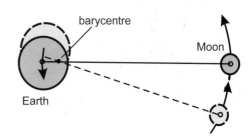

Earth–Moon system, the Earth will similarly trace an orbital path around the system's barycentre. The Earth's path has a radius also equal to the distance between its centre and the system's barycentre, some 4,700 km. The Earth's path is quite small and we are largely unaware of this motion relative to the barycentre. To someone standing on the Moon, the motion would appear as a small wobble in the Earth's position. As the Earth–Moon system revolves around the Sun, it is the system's barycentre that follows a smooth orbital path, while the individual paths of the Earth and Moon waver.

You should recall from Newton's Laws of Motion that a force can be defined as the product of a body's mass and acceleration. It follows then that acceleration can be defined as the ratio of force to mass, with the acceleration taking place in the direction of the force. We have a situation where the Earth and Moon are orbiting their barycentre. The acceleration every particle of mass on the Earth (and Moon) experiences in order to maintain its orbital motion is provided by a force. The necessary force in this case is the gravitational force. We call the gravitational force specifically required to keep the Earth–Moon system rotating in a stable manner the **centripetal force** $F_c$. It follows from Equation 3.1 that the centripetal force for the Earth–Moon system is

$$F_c = G\frac{m_E m_M}{R^2}$$

(3.2)

where $m_E$ and $m_M$ are the masses of the Earth and Moon, respectively. Since every particle of mass on the Earth follows the same orbit, and therefore experiences the same acceleration, the centripetal force must be acting with the same magnitude and direction on every particle. This is easily demonstrated by looking at the right-hand-side of Equation 3.2. Notice that none of the parameters would change in value from one location to another either on or inside the Earth. The direction in which the centripetal force acts is always parallel to the plane of rotation of the Earth–Moon system (Figure 3.2a).

The local **gravitational force**, on the other hand, does depend on location. It follows from Equation 3.1 that the local gravitational force $F_{lg}$ is

$$F_{lg} = G\frac{m_E m_M}{(R \pm r)^2}$$

(3.3)

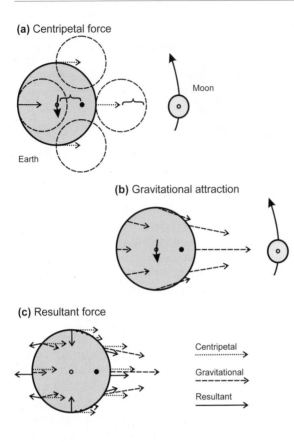

**(a)** Centripetal force

Earth

Moon

**(b)** Gravitational attraction

**(c)** Resultant force

Centripetal

Gravitational

Resultant

**Figure 3.2** – The view in all three panels is looking down upon the Earth's north pole. (a) The dashed circles trace the paths of positions on the Earth's surface as the Earth–Moon system completes one full rotation. The radius of the circles is equal to the distance between the Earth's own centre-of-mass and the barycentre (solid dot). The magnitude and direction of the centripetal force is indicated by the length and direction of the dotted arrows. (b) The magnitude and direction of the gravitational attraction is indicated by the length and direction of the dashed arrows. (c) The resultant tide-generating force (vector addition of the centripetal and local gravitational forces) is indicated by the length and direction of the solid arrows.

where $r$ is the distance between the Earth's centre and the point of interest on the Earth's surface ($-r$ for points on the Moon-facing side of the Earth and $+r$ for points on the opposite side). Locations on the Moon-facing side of the Earth will experience a local gravitational force that is larger than that experienced at locations on the opposite side, due to their closer proximity to the Moon (Equation 3.3). The direction in which the local gravitational force acts is towards the centre of the Moon's mass, which in most cases is at an angle to the centripetal force (Figure 3.2b).

In order for the rotation of the Earth–Moon system to remain stable, the average centripetal force per unit of mass must equal the average gravitational force. If these forces were not equal then the Earth and Moon would either accelerate towards each other or away from each other. Let us now move from considering the total system to considering the forces acting locally on the Earth's surface. It should be evident from Equations 3.2 and 3.3 that the centripetal force per unit mass is the same everywhere on or inside the Earth, but the gravitational force per unit mass varies locally. If $F_c$ and $F_{lg}$ were equal everywhere then there would be no tide-generating force. It is the local differences in magnitude of $F_c$ and $F_{lg}$ that are ultimately responsible for the tides. The ocean on the Moon-facing side of the Earth experiences a small acceleration due to the fact that $F_{lg} > F_c$, whereas on the other side of the Earth it experiences a small acceleration due to the fact that $F_c > F_{lg}$. Since the ocean mass is accelerated a force must be involved. This is the **tide-generating force** $F_t$ and it is the resultant vector addition of $F_c$ and $F_{lg}$

$$F_t = \frac{(\pm r)2Gm_Em_M}{R^3} \qquad (3.4)$$

On the Moon-facing side of the Earth, the resultant tide-generating force is positive, which means that it is directed towards the Moon (Figure 3.2c). Conversely, on the other side of the Earth it is negative and thus directed away from the Moon.

At this stage it is tempting to draw the conclusion that the local variation in $F_t$ shown in Figure 3.2c causes the ocean water on the Earth's surface to be drawn towards two points on opposite sides of the Earth. This is not completely correct, because it ignores the fact that $F_t$ is only a very small fraction of the Earth's own attractive force acting upon the ocean, which is every where directed towards the centre of the Earth. To demonstrate this fact let us consider a unit mass (1 kg) of ocean water on the Earth's surface located directly beneath the Moon. Using Equation 3.1 and the data in Table 3.1 we can calculate the Earth's gravitational attraction $F_g$ on this water mass to be approximately 9.68 N (note that in this case $m_1$ is the Earth's mass, $m_2$ is 1 kg and $R$ is the radius of the Earth). From Equation 3.4 the lunar tide-generating force acting on the water mass is approximately $1.09 \times 10^{-6}$ N. We have now determined that the Earth's gravitational attraction acting on the ocean is a factor $10^7$ larger than the lunar tide-generating force.

It turns out that it is actually the vector component of the tide-generating force that is tangential to the Earth's surface that draws the ocean into two bulges on either side of the Earth (Figure 3.3). This tangential vector component is also small in magnitude, but is not directly opposed by the Earth's attractive force and so is comparable to other forces acting in the ocean. The tangential component of the tide-generating force is known as the **tractive force** $F_T$ and is

$$F_T = \frac{3m_Mr^3}{2m_ER^3} g \sin2\theta \qquad (3.5)$$

where $\theta$ is the angle between the point of interest on the Earth's surface and the line joining the centres of the Earth and Moon (Figure 3.3). The tractive force is smallest for $\theta$ approaching $0°$ and $90°$, which is where the Earth's gravitational attraction directly opposes the lunar tide-generating force, and largest for $\theta = 45°$ where it does not.

To summarize up to this point, in order for the Earth–Moon system to be in stable rotation around its barycentre, the average centripetal force per unit mass in the system must equal the average gravitational attractive force between the two bodies. Because the distance to the Moon varies with location on the Earth's surface there are local differences between the centripetal

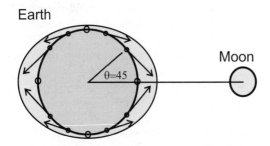

Earth

Moon

$\theta=45$

**Figure 3.3** – The magnitude and direction of the tractive force responsible for causing two tidal bulges on opposite sides of the Earth that are aligned with the Moon. The tractive force is maximum at an angle of $\theta$ = 45° and decreases towards zero as $\theta$ approaches both 0° and 90°.

## Box 3.1 – Derivation of a Pressure-Gradient Force

The hydrostatic pressure $P$ at any given depth is the weight of the overlying water acting on a unit area, which can be written as

$$P = \rho g h$$

where $\rho$ is the water density, $g$ is the gravitational acceleration and $h$ is the height of water above the depth of interest. A non-zero water surface gradient (i.e., a sloping water surface) is effectively the same as a pressure gradient. Consider the sloping water surface in Figure 3.4 and assume that the water density is uniform. The hydrostatic pressure at Site A and Site B is, respectively

$$P_A = \rho g h \quad \text{and} \quad P_B = \rho g (h + \Delta h)$$

The pressure gradient between the two sites is then

$$\frac{\Delta P}{\Delta x} = \frac{P_B - P_A}{\Delta x} = \rho g \frac{\Delta h}{\Delta x} = \rho g \, tan\theta$$

where $\theta$ is the slope of the water surface. We have thus derived the horizontal pressure-gradient force per unit volume of water (i.e., $\rho g tan\theta$). It is this pressure-gradient force that opposes the tractive force (Equation 3.5), and thus maintains a steady sloping water surface that is the tidal bulge in the Equilibrium Theory. It is also this pressure gradient that drives coastal tidal currents, as water is forced to flow horizontally from areas of high to low pressure (Section 7.3).

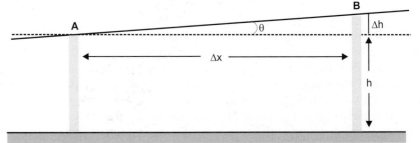

**Figure 3.4** – Schematic diagram of a sloping water surface, which defines the terms used to derive the pressure-gradient force. [Modified from Open University, 1993.]

and gravitational forces. The vector sum of these two forces acting on the ocean is the tide-generating force. The small tangential component of the tide-generating force (*i.e.*, the tractive force), which is largely un-opposed by the gravitational attraction of the Earth, causes water to move around the Earth to form two 'tidal bulges' on opposite sides of the Earth and aligned with the Moon. The sloping water surface of these bulges is maintained through a balance between the tractive force and a horizontal **pressure–gradient force** (Box 3.1).

## 3.3 THE EQUILIBRIUM THEORY OF TIDES

The three principal assumptions of the **Equilibrium Theory of tides** are: (1) the Earth is covered entirely by an ocean of uniform depth (*i.e.*, there are no continental land masses); (2) there is no inertia in the system (*i.e.*, the oceans respond immediately to the tide-generating force); and (3) the Coriolis and friction effects can be neglected. The basis of the Equilibrium Theory is the tidal bulges shown in Figure 3.3. If there is no inertia in the system then these bulges will follow the Moon around the Earth.

### 3.3.1 Earth rotation and the semi-diurnal tide

In the previous section we discussed the Earth and Moon rotating as a system around their common centre-of-mass or barycentre. In addition to that rotation, the Earth also rotates (or spins) in an anticlockwise direction on the axis through its poles (Figure 3.5). The time it takes to spin through one rotation is a **solar day** (24 hours). Consider an observer at Point A on the Earth's surface. The observer would experience two high tides separated by

**Figure 3.5** – The view in the top panel is looking down upon the Earth's north pole. This panel shows the Earth completing one rotation on its polar axis while the Moon (and tidal bulges) moves along an arc around the barycentre during the same time period. The numbers on the Earth indicate the time in hours required for Point A on the Earth's surface to rotate around to that location. The bottom panel shows the tidal water level observed at Point A as the Earth rotates beneath the tidal bulges. The solid line shows the tide record if the Moon did not change its position relative to the Earth, and the dashed line shows the true tide record.

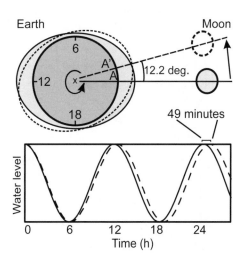

two low tides as the Earth spins through one rotation beneath the two tidal bulges. While the Earth is rotating on its polar axis the Moon is also moving in the same direction around the barycentre. During one rotation of the Earth on its polar axis the Moon has advanced along its path by about 12.2 degrees, thus shifting the location of the tidal bulges by the same amount. It therefore takes a little longer than a solar day for Point A to reach the first bulge again (now located at A'). In fact it takes an additional 49 minutes. That is why the times of high and low water are approximately an hour later than the times that they occurred on the previous day.

## 3.3.2 Lunar declination and the diurnal inequality (tropical and equatorial tides)

Up until now we have assumed that the Moon is aligned with the Earth's equator. In fact the orbital plane of the Earth–Moon system is tilted up to an angle of 28.5° to the Earth's equatorial plane (Figure 3.6). This tilt is known as the **lunar declination**. As the Moon revolves around the barycentre on the tilted orbital plane, its position above the Earth varies between latitudes 28.5° North and 28.5° South. When the Moon is above either 28.5° North or South, the tidal bulges are tilted relative to the equatorial plane. When the Moon is positioned above the equator, the bulges are in line with the equatorial plane.

As the Earth spins on its polar axis, an observer at Point A in Figure 3.6 would experience a diurnal (daily) variation in the magnitude of high and low tide. Initially Observer A experiences a high-high tide, but when the observer reaches the tidal bulge facing away from the Moon a low-high tide

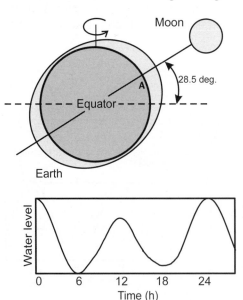

**Figure 3.6** – The view in the top panel is looking side-on to the Earth from a position level with the Equator. Notice that rather than showing the orbital plane of the Earth–Moon system (solid line connected to Moon) as horizontal, the equatorial plane (dashed line) is horizontal. The angle of declination shown between the two planes is 28.5°. The bottom panel shows the tide record that would be observed as Point A rotates beneath the tidal bulges when they are at maximum declination.

occurs. When the observer returns to the starting point approximately one day later, a high-high tide occurs again. Similarly, diurnal variation in the low tides also occurs, but this is more difficult to visualize in the figure because they occur when Observer A moves into and out of the page as the Earth spins. The maximum diurnal variation or inequality in the tide range occurs at those times during the month when the Moon is positioned roughly over the tropics (either Capricorn or Cancer), hence tides that display a diurnal inequality are referred to as **tropical tides**. There is minimal diurnal variation in the tide during those times of the month when the Moon is positioned over the Equator, hence tides that display virtually no diurnal inequality are referred to as **equatorial tides**. We often observe two periods

**Table 3.2** Definition of astronomical terms and their relationship to the ocean tide

| Terms | Related tide behaviour |
| --- | --- |
| **Anomalistic month** is 27.6 days and is the period associated with the elliptical orbit of the Moon <br> **Apogee** is the position in the Moon's orbit that is furthest from the Earth <br> **Perigee** is the position in the Moon's orbit that is closest to the Earth | Variation in the Moon's distance to the Earth causes the tides to be larger at perigee and smaller at apogee |
| **Anomalistic year** is 366.5 days and is the period associated with the elliptical orbit of the Earth–Moon system around the Sun <br> **Aphelion** is the position in the Earth–Moon's orbit that is furthest from the Sun <br> **Perihelion** is the position in the Earth–Moon's orbit that is closest to the Sun | Variation in the Sun's distance from the Earth causes the tides to be larger at perihelion and smaller at aphelion |
| **Sidereal month** is 27.3 days and is the time required for the Earth–Moon system to complete one orbit around their **barycentre** | Lunar declination causes the two tides each day to alternate between being equal (equatorial) and unequal (tropical) twice over the month |
| **Sidereal year** is 365.3 days and is the time required for the Earth–Moon system to complete one orbit around the Sun | Solar declination causes diurnal inequality in the tides to be most pronounced in June and December |
| **Synodic month** is 29.5 days and is the time between successive conjunctions of the Earth, Moon and Sun | Relative positions of the Moon and Sun cause larger (smaller) than average spring (neap) tides to occur twice a month |

of tropical and two periods of equatorial tides in a sidereal month (for reference purposes many of the astronomical terms used in this chapter are listed in Table 3.2).

### 3.3.3 Solar-lunar interaction and the spring-neap tide cycle

The same line of reasoning presented in Section 3.2 to explain the tide-generating force of the Moon is equally applicable to understanding the tide-generating force of the Sun. If we make the appropriate substitutions from Table 3.1 into Equation 3.4 we find that the tide-generating force of the Sun is about 46% that of the Moon. This is due to the fact that the magnitude of the tide-generating force varies inversely with the distance cubed, so the close proximity of the Moon more than compensates for the enormous mass of the Sun. Nevertheless, the Sun has a noticeable moderating and amplifying effect on the tide.

Figure 3.7 shows the relative magnitudes of the lunar and solar tidal bulges and their positions for different phases of the Moon. When the Moon is new it is located in a line between the Earth and Sun, and it is barely visible from Earth (Figure 3.7a). This is because the Earth-facing side of the

**(a)** Spring Tides (syzygy)

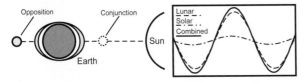

**(b)** Neap Tides (quadrature)

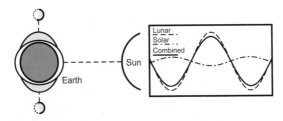

**(c)** Tide record with corresponding phases of the moon

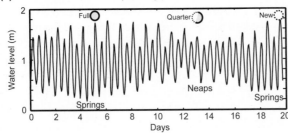

**Figure 3.7** – The view in the top two panels is looking down upon the Earth's north pole. (a) When the Moon is in conjunction (new Moon phase) or opposition (full Moon phase) with the Sun the lunar (grey) and solar (white) tidal bulges are aligned. The panel on the right shows the tide record resulting from each of the lunar and solar bulges as the Earth rotates beneath them. (b) When the Moon is in quadrature (quarter Moon phase) the lunar and solar tidal bulges are at 90° to one another. (c) A tide record covering nearly one and a half spring-neap tide cycles, demonstrating the relationship with the phases of the Moon.

Moon is not being illuminated directly by the Sun – it is being illuminated only by sunlight reflected back from Earth. When the Moon is full it is located on the opposite side of the Earth from the Sun and it is being illuminated directly by the Sun. When the Earth, Moon and Sun are all aligned, during either a full or new moon, the three bodies are said to be in **syzygy**. When the Moon is at a right-angle to the Earth with respect to the Sun, the portion of the Moon that is illuminated, when viewed from the Earth, appears to be one-quarter of the Moon's surface. In this arrangement the Earth, Moon and Sun are said to be in **quadrature** (Figure 3.7b).

When the Sun and Moon are in syzygy their respective tidal bulges are aligned. This yields a combined tidal bulge that is the sum of the individual contributors. The panel on the right-hand-side of Figure 3.7a shows that the corresponding lunar- and solar-tide records are in phase (crests and troughs are aligned), thus the respective tidal bulges constructively interfere with each other. The result is an amplified combined-tide range, in which the high tide is higher and the low tide is lower than the lunar-tide alone. When the Sun and Moon are in quadrature their respective bulges are at right-angles to each other (Figure 3.7b). It is important to realize that this does not mean there are four tides a day as the Earth rotates beneath the individual bulges. The tidal bulges do not behave independently, rather they are summed together. During quadrature the lunar- and solar-tide records are out of phase, thus the respective tidal bulges destructively interfere with each other to produce a combined-tide range that is reduced.

In summary, tides during syzygy (either conjunction or opposition) are largest and are called **spring tides**, whereas tides during quadrature are smallest and are called **neap tides**. One entire sequence of lunar phases (*i.e.,* new Moon, quadrature, full Moon, quadrature and back to new Moon) takes 29.5 days – a **synodic month**. A set of spring-neap tides therefore occurs approximately every 15 days (Figure 3.7c).

### 3.3.4 Elliptical orbits and their effect on the tide

Until now we have been considering a circular path for the Earth and Moon about their barycentre. It makes little difference in the case of the Earth, because the diameter of the path is so small, but a more realistic elliptical path for the Moon is shown in Figure 3.8a. The Moon is closest to the Earth at **perigee** (357,000 km) and furthest from the Earth at **apogee** (407,000 km). We therefore expect from Equation 3.4 that the tide-generating force will be larger at lunar perigee than at lunar apogee. This variation in the Moon's orbit has a period of 27.6 days – an **anomalistic month**. For roughly half this time, the lunar tide-generating force will be larger than average and for the other half it will be smaller than average. As a result, one set of spring-neap tides is usually larger than the other in any given month. There will be one occasion during the year when the new Moon and one when the full Moon coincides with lunar perigee. These occasions occur in March and September, which is when maximum spring tides occur.

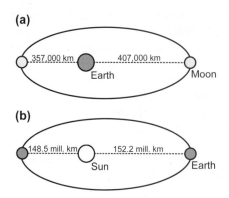

**Figure 3.8** – The elliptical orbit of (a) the Moon around the Earth–Moon barycentre and (b) the Earth around the Sun. Maximum and minimum distances from the Earth are also shown. [Modified from Thurman and Burton, 2001.]

The orbit of the Earth–Moon system around the Sun is also elliptical (Figure 3.8b). The Earth is closest to the Sun at **perihelion** (148,500,000 km) and furthest from the Sun at **aphelion** (152,200,000 km). This variation in the orbit of the Earth–Moon system has a period of 366.5 days – an **anomalistic year**. For roughly half of this time the solar tide-generating force will be larger than average and for the other half it will be smaller than average. Thus the tides are marginally larger in the 6 months of the year centred on January and marginally smaller in the 6 months centred on July.

### 3.3.5 Solar declination and the seasonal inequality of the tide

The plane of the Earth–Moon system's orbit around the Sun is tilted at an angle of 23.5° to the Earth's equatorial plane. The effect of this **solar declination** is similar to that described for the lunar declination (Figure 3.5). The Sun's position over the Earth varies between the Tropic of Cancer (latitude 23.5° North) and the Tropic of Capricorn (latitude 23.5° South) and back again in the course of 365.25 days – **sidereal year**. It is over the tropics during the solstices (21 June and 21 December) and over the equator during the equinoxes (21 March and 21 September). During the solstices, the solar tidal bulge will add a small amount to the diurnal inequalities in tide range produced by the lunar declination, whereas during the equinoxes it will not. The solar declination has a more profound effect on the tide when we consider the interaction of the solar and lunar declinations through time. There is a precession of the lunar declination with respect to the solar declination, which produces an 18.6-year periodicity in the tides (see Thurman and Burton, 2001 for details).

### 3.3.6 Shortcomings of the Equilibrium Theory

While it appears from our preceding discussion that the Equilibrium Theory can explain most of the observed features of the tide there are four significant shortcomings: (1) the predicted tide range is typically smaller than the observed range; (2) the predicted tidal range is not constant, but varies with location around the globe; (3) the timing of high water is generally several hours before or after the time of transit of the Sun and Moon; and (4) the timing of spring and neap tides does not always coincide with syzygy or quadrature, but is typically a day or more different. The first two shortcomings suggest that there is some sort of preferential response to the tide-generating force, determined by local characteristics, and the second two suggest that the assumptions of zero inertia and friction are too restrictive. For these reasons the Equilibrium Theory cannot be used in any precise way to predict the tide at a given location. The fact that the theory can explain most of the periodicities in the tide indicates that it adequately accounts for the causative astronomical processes, but we now need to consider more carefully the Earth-based factors that influence tide behaviour.

## 3.4 THE DYNAMIC THEORY OF TIDES

The basic premise of the **Dynamic Theory of tides** is that the two tidal bulges discussed in Section 3.3 actually behave as waves. Since the bulges are on opposite sides of the Earth, the wavelength is equal to half the Earth's circumference, or some 20,000 km. Because the wavelength is very large compared with the water depth, the tidal bulges behave as **shallow-water waves** or **long waves** (you will learn more about the properties of long waves in Section 4.4). Another example of long waves are tsunami, which are sometimes incorrectly referred to as tidal waves (Box 3.2). In the Dynamic Theory there is a long wave associated with each contributor to the tide-generating force (principally the Moon and Sun), just as there was a tidal bulge associated with each contributor in the Equilibrium Theory. Since the Earth is spinning in an anticlockwise direction, the relative movement of the long waves is from east to west around the globe. The tide-generating force is always present, thus the long wave is always being driven by it. We refer to such waves as **forced waves**.

Because the tide is a long wave forced largely by the Moon we might expect the crest of the tide to always be located beneath the Moon, just like the tidal bulges in the Equilibrium Theory. For this to occur, the crest of the long wave must be capable of travelling at the same linear velocity that the Earth's surface is spinning relative to the Moon's position. The linear velocity of the Earth's surface is at a maximum of 449 m s$^{-1}$ at the equator and diminishes to zero at the poles (Figure 3.9). The velocity of a long wave is equal to $\sqrt{gh}$ (Equation 4.10), where $g$ is the gravitational acceleration (9.81 m s$^{-2}$) and $h$ is the water depth. For an average water depth in the Earth's oceans of $c.$ 4,000 m, the long wave velocity is approximately 198 m s$^{-1}$. This means that the forced long wave that is the tide can only remain in equilibrium with the lunar forcing at latitudes of 65° or higher. At lower latitudes the long wave cannot travel fast enough to keep up with the lunar-forcing. The limited ocean depth (which restricts the long wave velocity) and the

## BOX 3.2 – WHEN A TIDAL WAVE IS NOT A TIDAL WAVE

It is common in the popular press for a **tsunami** to also be referred to as a tidal wave. It is true that they are both examples of shallow-water waves, but they differ from each other in many respects, most importantly in the way that they are generated. Tide waves are generated by astronomical forces involving the Earth, Moon and Sun. A tsunami can be generated by any of the following three mechanisms:

■ A displacement of the sea bed by a submarine earthquake (e.g., 1998 Papua New Guinea earthquake produced a tsunami that killed more than 2,000 people).

■ An impulse generated by a large landslide into the ocean, perhaps during a volcanic eruption (e.g., 1883 Krakatau eruption produced a tsunami that killed more than 36,000 people).

■ An impulse generated by a meteorite impact striking the ocean. This generation mechanism has not occurred in historic time, but mega-tsunami events recorded in the geological record have been attributed to meteorite impact (See Bryant, 1997 for a comprehensive review).

In the open ocean, tsunami typically have a wavelength of a few hundred kilometres and a height less than a couple of metres. This results in very small water surface slopes so that they can travel across the ocean basin at speeds of around 800 km hour[-1] and hardly be noticed. When they cross the edge of a continental shelf, however, they rapidly begin to shoal, which means they slow down, become shorter in wavelength and larger in height (see Section 4.5.2 for a discussion of shoaling). By the time tsunami reach the coastline they can be several tens of metres high and in a period of only a couple of hours they can cause enormous property damage and loss of life. Tsunami and tide waves are distinctly different phenomena and the terms should not be confused.

---

fact that the ocean is not continuous, but broken up into deep basins separated by shallow shelves and continental land masses, means that the forced long waves are effectively broken up into smaller systems. These systems are called amphidromes.

An idealized representation of how **amphidromes** operate is shown in Figure 3.10. The forced long wave rotates around the central point of the

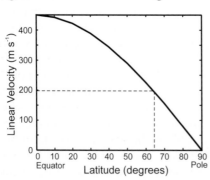

Figure 3.9 – Graph showing the linear velocity of the Earth's surface as a function of latitude. The dashed line indicates the speed of a shallow-water wave in a water depth of 4,000 m (typical ocean depth).

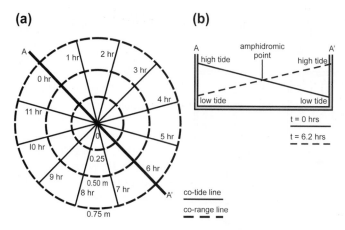

**Figure 3.10** – (a) Plan view of an idealized amphidromic system in the Southern Hemisphere. Clockwise rotation is indicated by co-tide lines, which mark arrival times of the wave crest (high tide) as it rotates around the system. The co-range lines show that minimum and maximum tide range occurs at the centre and outer edge of the system, respectively. (b) Cross-section through the amphidrome along the line A–A', and showing the system to be confined by solid boundaries representing the edge of the ocean basin. When the wave crest (high tide) is on one side of the basin the wave trough (low tide) is on the opposite side of the basin. [Modified from Pinet, 2000.]

amphidrome (node), achieving one circuit in a time period that corresponds with the astronomical forcing. The sense of rotation is clockwise in the Southern Hemisphere and anticlockwise in the Northern Hemisphere. The sense of rotation can be determined from the **co-tide lines**, which show the location of the wave crest at successive times. When the wave crest (high tide) is at A, the wave trough (low tide) is on the directly opposite side of the amphidrome at A'. The minimum tide range is zero at the centre of the amphidrome and the tide range achieves its maximum at the outer edge as indicated by the **co-range lines**.

In order to understand how amphidromic systems form in a fully enclosed ocean basin, recall that the tidal long wave is forced and travels from east to west. This causes sea level to be elevated against the western margin of the basin (Figure 3.11). The resultant sloping surface produces a pressure-gradient force that causes the water to flow eastward (to understand why refer to Box 3.1). For an ocean basin in the Southern Hemisphere, the eastward flow is deflected to the left (north) by the coriolis force. This ultimately causes the water level to become elevated against the northern margin of the basin. Again the resultant sloping surface produces a pressure gradient force that drives water southward, but which is ultimately deflected to the east and so on. In this way a wave crest (and trough located on the opposite side) rotates around the ocean basin – clockwise in the Southern Hemisphere and anticlockwise in the Northern Hemisphere. This wave is called a **Kelvin wave** named after its discoverer, Lord Kelvin.

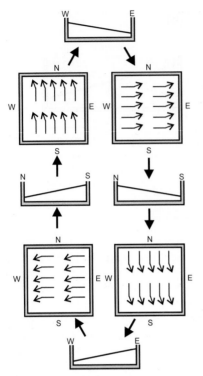

**Figure 3.11** – Schematic diagram showing how an amphidromic system develops a clockwise rotating Kelvin wave in the Southern Hemisphere. The letters N, S, E and W denote the compass directions north, south, east and west, respectively. The top panel shows an east-west cross-section of the ocean basin, the next panel to the right shows the basin in plan view, and so on. The sloping line in the cross-section view is the water surface and the arrows in the plan-view are the direction of water flow. See text for further discussion.

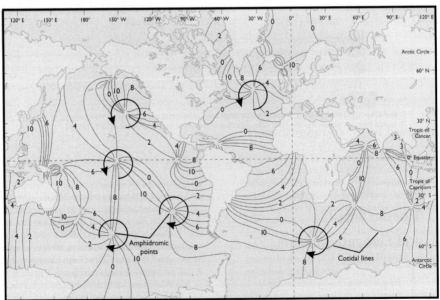

**Figure 3.12** – Amphidromic systems in the world's oceans identified by co-tidal lines. Notice that the sense of rotation is clockwise and anticlockwise for the Southern and Northern Hemispheres, respectively. [From Pinet, 2000; modified from Cartwright, 1969.] [Copyright © 2000 Jones and Bartlett Publishers, reproduced with permission.]

Figure 3.13 – Map showing amphidromic systems in the North Sea. Tides along the North Sea coast are a result of the complex interaction of three Kelvin waves with different heights (tide ranges). At a particular coastal location the observed tide range depends on which is the influential Kelvin wave and how far it is away from the amphidromic point. [From Pethick, 1984.] [Copyright © 1984 Edward Arnold, reproduced with permission.]

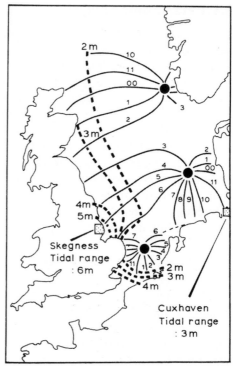

The major amphidromic systems of the Earth's oceans are shown in Figure 3.12. Many of the systems in the larger ocean basins are similar in character to the idealized system shown in Figure 3.10a. The similarity is less obvious for the smaller, convoluted basins that have an open connection to the larger basins. In reality the ocean tides are a complex interaction between co-oscillating forced Kelvin waves (amphidromic systems) occupying adjacent ocean basins, with an additional involvement from reflected forced- and free-waves that can be amplified to varying degrees by complex basin topography (Figure 3.13). By now you should have a clear appreciation of just how idealized a representation the Equilibrium and Dynamic Theories provide of what is an exceedingly complex situation.

## 3.5 PREDICTING TIDES

The periodicities of the tide-generating forces are obtainable from the Equilibrium Theory and they can be used in solving Laplace's equations (Dynamic Theory) to forecast realistic tide records for locations well away from the coastal margins of ocean basins. The accuracy of this approach diminishes markedly, however, as the coastline is approached. Since it is the coastal zone that we are primarily concerned with, we need to put theory aside and adopt an empirical approach that uses past tidal measurements in order to predict future tides.

## 3.5.1 Harmonic analysis

The fact that real tide records are highly periodic means that they can be modelled as the summation of several **partial tides**. This is illustrated graphically in Figure 3.14 where five partial tides are shown together with the combined tide. The latter is the numerical sum of the five partial tides shown. Unlike the real tide, partial tides have a single amplitude and frequency. There is some physical justification for these somewhat artificial partial tides in that their frequency can be linked to physical processes. For example, the tidal effect produced by the Earth–Moon system rotating on its barycentre (Figure 3.5) is represented by a partial tide with a frequency of 0.081 hour$^{-1}$ (*i.e.*, corresponding to the fundamental semi-diurnal tide). The partial tides are **harmonics** because the periodic motion of the water surface elevation is expressed in terms of a cosine function that includes time. A measured tide record can be expressed mathematically as

$$\eta = {}^{-}\eta + \sum_{i=1}^{n} a_i \cos(2\pi f_i t - G_i) + \eta_r(t) \qquad (3.6)$$

where $\eta$ is the tidal water level, $t$ is time, ${}^{-}\eta$ is the time-averaged water level (or mean sea level), and $a_i$, $f_i$, and $G_i$ are the amplitude, frequency and phase of the $i$-th partial tide, respectively. The last term $\eta_r$ is the residual water level, which we will discuss more fully in Section 3.5.2.

Equation 3.6 states that the measured tide record for a given locality can be broken down into a number of partial tides of specified amplitude, frequency and phase. In order to predict future tides at a given location using Equation 3.6 it is necessary to know the amplitudes, frequencies and phases of the partial tides. The frequencies of the partial tides are fixed, and can be obtained from the Equilibrium and Dynamic theories. The amplitudes and phases vary with location around the globe and must be obtained

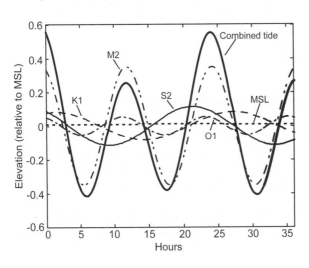

Figure 3.14 — Graph showing five partial tides with differing amplitudes, frequencies and phases that, when summed together, produce the combined tide (bold line). Brief descriptions of the partial tides are listed in Table 3.3.

**Table 3.3** List of the principal partial tides [After Pugh, 1987]

| Partial tide | Frequency, $h^{-1}$, period in h | Description |
|---|---|---|
| M2 | 0.081 (12.420) | Principal lunar |
| S2 | 0.083 (12.000) | Principal solar |
| N2 | 0.079 (12.658) | Elliptical lunar |
| K2 | 0.084 (11.976) | Lunar/solar declinations |
| K1 | 0.042 (23.928) | Principal lunar/solar |
| O1 | 0.039 (25.824) | Principal lunar |
| P1 | 0.042 (24.072) | Principal solar |

Note: The number in the symbol for each partial tide indicates roughly how many cycles occur each day. Diurnal partial tides always contain the number 1 and semi-diurnal partial tides contain the number 2.

empirically, *i.e.*, from measurements. Tidal prediction by the method of harmonic analysis typically involves a least-squares fitting of the principal partial tides to an existing tide record for the locality in order to obtain their amplitudes and phases. The amplitudes, frequencies and phases are then substituted into Equation 3.6 to forecast the tide into the future.

There are seven fundamental partial tides (Table 3.3), directly related to the astronomical tide-generating force, that will in most cases forecast the tide to within 10% of its true value for the coming month. The more complicated the shelf morphology at a particular location and the further into the future one wishes to forecast, the larger the number of partial tides required in the analysis. Often 20–30 partial tides are necessary to predict the future tide at coastal ports. This involves analysing a historic tidal record length spanning one month. In complicated situations such as estuaries up to 60 partial tides are used, which requires a record length spanning one year. The most accurate tidal predictions are based on partial tides calculated from historical records spanning 19 years.

## 3.5.2 The residual tide: Radiational tides and storm surge

It is often the case when predicting tides using harmonic analysis that there will be a small, but significant difference between the predicted and observed tidal records. This difference is referred to as the **residual tide** and is represented by the last term on the right-hand-side of Equation 3.6. The residual tide can display both periodic and random fluctuations in water level. In some cases the periodic fluctuations in the residual tide are due to an insufficient number of partial tides employed in the prediction. In other cases it is due to factors unrelated to the tide-generating force. One such factor is **radiational forcing**, which is where regular weather cycles cause the atmospheric pressure to fluctuate, thus loading and unloading the ocean surface. As a general rule-of-thumb sea level will rise 1 cm for every 1

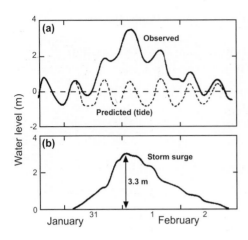

**Figure 3.15** – (a) Graph showing the predicted (dashed line) tidal water level and observed (solid line) water level at the Hook of Holland in January–February, 1953. (b) Graph showing that the residual tide, which is attributed to storm surge, reached 3.3 m. [Modified from Wemelsfelder, 1953.]

millibar fall in air pressure and *vice versa* – the so called '**inverse barometer effect**'. Diurnal and annual periodicities are commonly observed in the residual tide due to radiational forcing.

Non-periodic fluctuations in the residual tide are typically associated with the effect of wind on sea level. If the wind vector is directed shoreward it can pond water against the coast, causing an increase in the residual tide. The magnitude of the effect is related to the fetch-size, wind-strength and duration, so not surprisingly, the effect is strongest during storms, hence the term '**storm surge**' (Figure 3.15). The strongest winds are often associated with low-pressure systems, thus storm surge combines with the inverse barometer effect to produce a residual tide that at times can be responsible for widespread coastal inundation.

## 3.6 CLASSIFICATION OF TIDES

A simple scheme was introduced in Chapter 1 that classified tidal environments according to the spring tide range (Figure 1.4). It is now possible to introduce a more complex classification based on tidal period. Examples of tidal records from four coastal ports are shown in Figure 3.16. The two end-members of the classification scheme are the **semi-diurnal** and **diurnal tides**. On a continuum between these end-members lie the **mixed tides**, which display both diurnal and semidiurnal periodicities. Tidal records can be classified quantitatively using the **tidal form factor** $F$

$$F = \frac{a_{K1} + a_{O1}}{a_{M2} + a_{S2}} \tag{3.7}$$

where $a$ is the amplitude of the partial tides identified by the subscripts (Defant, 1958). The tidal form factor is effectively the ratio of the amplitudes of the major diurnal and semi-diurnal partial tides. The ranges of $F$ associated with each tidal type are:

**Figure 3.16** – Monthly tidal records from four coastal ports showing examples of semi-diurnal, mixed and diurnal tides. The tidal form factor *F* is defined in Equation 3.7. [Modified from Defant, 1958.]

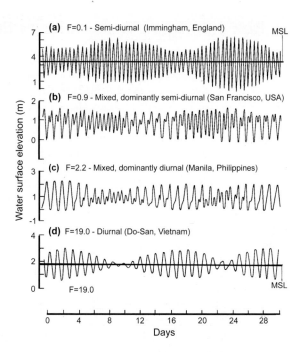

- $F = 0.00–0.25$   Semi-diurnal tide
- $F = 0.25–1.50$   Mixed, dominantly semi-diurnal tide
- $F = 1.50–3.00$   Mixed, dominantly diurnal tide
- $F > 3.00$          Diurnal tide

The type of tide depends on the local predominance of semi-diurnal versus diurnal partial tides, but how does this predominance arise? Box 3.3 demonstrates how a tide of given wavelength responds to ocean basins of variable length. There is a maximum amplification of the tide for basin lengths that correspond to integer multiples of a quarter of a tidal wavelength. It turns out that the major ocean basins of the world have a dimension that preferentially amplifies the semi-diurnal partial tides, which is why it is globally the most common tide. Occasionally, however, the morphology of continental shelves and coastal bays locally amplifies the diurnal partial tides.

## Box 3.3 – Tidal Resonance in Enclosed Basins

We have noted that most locations around the world experience a predominantly semi-diurnal tide, while some experience a predominantly diurnal tide (Figure 3.16a and 3.16d). The reason for this is that most ocean basins have dimensions that are so well matched to the wavelength of either the semi-diurnal or diurnal partial tides that they resonate.

Figure 3.17a shows a situation where the basin length $\lambda$ is half the tidal wavelength $L$. Or to put it another way, one half of one wavelength fits into the basin. The partial tide is of course forced, and upon encountering the basin margin it is reflected. The incident (solid curve) and reflected (dash-dot curve) wave interact to produce a standing wave. In Figure 3.17a the single node of the standing wave is at the centre of the basin and two antinodes occur at the basin margins. Notice that there is no tide range at the node, and the tide range at the antinodes is twice the range that it would be if there was no resonance.

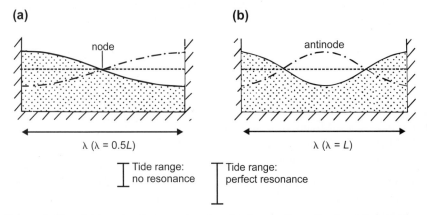

**(a)**    **(b)**

node    antinode

$\lambda$ ($\lambda = 0.5L$)    $\lambda$ ($\lambda = L$)

Tide range: no resonance    Tide range: perfect resonance

**Figure 3.17** – Schematic diagram showing the interaction of a forced wave (e.g., partial tide) and a free reflected wave, which leads to a resonant standing wave. Two different basin lengths are shown: (a) basin length = half a tidal wavelength; and (b) basin length = one tidal wavelength. [Modified from Pond and Pickard, 1983.]

In Figure 3.17b we have a similar situation, except this time the wavelength and basin length are equal. In this case there will be two nodes and three antinodes. Again the tide range at the margins of the basin will be twice that expected in the absence of resonance. Perfect resonance will occur with accentuated tides at the margins so long as the basin length is an integer-multiple of one-quarter of a tidal wavelength. The example shown here has been simplified considerably, but the concept applies equally to rotating Kelvin waves. It so happens that most of the ocean basins favour resonance of the semi-diurnal partial tides, which is why the semi-diurnal and the mixed-predominantly semi-diurnal tidal types are the most common. Nevertheless there are some ocean basins that resonate the diurnal partial tides, thus producing diurnal or mixed-predominantly diurnal tides.

# 3.7 SUMMARY

- The magnitude of the tide-generating force exerted by the Moon and Sun is directly related to their mass and inversely related to their distance from the Earth. The tide-generating force of the Sun is approximately 46% that of the Moon.
- The Equilibrium Theory is based on the premise that the tangential component of the tide-generating force (*i.e.,* the tractive force) causes the water to move around the Earth to form two tidal bulges on opposite sides of the Earth and aligned with the Moon and Sun. The Equilibrium Tide is the resulting rise and fall of the water level as the Earth spins on its polar axis beneath the bulges.
- The Equilibrium Theory adequately explains the various periods observed in the ocean tides, but does not account for lags between the occurrence of high tide and the transits of the Sun and Moon, nor does it account for variation in the tidal range observed across ocean basins.
- The Dynamic Theory is based on the premise that the tides are Kelvin waves rotating around amphidromic systems whose dimensions correspond to the size of ocean basins. The Dynamic Tide is the resulting rise and fall of the water as the crest and trough of the Kelvin wave pass by. The tidal period (time required for one rotation around the amphidrome) is determined by the size of the ocean basin, which favours the resonance of either diurnal or semi-diurnal partial tides.
- The Dynamic Theory adequately explains the lag between high tide and lunar/solar transit times, as well as the observed variation in tidal range across the ocean basins. However, the theory cannot deal with the effect of complex topography that is characteristic of coastal waters.
- In order to predict tides in coastal waters an empirical approach is required. The approach employed is harmonic analysis, which uses past tidal records from a location of interest to determine the amplitude and phase of partial tides. The frequencies of the partial tides are obtained from the Equilibrium Theory. Once the amplitude, frequency and phase of the partial tides have been determined for a location, the modelled partial tides are summed together to forecast the total tide into the future.

# 3.8 FURTHER READING

Cartwright, D.E., 1999. *Tides: A Scientific History.* Cambridge University Press. [A fascinating account of the historical developments leading to modern tidal theory and observation.]

Komar, P.D. 1998., *Beach Processes and Sedimentation.* 2nd Edition, Prentice Hall. [Chapter 4 of this text presents a very readable account of tides that goes into more depth than the level presented here.]

Pond, S. and Pickard, G.L., 1995. *Introductory Dynamical Oceanography*. 2nd Revised Edition, Butterworth-Heinemann Ltd. [A thorough presentation pitched at students with a junior-intermediate-level physics background.]

Pugh, D.T., 1987. *Tides, Surges and Mean Sea-Level*. John Wiley and Sons. [Simply the best book around on tidal processes, but certainly not an introductory text.]

Thurman, H.V. and Burton, E.A., 2001. *Introductory Oceanography*. 9th Edition, Prentice Hall. [A useful introductory text including a chapter on tides.]

# 4

# WAVES

## 4.1 INTRODUCTION

Along most coastlines, waves represent the dominant source of energy in the nearshore zone. For this reason, the global distribution of wave environments discussed in Chapter 1 (Figure 1.3) provides a useful first-order approach to classifying coastal environments. Part of the incoming wave energy is reflected at the shoreline and is propagated back to the open sea, very much the way light bounces off a mirror. Most of the incoming wave energy, however, is transformed to generate nearshore currents and sediment transport, and is ultimately the driving force behind morphological change. A sound knowledge of the dynamics of waves is therefore fundamentally important to understanding coastal morphology.

Ocean waves can be classified in a number of useful ways. One way to do this is by period $T$, or, what comes to the same thing, by frequency $f$, which

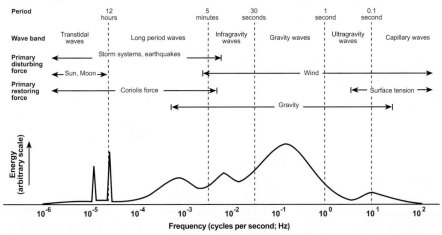

Figure 4.1 – Schematic representation of the energy contained in the surface waves of the oceans. [Modified from Kinsman, 1984.]

is the reciprocal of the period ($f = 1/T$). Alternatively, we can classify waves by the disturbing force that generates them, or by the restoring force that dampens the wave motion. Figure 4.1 shows a schematic wave spectrum, which plots wave energy as a function of frequency, with different types of ocean waves indicated. The type of wave that forms the subject of this chapter is the gravity wave, which is the most energetic of the wave types. **Gravity waves** have periods of 1–30 seconds and frequencies of 0.033–1 cycles per second (or Hertz). They are generated by wind and their main restoring force is gravity.

To achieve a solid understanding of wave processes and appreciate their role in shaping the coast, a certain amount of mathematical treatment is required. Therefore this chapter has a relatively strong theoretical flavour, perhaps at a level which may appear quite daunting at first sight. It should be borne in mind, however, that most of the equations look more complicated than they are and can be applied with relative ease. A large number of new terms and concepts will be introduced in this chapter. To help the reader along, Table 4.1 may serve as a useful reference while reading this chapter.

**Table 4.1** Overview of wave processes ($h$ = water depth, $L_o$ = deep water wave length, $H_o$ = deep water wave height)

| Process | Description of process | Types of waves |
|---|---|---|
| **Deep water: $h/L_o$ > 0.5 (waves are unaffected by sea bed):** | | |
| Wave generation | Waves are generated by wind. Wave height and period increase with increase in wind speed and duration | Sea |
| Wave dispersion | In deep water long-period waves travel faster than short-period waves. This results in a narrowing of the wave spectrum | Swell |
| **Intermediate and shallow water: $h/L_o$ < 0.5 (waves are affected by sea bed):** | | |
| Wave shoaling | The wave length shortens with decreasing water depth. This results in a concentration of the wave energy over a shorter distance and an increase in the wave height | Shoaling waves |
| Wave asymmetry | Shoaling waves become increasingly asymmetrical and develop peaked crests and flat troughs | Asymmetric waves |
| Wave refraction | Wave refraction causes wave crests to become more aligned with the coast, resulting in a decrease in the wave angle | Refracted waves |

| Wave diffraction | Leakage of wave energy along the wave crests into shadow areas | Diffracted waves |
|---|---|---|
| **Surf zone: $h < 1\ H_o$ (regular waves) or $h < 2\ H_o$ (irregular waves):** | | |
| Wave breaking | When the horizontal velocities of the water particles in the wave crest exceed the wave velocity, the water particles leave the wave form, and the wave breaks | Spilling, plunging, surging breakers |
| Energy dissipation | As breaking/broken waves propagate through the surf zone, they progressively decrease in height as a result of wave energy dissipation | Surf zone bores |
| Wave reflection | On steep beaches a significant part of he incoming wave energy is not dissipated in the surf zone, but is reflected back to sea | Standing waves |
| Infragravity wave energy | A significant part of the wave energy in the surf zone is at very low (infragravity) frequencies. Infragravity waves are particularly energetic during storms | Infragravity waves, long waves, edge waves |
| **Swash zone: $h = 0$:** | | |
| Runup | The maximum water level attained on a beach is higher than the still water level. This vertical displacement of the water level is known as runup and consists of a steady component (wave set-up) and a fluctuating component (swash) | Wave runup, swash, wave set-up |

## 4.2 CHARACTERISTICS AND ANALYSIS OF NATURAL WAVES

It is important at the outset to make the distinction between regular (or monochromatic) and irregular (or random) waves. This is because theoretical discussions on wave processes most commonly assume regular wave motion, whereas natural waves are characteristically irregular. The motion of **regular waves** is periodic, that is, the motion is repetitive over space or through fixed periods of time such as shown in Figure 4.2. Regular waves can be described in terms of a single representative wave height $H$, wave length $L$ and wave period $T$. The **wave height** is the difference in elevation between the crest and the trough of the wave, the **wave length** is the

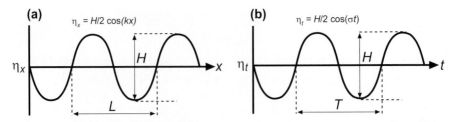

**Figure 4.2** – Schematic showing a regular wave train. (a) The spatial variation in water level $\eta_x$ is measured at a single moment in time along the direction of wave travel. From such data the wave length $L$ can be derived. (b) The temporal variation in water level $\eta_t$ is measured at a single location in space over a representative time period. Such data enable the determination of the wave period $T$. The wave height $H$ can be derived from both types of wave data.

distance between successive crests (or troughs) and the **wave period** is the time it takes for the wave to travel a distance equal to its wave length. An important characteristic of natural waves is that they are highly irregular and that a range of wave heights and periods are present (Figure 4.3a). To properly describe the wave conditions of **irregular waves** in quantitative terms, statistical techniques are required.

**Wave-by-wave analysis** is one of the most common approaches to describe irregular waves. This technique consists of identifying the individual waves in the wave record and determining representative wave parameters from the sub-set of wave heights and periods (Figure 4.3b, c). The following wave parameters are most commonly used to describe an irregular wave field:

- $H_s$ (or $H_{1/3}$) – The **significant wave height** represents the average wave height of one-third of the highest waves in a wave record. The significant wave height approximately corresponds to visual estimates of wave heights and has been found by coastal engineers to be particularly useful for practical design purposes.
- $H_{rms}$ – The **root-mean-square wave height** is obtained by taking the square root of the mean squared wave height, using all the waves in the wave record. As a rule of thumb, $H_s = 1.41H_{rms}$.
- $T_z$ – The **mean wave period** is the mean period of all the waves in the wave record. It is also referred to as the zero-crossing wave period.
- $T_s$ (or $T_{1/3}$) – The **significant wave period** is the mean wave period of one-third of the highest waves in the wave record.

An alternative method to describe the properties of irregular waves is **spectral analysis**. This method can be used to identify the dominant wave frequencies (or periods) present in the wave record. Spectral analysis produces a **wave spectrum** (Figure 4.3d) which is a plot of wave energy versus wave frequency. The maximum wave energy in the spectrum is referred to as the spectral peak and the wave period associated with this peak is the

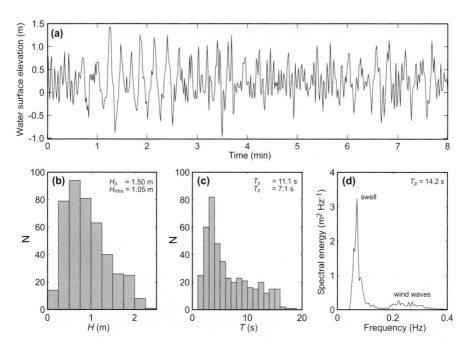

**Figure 4.3** – Analysis of wave data collected in 48-m water depth off the coast of Perth, Western Australia: (a) time series of 8 minutes of water surface elevation data; (b) frequency distribution of the wave heights in the data record; (c) frequency distribution of the wave periods in the data record; and (d) wave spectrum. The wave data represent a mixture of short-period wind waves and long-period swell waves, accounting for the very irregular nature of the time series, the wide distribution of wave periods and the bi-modal wave spectrum.

**peak spectral period** $T_p$. The wave spectrum is also useful for partitioning wave energy over distinct frequency bands. For example, the wave spectrum shown in Figure 4.3d indicates the presence of two distinct wave fields: swell with a peak spectral period of 14 s ($f \approx 0.07$ Hz) and wind waves with periods ranging from 3–5 s ($f \approx 0.25$ Hz). An irregular wave field is often made up of more than one wave source and it is quite common for wave spectra to be characterized by more than one peak. This brings into question the use of a single parameter to describe the wave period of the wave field.

Typical significant wave heights and periods along swell-dominated coastlines are 1–2 m and c. 10 s, respectively. Along protected sea environments locally-generated wind waves dominate, and wave heights and periods are generally 0.5–1 m and c. 4 s, respectively. During storms, wave heights can be substantially larger and waves exceeding heights of 10 m are routinely reported from offshore oil rigs and vessels (Box 4.1).

# BOX 4.1 – EXTREME HEIGHT OF NATURAL WAVES

Extremely high waves are, not surprisingly, generated by strong winds that blow over long periods and large distances. The largest wave ever reliably reported had an estimated height of 34 m and was encountered on 7 February 1933 by the U.S.S. *Ramapo*, a 146-m long naval tanker on route from Manila to San Diego (Bascom, 1980). Wave heights in excess of 15 m do occur, but they are not common, and even in the most severe storms, wave heights are generally smaller. According to Kinsman (1984), 45% of ocean waves are less than 1.2 m high, 80% are less than 4 m high, and only 10% are greater than 7 m. Figure 4.4 shows a map of the global distribution in significant wave height. The most energetic wave regions are the southern Indian and Pacific Oceans and the northern Atlantic and Indian Oceans, where a wave height of 5 m is exceeded 10% of the time.

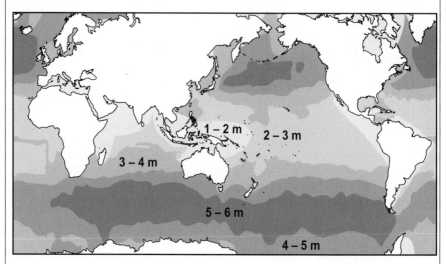

**Figure 4.4** – Global values for significant wave height which will be exceeded 10% of the time. Based on Geosat altimetry. [From Short, 1999; modified from Young and Holland, 1996.] [Copyright © 1999 John Wiley & Sons, reproduced with permission.]

Extreme wave conditions may also occur in certain locations due to constructive interference involving different wave fields. For example, wave refraction around an island results in two wave trains wrapping around the island. There will be a narrowly-defined region in the lee of the island where the two wave trains meet and where their addition results in relatively rough wave conditions with large waves. Large singular waves, sometimes referred to as 'king waves', 'freak waves' or 'rogue waves' also occur and these waves are often held responsible for sweeping sea fishermen off rock ledges. These extreme waves develop when two (or more) large waves from different wave fields (e.g., from different directions) coincide, resulting in a new wave with a much increased height. The occurrence of such waves is virtually unpredictable and catches many unaware.

# 4.3 WAVE FORMATION AND PREDICTION

It is obvious even to a casual observer that waves are generated by wind acting on a water surface. In physical terms, the formation of waves constitutes a transfer of energy from wind to waves. The generally accepted theory to account for the growth of wind waves is the combined Miles-Phillips mechanism. The mechanism incorporates two distinct phases of energy transfer from wind to waves, namely an initial slow growth phase that accounts for the formation of waves on a calm water surface, and an ongoing rapid growth of the waves due the interactive coupling between wind and waves. In a developing wave field, wave height and wave period steadily increase.

Wave growth is not infinite and in the open ocean is limited by the ratio between the wave height and the wave length. This ratio is the **wave steepness** $H/L$ and increases progressively during wave formation. When waves reach their limiting steepness ($H/L \approx 1/7$ in deep water) they break in the form of white caps. An equilibrium can eventually be achieved whereby the energy losses by wave breaking are balanced by the addition of new energy being transferred from the wind to the waves. Such an equilibrium wave field is referred to as a **fully arisen sea**. Its development may take several days or hundreds of kilometres of wave travel under strong wind conditions, but a fully arisen sea may develop relatively quickly (less than 10 hours) under mild wind conditions.

The characteristics of locally-generated waves can be predicted from a knowledge of the existing wind conditions (it is less straightforward to predict the characteristics of waves from a distant source). This technique is known as **wave forecasting** when predicted wind data are used, or **wave hindcasting** when historical wind data are used. The wind parameters used for wave prediction are the wind velocity $U$, the duration $D$ that the wind blows and the distance $F$ over which the wind blows. The latter parameter is known as the **fetch** or **fetch length**. Wind duration and fetch length may restrict the time during which individual waves are moving under the action of the wind. They therefore govern the time during which energy can be transferred from wind to waves. For the wave field to develop into a fully arisen sea, duration and fetch will have to be relatively long and large, respectively. If the fetch is too short, for example in the case of a small lake or a sheltered coastal environment, the wave conditions are said to be **fetch-limited**.

The simplest method available for wave prediction is the significant wave approach. This method has acquired the abbreviated name *S-B-M* method after the names of its developers (Sverdrup and Munk, 1946; Bretschneider, 1952). Using the *S-B-M* method, the significant wave height $H_s$ and peak spectral period $T_p$ of a wave field can be predicted from $U$, $D$ and $F$ using a nomogram (Figure 4.5). For example, for a wind speed of 10 m s$^{-1}$, unlimited fetch and a wind duration of 10 hours, the resulting significant

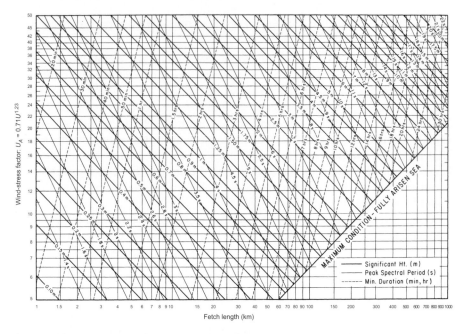

**Figure 4.5** – Nomogram to predict significant wave height $H_s$ and peak spectral period $T_p$ from wind speed $U$, wind duration $D$ and fetch $F$. Note that before using the diagram, the measured or estimated wind speed $U$ has to be converted to a wind-stress factor $U_A$ given by $U_A = 0.71U^{1.23}$. The diagonal line at the right-hand side of the nomogram indicates fully arisen sea conditions. [From CERC, 1984.] [Reproduced with permission from U.S. Army Corps of Engineers.]

wave height and peak spectral period are 2.2 m and 7.3 s, respectively. For the same wind speed, unlimited wind duration and a fetch length of 20 km, the resulting significant wave height and peak spectral period are 0.87 m and 3.9 s, respectively.

We have already seen that the wave spectrum provides a more comprehensive description of a wave field than a single measure of wave height and period. Therefore, a more sophisticated method to predict wave characteristics from wind conditions is the wave spectrum approach, which characterises the growth of a wave field by its spectrum. One of the most commonly used spectral formulations is the **JONSWAP spectrum** based on extensive measurements conducted in the North Sea. Wave spectra derived from a developing wave field indicate a progressive increase in the wave energy level and period (Figure 4.6). In addition, the wave field becomes more narrow-banded, meaning that the range of wave frequencies (periods) present in the wave field decreases. In other words, the waves become more organized and regular, the longer the wind blows.

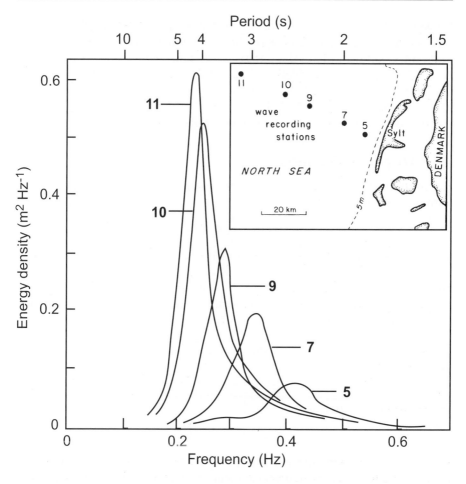

**Figure 4.6** – A series of wave spectra measured at recording stations offshore from the island of Sylt on the North Sea coast of Denmark during the JONSWAP experiments (Hasselmann et al., 1976). Offshore winds prevailed during the measurements, resulting in the growth of spectral energy from stations 5 through to 11 and an increase in the peak spectral period. [Modified from Komar, 1998.]

## 4.4 LINEAR WAVE THEORY

**Wave theories** are mathematical formulations that predict the change in wave properties, such as water particle velocity, wave height and wave energy, with depth. They allow the derivation of nearshore wave characteristics from offshore wave conditions and are therefore an invaluable tool in coastal engineering applications. A large number of wave theories have been developed with varying levels of complexity and accuracy. Generally speaking, the shallower the water, the more complex the applicable theory becomes. The simplest wave theory available is **linear wave theory**. This theory is also known as **Airy wave theory**, named after the British mathematician George Biddell Airy (1801–1892), and is the most widely used wave theory. Despite its relative simplicity, linear wave theory adequately describes important wave processes such as shoaling and refraction. Since these latter processes are of fundamental importance to nearshore processes, linear wave theory is discussed in some detail here.

## 4.4.1 Dispersion equation

According to linear wave theory, the fluctuation of water surface elevation with time $\eta(x,t)$ is described by

$$\eta(x,t) = \frac{H}{2}\cos(kx - \sigma t) \tag{4.1}$$

where $x$ is the coordinate axis in the direction of wave propagation, $t$ is time, $H$ is the wave height, $k = 2\pi/L$ is the wave number ($L$ is the wave length) and $\sigma = 2\pi/T$ is the wave radian or angular frequency ($T$ is the wave period). Linear wave theory considers natural waves to be simple sinusoids such as shown in the Figure 4.2.

An important relationship derived from linear wave theory is the **dispersion equation**, which expresses the relation between wave length and wave period

$$\sigma^2 = gk\,\tanh(kh) \tag{4.2}$$

which can be rewritten as

$$L = \frac{g}{2\pi} T^2 \tanh\left(\frac{2\pi h}{L}\right) \tag{4.3}$$

where $h$ is the water depth and $g$ is the gravitational acceleration. The tanh term is referred to as the hyperbolic tangent. Over a single wave period $T$, a wave travels a distance equal to its wavelength $L$. The speed at which the wave travels, the **wave phase velocity** $C$, is therefore given by the ratio $L/T$ according to

$$C = \frac{L}{T} = \frac{g}{2\pi} T \tanh\left(\frac{2\pi h}{L}\right) \tag{4.4}$$

The wave phase velocity is also referred to as the **wave celerity**.

The dispersion equation can not be solved directly because it contains $L$ on either side of the equation and therefore has to be solved using an iterative process. However, Hunt (1979) derived a convenient, albeit slightly daunting, function which is accurate to 0.1% and is given by

$$(kh)^2 = y^2 + \frac{y}{1 + 0.666\,y + 0.355\,y^2 + 0.161\,y^3 + 0.0632\,y^4 + 0.0218\,y^5 + 0.00654\,y^6} \tag{4.5}$$

where

$$y = \frac{\sigma^2 h}{g} = \frac{4\pi^2 h}{gT^2} = 4.03\,\frac{h}{T^2} \tag{4.6}$$

Using Equations 4.5 and 4.6, the wave number $k$ can be determined for any given water depth $h$ and period $T$, and the wave length $L$ can then be derived using $L = 2\pi/k$.

## 4.4.2 Deep and shallow water approximations

The use of computers makes computations involving the general dispersion relationship (Equations 4.2 and 4.3) relatively straightforward, especially if Equations 4.5 and 4.6 are used. In the past, however, simpler approximations to the dispersion relationship were desirable and these were made possible due to the characteristics of the tanh function shown in Figure 4.7. If $kh = 2\pi h/L$ becomes large (in relatively deep water), $\tanh(kh) \approx 1$ and therefore Equations 4.3 and 4.4 can be reduced to

$$L_o = \frac{gT^2}{2\pi} \tag{4.7}$$

and

$$C_o = \frac{gT}{2\pi} \tag{4.8}$$

Equations 4.7 and 4.8 are known as the **deep water approximations** and are valid for $kh > \pi$ (or $h/L_o > 0.5$, *i.e.*, the water depth is greater than twice the wave length). The subscript 'o' is used here to denote deep water conditions. If, on the other hand, $kh = 2\pi h/L$ approaches zero (in relatively shallow water), $\tanh(kh) \approx kh$ and Equations 4.3 and 4.4 reduce to

$$L_s = T\sqrt{gh} \tag{4.9}$$

and

$$C_s = \sqrt{gh} \tag{4.10}$$

Equations 4.9 and 4.10 are known as the **shallow water approximations**, hence the use of the subscript 's', and are valid for $kh < 0.1\,\pi$ (or $h/L_o < 0.05$, *i.e.*, the water depth is less than 1/20th of the wave length).

**Figure 4.7** – Properties of the tanh function. See text for explanation.

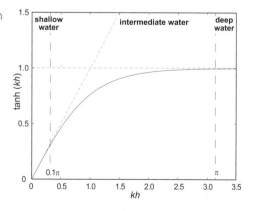

Intermediate depth conditions prevail for $0.05 < h/L_o < 0.5$ and here the general wave equations (Equations 4.3 and 4.4) must be used.

The deep and shallow water approximations are straightforward to apply to determine various characteristics of waves in deep and shallow water. For example, consider a wave with a period of 10 seconds. In deep water, the wave length and wave velocity are only dependent on the wave period and application of Equations 4.7 and 4.8 gives $L_o = 156$ m and $C_o = 15.6$ m s$^{-1}$. In shallow water, the water depth also needs to be considered. For a water depth of 1 m and a wave period of 10 s, Equations 4.9 and 4.10 yield $L_s = 31$ m and $C_s = 3.1$ m s$^{-1}$. The above example highlights a fundamental characteristic of waves – waves get closer together and slow down as they propagate from deep to shallow water.

### 4.4.3 Wave orbital motion

The distinction between deep, intermediate and shallow water is not simply an artefact of the characteristics of linear wave theory and in turn the tanh function, but has a physical explanation that becomes apparent when the **water particle motion** under waves is considered. As waves propagate across the sea surface, the water particles beneath undergo an almost closed circular path. The water particles move forward under the crest of the wave and move seaward under the trough of the wave. Because the water particle velocities decrease with depth, the forward velocity at the top of the orbit is slightly greater than the seaward velocity at the bottom. Consequently, there is a net drift of water in the direction of wave travel, known as **Stokes drift**. For our purpose, we can ignore Stokes drift and focus on how the characteristics of the water particle motion vary with water depth according to linear wave theory (Figure 4.8).

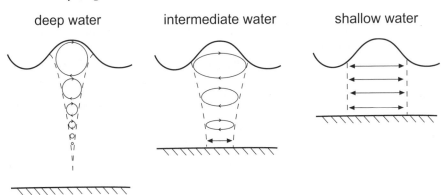

deep water          intermediate water          shallow water

**Figure 4.8** – Motion of water particles under waves according to linear wave theory. In deep water, water particles follow a circular motion with the radius of the orbits decreasing with depth beneath the surface. In intermediate water, the orbits are elliptical and become flatter as the sea bed is approached. In shallow water, all water motion consists of horizontal movements to-and-fro which are uniform with depth.

In deep water the water particles under the wave move in a circular path with the diameter of the circles decreasing with increasing depth beneath the water surface according to

$$d = He^{-kz} = He^{-2\pi z/L} \tag{4.11}$$

where $d$ is the diameter of the circular orbit and $z$ is the depth beneath the water surface. The maximum velocity of the rotating water particle $u_m$ also decreases with increasing depth beneath the water surface and is given by

$$u_m = \frac{\pi H}{T} e^{-kz} = \frac{\pi H}{T} e^{-2\pi z/L} \tag{4.12}$$

According to Equations 4.11 and 4.12, at depths greater than half the wave length ($h/L > 0.5$; the deep water region), the orbital diameter and velocities are less than 5% of those at the sea surface. Therefore, it can be inferred that the wave motion in deep water is hardly experienced at the sea bed, and conversely, deep water waves are not significantly affected by the presence of the sea bed.

In contrast, at intermediate water depths the wave motion extends to the sea bed and the surface waves 'feel' the presence of the sea bed. As a result, the water particles now follow an elliptical path with the ellipses becoming flatter and smaller as the sea bed is approached. At the sea bed, the water particles merely undergo a horizontal to-and-fro motion. The excursion of this to-and-fro motion at the bed $d_0$ is given by

$$d_0 = \frac{H}{\sinh(kh)} \tag{4.13}$$

and the associated maximum flow velocity $u_0$ is

$$u_0 = \frac{\pi}{T} d_0 = \frac{\pi H}{T \sinh(kh)} \tag{4.14}$$

where the sinh term is referred to as the hyperbolic sine.

In shallow water all water motion consists of horizontal movements to-and-fro which are uniform with depth. Because in shallow water $\sinh(kh) = kh$, Equations 4.13 and 4.14 reduce to

$$d_0 = \frac{H}{kh} = \frac{HT}{2\pi} \sqrt{\frac{g}{h}} \tag{4.15}$$

and

$$u_0 = \frac{\pi}{T} d_0 = \frac{H}{2} \sqrt{\frac{g}{h}} \tag{4.16}$$

As $kh$ always has values of 0.6 or less in shallow water, the excursion of the water motion will always be greater than the wave height.

### 4.4.4 Wave energy and wave energy flux

We have already seen that there is only limited net movement of water associated with wave motion because the water particles follow almost-closed orbits in deep and intermediate water, and undergo a to-and-fro motion in shallow water. However, the propagation of the wave form itself constitutes a transfer of energy over the sea surface. Waves have potential energy associated with the deformation of the water surface (*i.e.*, trough/crest) and kinetic energy due to the orbital motion of the water particles. Applying linear wave theory, the two forms of energy are equal and the total **wave energy** $E$ is given by

$$E = \frac{1}{8} \rho g H^2 \qquad (4.17)$$

where $\rho$ is the density of water. The wave energy is expressed as the amount of energy per unit area ($N\ m^{-2}$) and is more appropriately referred to as the wave energy density. Wave energy depends on the square of the wave height, therefore, a doubling of the wave height will result in a fourfold increase in wave energy. Because $E$ is the amount of energy per unit area, the total amount of energy associated with a long-period wave is greater than that of a short-period wave because the long-period wave has a larger wave length.

The rate at which wave energy is carried along by the moving waves, known as the **wave energy flux** $P$, is given by

$$P = EC_g \qquad (4.18)$$

where $C_g$ is the speed at which the wave energy is carried along and is known as the **wave group velocity**, because it represents the speed at which wave groups travel (Box 4.2). Wave groups do not travel at the same speed as individual waves and the wave group velocity $C_g$ is related to the speed of individual waves $C$ according to

$$C_g = Cn \qquad (4.19)$$

where $n$ is given by

$$n = \frac{1}{2}\left[ 1 + \frac{2kh}{\sinh(2kh)} \right] \qquad (4.20)$$

The parameter $n$ increases from 0.5 to 1 from deep to shallow water, in other words, deep water waves travel at twice the speed of the wave groups ($n = 0.5$), whereas shallow water waves propagate at the same speed as the wave groups ($n = 1$). Equation 4.18 can also be expressed as

$$P = ECn \qquad (4.21)$$

which is more commonly used than Equation 4.18.

# Box 4.2 – Formation of Wave Groups

If two sets of waves (or wave trains) with slightly different wave lengths (and frequencies) are present at the same time, they will interfere and produce a single set of resultant waves (Figure 4.9). Where the sets of waves are in phase (*i.e.*, crests and troughs of both wave trains coincide), the wave amplitudes are added, and the resultant wave has twice the amplitude of the two original waves. Where the two wave trains are out of phase (*i.e.*, crests of one wave train coincides with the troughs of the other, and *vice versa*), the amplitudes cancel out and the water surface displacement is minimal. The two wave trains thus interact, each losing its individual identity, and combine to form a series of **wave groups**, separated by regions almost free of waves.

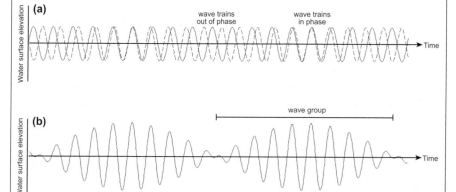

**Figure 4.9** – The merging of two wave trains of slightly different wave lengths, but the same amplitude, to form wave groups. [Modified from Open University, 1994.]

The velocity of the wave group can be derived from the properties of the two sets of waves that generated the group (Open University, 1994). The wave group speed $C_g$ is given by

$$C_g = \frac{\sigma_1 - \sigma_2}{k_1 - k_2}$$

where $\sigma_1 - \sigma_2$ is the difference between the radian frequencies of the two wave trains and $k_1 - k_2$ is the difference between the two wave numbers. The equations in Section 4.4.1 demonstrate that both $T$ and $L$, and hence $\sigma$ and $k$, can be expressed in terms of the wave speed $C$. Using the deep water wave equations and some manipulation, $C_g$ can be expressed in terms of the respective speeds, $C_1$ and $C_2$ of the wave trains according to

$$C_g = \frac{C_1 C_2}{C_1 + C_2}$$

If $C_1$ is nearly equal to $C_2$ then the simple result $C_g = C^2/2C = C/2$ is obtained, where $C$ is the average speed of the wave trains. In other words, the wave group travels at half the speed of the individual waves.

## 4.5 WAVE PROCESSES OUTSIDE THE SURF ZONE

A number of wave processes take place outside the surf zone, including wave dispersion, wave shoaling, development of wave asymmetry, wave refraction and wave diffraction. These processes modify the characteristics of offshore waves during their travel across the ocean or sea and will be discussed below.

### 4.5.1 Wave dispersion

A developing wave field is characterized by a broad-banded wave spectrum, indicating that a wide range of wave periods is represented in the wave field. Such a wave field is referred to as **sea**. In deep water, wave velocity increases with wave period (Equation 4.8). In a broad-banded wave field, therefore, waves propagate at a range of wave velocities with the long-period waves travelling faster than the short-period waves. Given sufficient time, the long-period waves will outrun and leave behind the short-period waves. This sorting of the waves by period is termed **wave dispersion** and results in the transformation of broad-banded sea into a regular wave field known as **swell**. The longer the distance of travel from the area of wave generation, the more effective the wave sorting process and the narrower the wave spectrum becomes.

### 4.5.2 Wave shoaling

Wave motion in intermediate and shallow water depths extends to the sea bed. The presence of the sea bed is felt by the waves and this has a significant effect on wave motion. The change in wave characteristics from deep to intermediate and shallow water can be predicted reasonably well using linear wave theory. Some of the variations in wave properties are summarized in Figure 4.10. It is worth noting that wave period is the only wave property that remains constant as waves travel from deep to shallow water.

The changes in wave length, wave group velocity, wave velocity and $n$ that occur when waves propagate from deep to shallow water have a profound effect on the wave height. The variation in the height of the shoreward-propagating waves can be calculated from a consideration of the wave energy flux $P$. We assume hereby that energy losses due to bed friction can be ignored, therefore the wave energy flux $P = ECn$ remains constant during wave propagation. This can be expressed as

$$P = (ECn)_1 = (ECn)_2 = \text{constant} \tag{4.22}$$

where the subscripts '1' and '2' indicate two different locations along the path of wave travel ($h_1 > h_2$). Substituting the wave energy $E = 1/8\rho g H^2$ in Equation 4.22 and re-arranging the result yields

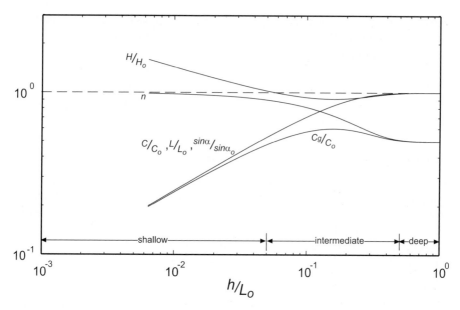

**Figure 4.10** – Shoaling transformations for linear waves as a function of the ratio of the water depth $h$ and the deep-water wave length $L_o$.

$$H_2 = \left(\frac{C_1 n_1}{C_2 n_2}\right)^{1/2} H_1 \qquad (4.23)$$

The ratio of the local wave height $H$ to the deep water wave height $H_o$ can now be derived from Equation 4.23

$$\frac{H}{H_o} = \left(\frac{1}{2n}\frac{C_o}{C}\right)^{1/2} = K_s \qquad (4.24)$$

where $K_s$ is referred to as the **wave shoaling coefficient**.

Equation 4.24 is plotted in Figure 4.10 and illustrates that the wave height initially decreases while entering intermediate water depth followed by a rapid increase. This increase in wave height is known as **wave shoaling** and is particularly pronounced just before wave breaking at the seaward edge of the surf zone. Due to the dependence of the wave velocity in deep water on the wave period (Equation 4.8), $K_s$ increases with wave period. Therefore, long-period waves shoal more than short-period waves.

Wave energy losses due to **bed friction** were ignored in the development of Equations 4.22–4.24, but may have a significant impact on the wave height development during shoaling. The two main factors that influence the amount of wave energy lost due to bed friction are the width and gradient of the continental shelf, because these control the distance over which shoaling, and hence frictional losses, occur. The roughness of the sea bed is important as well with rippled sea beds (Chapter 5) inducing greater energy losses than smooth beds. Friction-induced energy losses during shoaling are most pronounced over wide, low-gradient and rippled continental shelves. In these coastal settings, the decrease in wave height due to energy losses may exceed the increase in wave height due to shoaling, resulting in breaking wave conditions that are less energetic than deep water wave conditions. Along coastlines fronting steep and narrow continental shelves, wave energy losses due to bed friction are limited and here the breaker height is generally larger than the deep water wave height.

## 4.5.3 Development of wave asymmetry

In deep water, waves are characterized by a sinusoidal shape and the water particle velocities associated with the wave motion are symmetrical, meaning that the onshore velocities are of equal strength and duration as the offshore velocities. However, as the waves enter intermediate water, they become increasingly asymmetrical and develop peaked crests and flat troughs (Figure 4.11). The associated flow velocities also become asymmetric with the onshore stroke of the wave being stronger, but of shorter duration than the offshore stroke of the wave. Linear wave theory can no longer be used to accurately describe the dynamics of such waves and higher-order wave theories are necessary if information is required on the development of **wave asymmetry**. Examples of such wave theories are Stokes and Cnoidal wave theory (Komar, 1998).

Figure 4.11 – Asymmetric waves in shallow water characterized by peaked wave crests and broad wave troughs. [Photo M.G. Hughes.]

Figure 4.12 shows the wave profile and associated water velocities at the sea bed for a wave in intermediate water depth according to linear wave theory and the more advanced second-order **Stokes wave theory**. It is evident that the wave form and the associated velocity field of the Stokes wave significantly departs from that of the linear wave. Most significantly, the velocity field under the linear wave is symmetrical, whereas the maximum onshore water velocity under the Stokes wave is twice that of the maximum offshore water velocity. The development of wave asymmetry is very important from a sediment transport point of view because it promotes the onshore transport of sediment particles. This will be discussed in more detail in Chapter 8.

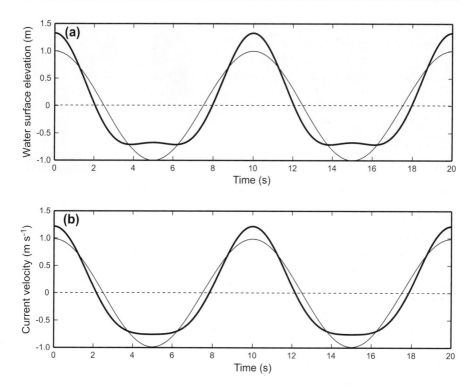

**Figure 4.12** – Comparison of the theoretical wave motion according to linear wave theory (thin line) and second-order Stokes wave theory (thick line): (a) wave form and (b) orbital velocity at the sea bed. The wave motion is calculated using $H = 2$ m, $T = 10$ s and $h = 8$ m. The linear wave describes a sinusoidal variation in the water level and flow velocities, whereas the Stokes wave is highly asymmetric and is characterized by a peaked crest and a flat trough.

## 4.5.4 Wave refraction

When a wave approaches the coast with its crest at an angle to the bottom contours, the water depth will vary along the wave crest. If the wave is in intermediate or shallow water, the wave velocity $C$ will also vary along the wave crest with the part of the wave in deeper water propagating at a faster rate than the part of the wave in shallower water (Equations 4.4 and 4.10). This results in a rotation of the wave crest with respect to the bottom contours, or in other words a bending of the wave rays. This process is known as **wave refraction** and is of great relevance to nearshore currents, sediment transport and coastal morphology. Two examples of wave refraction patterns are shown in Figure 4.13. The examples represent distinctly different coastal settings, but in both cases refraction causes the waves to become aligned more parallel to the shoreline as they propagate into shallow water.

**Figure 4.13** – (a) Aerial photograph showing the refraction pattern of waves approaching a straight beach under a small angle to the shoreline ($\alpha \approx 20°$). The waves clearly bend so that they become more closely parallel to the shoreline as they enter shallow water. [From Komar, 1998.] (b) Oblique photograph illustrating the wave refraction pattern of waves approaching a gravel barrier under a large angle with the shoreline ($\alpha \approx 45°$). The bending of the waves is not complete and as a result the angle between the shoreline and the crests of the waves remains relatively large, in particular along the right margin of the barrier. [From Orford et al., 1996; photo J.D. Orford] [Copyright © 1996 Coastal Education & Research Foundation, reproduced with permission.]

The change in wave direction by wave refraction is related to a change in wave velocity $C$ and can be described in a similar way to the bending of light rays according to Snell's law

$$\frac{\sin\alpha_1}{C_1} = \frac{\sin\alpha_2}{C_2} = \text{constant} \tag{4.25}$$

where $\alpha$ refers to the angle between the wave crest and the bottom contours and the subscripts '1' and '2' are used to indicate two different locations along the path of wave travel ($h_1 > h_2$) (Figure 4.14). For a straight coast with parallel bottom contours, the angle at a given depth can be related to the angle of wave approach in deep water $\alpha_o$ according to

$$\sin\alpha = \frac{C}{C_o}\sin\alpha_o \tag{4.26}$$

**Figure 4.14** – Schematic showing wave refraction. See text for explanation.

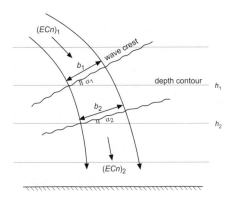

The expression shows that as the wave velocity decreases in shallow water, the angle between the wave crest and the bottom contour also decreases.

The refractive bending of the wave rays also causes the wave rays to spread out, *i.e.*, the distance between rays increases as the waves are being refracted (Figure 4.14). If $b$ is the spacing between wave rays, then the energy flux between the wave rays at two different depths should be constant so that

$$P = (ECnb)_1 = (ECnb)_2 = \text{constant} \tag{4.27}$$

Inserting $E = 1/8\rho g H^2$ in Equation 4.27 then yields

$$H_2 = \left(\frac{n_1 C_1}{n_2 C_2}\right)^{1/2} \left(\frac{b_1}{b_2}\right)^{1/2} H_1 \tag{4.28}$$

where

$$\frac{b_1}{b_2} = \frac{\cos\alpha_1}{\cos\alpha_2} \tag{4.29}$$

for a straight coast with parallel contours. The ratio of the local wave height $H$ to the deep water wave height $H_o$ can be derived from Equation 4.28

$$\frac{H}{H_o} = \left(\frac{1}{2n}\frac{C_o}{C}\right)^{1/2} \left(\frac{b_o}{b}\right)^{1/2} = K_s K_r \tag{4.30}$$

where $K_s$ is the shoaling coefficient discussed earlier (Equation 4.24) and $K_r$ is referred to as the **wave refraction coefficient**. The wave refraction process

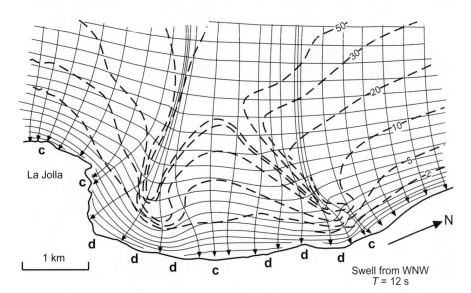

**Figure 4.15** – Wave refraction over submarine canyons and along the headland of La Jolla, California. The letters 'd' and 'c' refer to wave divergence and convergence, respectively. [Modified from Munk and Traylor, 1947.]

is not affected by the wave height, but it does depend on the wave period. Long-period waves feel the bottom earlier than short-period waves. They therefore refract more than short-period waves and become more aligned parallel to the coast.

Irregular bottom topography can cause waves to be refracted in a complex way and produce significant variations in wave height and energy along the coast. Spreading of the wave rays, referred to as **wave divergence**, occurs when waves propagate over a localized area of relatively deep water (*e.g.*, depression in the sea floor). Wave divergence is characterized by an increase in the spacing of the wave rays ($K_r < 1$) and this causes a reduction in the wave energy and wave height. Focusing of the wave rays is known as **wave convergence** and occurs when waves travel over a localized area of relatively shallow water (*e.g.*, shoal on the sea floor). The resulting decrease in the spacing of the wave rays ($K_r > 1$) causes an increase in the wave

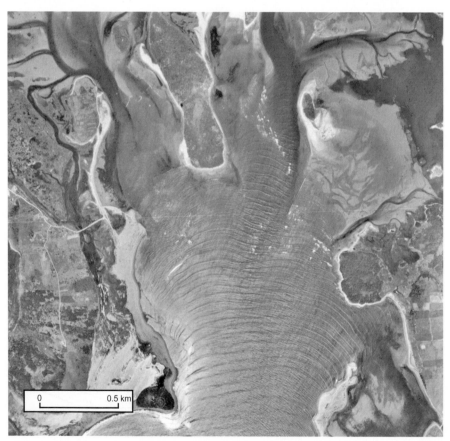

0    0.5 km

Figure 4.16 – Aerial photograph showing the combined refraction/diffraction pattern associated with waves entering an irregular bay. [From Orford *et al.*, 1996.] [Copyright © 1954 Her Majesty the Queen in Right of Canada (airphoto A14288-148). Reproduced with permission from Natural Resources Canada.]

energy and wave height. An example of wave refraction over complex bottom topography, including examples of both wave divergence and convergence, is given in Figure 4.15.

### 4.5.5 Wave diffraction

**Wave diffraction** is the process of wave energy transfer along the wave crest rather than in the direction of wave propagation, and occurs irrespective of water depth. Wave diffraction occurs when an otherwise regular train of waves encounters a feature such as an island, breakwater or offshore reef. A wave shadow zone is created behind the obstacle, but wave diffraction causes wave energy to spread into the shadow zone. Wave diffraction also enables wave energy to enter into narrowly-confined bays and harbours. Wave diffraction is a fundamentally different process from wave refraction, but both mechanisms often operate in concert. This is clearly illustrated in Figure 4.16, showing the combined refraction/diffraction pattern associated with waves entering an irregular bay. The wave crests in the bay are bent due to wave refraction and indicate that the fastest wave celerities and hence the greatest water depths are encountered in the centre of the bay. The margins of the bay are in the wave shadow zone, but wave diffraction results in some wave energy penetrating into these sheltered regions.

## 4.6 WAVE PROCESSES IN THE SURF AND SWASH ZONE

### 4.6.1 Wave breaking

At some point during wave shoaling, the water depth becomes too shallow for a stable wave form to exist and the wave will break. **Wave breaking** occurs when the horizontal velocities of the water particles in the wave crest exceed the velocity of the wave. Consequently, the water particles leave the wave form, resulting in a disintegration of the wave into bubbles and foam. Wave breaking is an important process, because when waves break, their energy is released and can result in the generation of nearshore currents and the transport of sediment. The reduction of wave energy due to wave breaking is referred to as **wave energy dissipation**.

The depth of water in which waves break is related to the height of the breaking wave according to

$$H_b = \gamma h_b \tag{4.31}$$

where $H_b$ is the breaker height, $h_b$ is the mean water depth at the point of breaking and $\gamma$ is referred to as the **breaker index** or **breaking criterion**. When the height-to-depth ratio of the wave exceeds the breaker index, breaking will occur. A useful first-order approximation for the breaker index is $\gamma = 0.78$.

**Spilling**

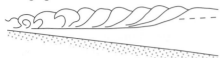

**Plunging**

**Surging**

**Figure 4.17** – The three main types of breakers: spilling, plunging and surging. [Modified from Galvin, 1968.]

A continuum of breaker shapes occurs in nature. However, three main breaker types are commonly recognized (Figure 4.17):

- **Spilling breakers** are associated with gentle beach gradients and steep incident waves (wave height large relative to the wave length). They are characterized by a gradual peaking of the wave until the crest becomes unstable, resulting in a gentle forward spilling of the crest.
- **Plunging breakers** tend to occur on steeper beaches than spilling breakers, with waves of intermediate steepness. They are distinguished by the shoreward face of the wave becoming vertical, curling over, and plunging forward and downward as an intact mass of water. Plunging breakers are the most desirable type of breaker for surfers, because they offer the fastest rides and produce 'tubes'.
- **Surging breakers** are found on steep beaches with low steepness waves. In surging breakers, the front face and crest of the wave remain relatively smooth and the wave slides directly up the beach without breaking. In surging breakers a large proportion of the incident wave energy is reflected at the beach.

Dimensionless parameters have been developed to predict the type of breaker and the most widely-used formulation is the **Iribarren Number** $\xi$ (Battjes, 1974)

$$\xi = \frac{\tan \beta}{\sqrt{H_b/L_o}} \tag{4.32}$$

where $\tan\beta$ is the gradient of the beach and the subscripts 'b' and 'o' indicate breaker and deep water conditions, respectively. Small values for $\xi$ are attained when the beach has a gentle gradient and the incident wave field is characterized by a large wave height and a short wave length (or a short wave period). Such conditions can be parameterized by $\xi < 0.4$ and promote the formation of spilling breakers. Large values of ξ are found when the beach is steep and the incident wave field is characterized by a small wave height and a long wave length (or a long wave period). Such conditions can be parameterized by ξ > 1 and favour the formation of surging breakers. Plunging breakers prevail when ξ = 0.4–1.

## 4.6.2 Wave reflection

Under certain conditions, waves do not break in shallow water, but reflect at the shoreline. The interaction of the reflected wave and the incoming wave results in the formation of a standing wave. **Wave reflection** occurs when ξ > 1 and is associated with surging breakers. Swell and wind waves only reflect off very steep beaches and vertical shorelines formed by sea walls and cliffs, but infragravity waves may even reflect off low-gradient sandy beaches (Box 4.3). Wave reflection is an important processes and standing wave motion has been implicated in the formation of nearshore bar morphology (Section 8.3.3).

There are some fundamental differences in the fluid motion between **progressive waves** (*i.e.*, waves that propagate) and **standing waves**. Figure 4.18a shows the successive stages or phases of a progressive wave as it travels from left to right. It is customary to indicate the different phases of the wave motion on a scale from 0 to 360° (or 0 to 2π radians). Thus, if half the wave cycle is completed, the phase of the wave is 180° (or π). In a progressive wave, the wave shape remains constant and travels a distance equal to one wave length over one wave period. In contrast, the wave shape associated with the standing wave, shown in Figure 4.18b, does not propagate and the water level simply moves up and down over one complete wave period. The excursion of the water level is maximum at the antinodes, while at the nodes the water level remains stationary at MSL.

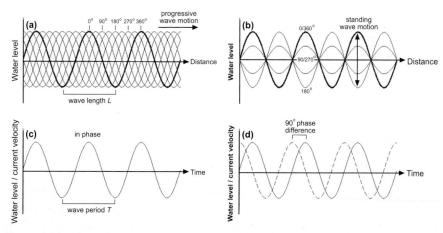

**Figure 4.18** – Progressive and standing wave motion. See text for explanation.

The time series of water level and cross-shore current for a progressive wave is shown in Figure 4.18c. Only one wave signal is apparent and this is because the two time series are in phase, *i.e.*, the crest (trough) of the progressive wave coincides with the maximum onshore (offshore) directed velocity. The correlation between water level and cross-shore current associated with standing wave motion is more complicated. At nodes, the cross-shore current velocity is maximum, but the water level does not change. In contrast, at antinodes, the excursion of the water level is maximum, while the cross-shore current is zero. When water level and cross-shore current are measured between a node and antinode, they display a 90° ($\pi/2$) phase difference as shown in Figure 4.18d. Such a phase difference, which is referred to as quadrature, implies that the strongest cross-shore current occurs when the water level crosses mean sea level. Whether the water level leads or lags the cross-shore current depends on the location of the measurements in relation to the position of the antinode.

### 4.6.3 Broken waves in the surf zone

A natural wave field is characterized by a range of wave heights and periods. Therefore, not all the waves arriving at the shore will break at the same location. The largest waves will break furthest offshore, while the smaller waves approach more closely to the shore before they break in shallower water. Thus, at any position within the surf zone, some waves are breaking, while other waves are still undergoing their transformations leading to initial breaking. The proportion of breaking waves progressively increases in a shoreward direction across the surf zone.

Field measurements have indicated that in the inner part of the surf zone, where most waves are broken, the wave height is limited by the water depth (Thornton and Guza, 1982). The wave height becomes a direct function of the local water depth and can be expressed by

$$H_s = \langle \gamma \rangle h \tag{4.33}$$

The coefficient $\langle \gamma \rangle$, which is sometimes referred to as the relative wave height, ranges from 0.4 to 0.6 and increases with beach slope (Sallenger and Holman, 1985). A surf zone in which the wave height is limited by the local water depth, as prescribed by Equation 4.33, is referred as a **saturated surf zone**.

### 4.6.4 Infragravity wave motion

Wave and current spectra derived from hydrodynamic measurements collected in the surf zone are usually characterized by the presence of energy at frequencies lower than those of the incident swell and wind waves (Figure 4.19). This motion is referred to as infragravity wave motion and is generally characterized by periods ranging from 20 seconds to several minutes.

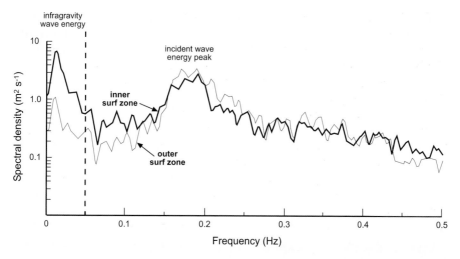

**Figure 4.19** – Examples of cross-shore velocity spectra from the surf zone demonstrating the presence of large amounts of infragravity energy at frequencies below 0.05 Hz (periods larger than 20 s). The two spectra were recorded simultaneously and indicate that infragravity waves in the inner surf zone are more energetic than in the outer surf zone. [From Aagaard and Masselink, 1999.] [Copyright © 1999 John Wiley & Sons, reproduced with permission.]

Especially in the 1980s, a large amount of nearshore research was aimed at understanding infragravity wave motion. Interest in **infragravity waves** was partly motivated by the theoretical finding that certain types of infragravity waves (**edge waves**) may be responsible for the formation of many nearshore morphological features, such as rip channels and multiple bars (Holman and Bowen, 1982). Despite concerted research efforts to confirm this theoretical result, field evidence remains largely inconclusive, although the presence of edge waves on beaches has been convincingly demonstrated (Huntley *et al.*, 1981).

Infragravity wave motion is also important from another point of view. Infragravity waves are forced directly or indirectly by the incoming waves and therefore the energy associated with infragravity wave motion is proportional to the incident wave energy level. Typically, the height of the infragravity waves in the surf zone is 20–60% of the offshore wave height (Guza and Thornton, 1985). The amount of infragravity wave energy increases significantly across the surf zone in the onshore direction. This increase is for the large part ascribed to shoaling of the infragravity wave as it enters shallow water. At the same time, the incident wave energy level decreases in the onshore direction across the surf zone due to wave breaking.

The dependence of the infragravity wave height on the incident wave height, and the opposing cross-shore trends in infragravity and incident wave energy levels have major ramifications for the behaviour of beaches during storm conditions. These are schematically illustrated in Figure 4.20. During storm conditions, the surf zone will become wider, but the conditions at the shoreline will not become more energetic due to incident wave energy dissipation. In other words, the larger incident storm waves would

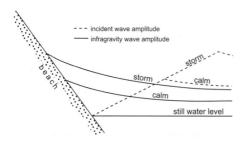

**Figure 4.20** – Schematic representation of the variation in incident and infragravity wave energy across the nearshore zone during storm and calm weather conditions. [Modified from Holman, 1983.]

## BOX 4.3 – WAVE REFLECTION OF INFRAGRAVITY WAVES

Wave energy can be reflected at the shoreline similar to the reflection of light off a mirror, or that of sound off a large wall in the form of an echo. Reflection of wave energy is an important process because the reflected wave energy will propagate back to the sea and is therefore not available for generating nearshore currents. Reflection of waves off beaches is difficult to perceive because it is commonly not possible to see evidence of the outgoing waves. However, if barred morphology is present, reflected waves can sometimes be seen breaking on the landward face of the bar. If such reflected breaking waves meet incoming breaking waves their collision produces a line of rising foam that can be a spectacular sight if the associated waves are large.

Wave reflection produces standing wave motion in the surf zone, characterized by nodes and antinodes. If the cross-shore currents together with water depth are measured at the same position, the two time series would show a 90° ($\pi/2$) phase difference. The Iribarren Number $\xi$ can be used to indicate the importance of wave reflection

$$\xi = \frac{\tan \beta}{\sqrt{H_b/L_o}}$$

Wave reflection is dominant when $\xi > 1$ and is promoted by steep beach gradients, long wave periods, and to a lesser extent, small wave heights. Therefore on a given beach, the potential for wave reflection and hence standing wave generation, is greatest for longer period waves, in particular, infragravity waves. The latter type of waves generally do not break on natural beaches. For example, if an infragravity wave has a wave height of 0.5 m and a wave period of 40 s, then the beach slope required to cause this wave to break rather than reflect can be obtained by re-arranging the above Equation and inserting the appropriate numbers

$$\tan \beta = \xi \sqrt{H_b/L_o} = \sqrt{0.5/2496} = 0.014$$

Thus, the infragravity wave under question will only break if the beach gradient is less than 0.014 which is not typical of natural beaches.

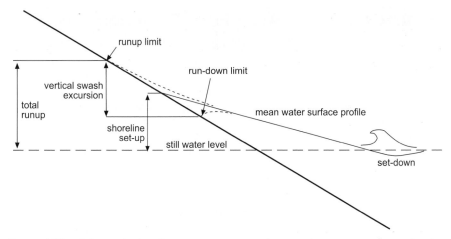

**Figure 4.21** – Schematic showing wave runup, swash excursion, wave set-up and wave set-down. [Modified from Komar, 1998.]

simply break further offshore, dissipating their energy through the resulting wider surf zone. Infragravity energy, on the other hand, is not dissipated in the surf zone because the long wave length of the infragravity wave inhibits wave breaking (Box 4.3). Because the infragravity energy level is proportional to the offshore wave height, infragravity energy may dominate the water motion in the inner surf zone during storm conditions. Furthermore, infragravity waves periodically raise the mean water level at the beach, thus enabling the incident waves to reach the back beach and the toe of the dunes.

## 4.6.5 Wave set-up, swash and runup

When waves break on the beach, they produce **wave set-up**, which is a rise in the mean water level above the still water-level elevation of the sea (Figure 4.21). In popular terms, wave set-up is conceived as a piling up of water against the shoreline due to the waves and is caused by the breaking waves driving water shoreward. However, it should be noted that such an explanation is somewhat simplistic and does not take into account the forces required to drive wave set-up (Box 4.4). The wave set-up at the shoreline can be defined as the intersection of the sloping mean water surface (set-up gradient) with the beach. As a general rule of thumb, the shoreline set-up is 20% of the offshore significant wave height.

According to the saturated surf zone concept, the wave height at the shoreline should be reduced to zero to due to wave breaking (Equation 4.33). However, in reality there will always be a residual wave at the shoreline and this wave will propagate onto the 'dry' beach in the form of swash. **Swash** motion consists of an onshore phase with decelerating flow velocities (uprush or swash) and an offshore phase characterized by

## Box 4.4 – Momentum Flux, Radiation Stress and Wave Set-up

Wave set-up is the super elevation of the nearshore water level due to the presence of waves. A theoretical explanation for wave set-up was provided by Longuet-Higgins and Stewart (1962, 1964) using the **radiation stress** concept. This is not an easy concept to explain, but since wave set-up is an important process it deserves extra attention.

Associated with wave propagation, there is a transport of momentum due to the pressure and velocity fluctuations under waves. Longuet-Higgins and Stewart introduced the term 'radiation stress' and defined this as the 'excess flow of momentum due to the presence of the waves'. It is difficult to perceive waves as possessing momentum, but fortunately the **momentum flux** can be readily computed from linear wave theory, because it is directly related to the wave energy density $E$

$$S_{xx} = \frac{3}{2} \, E = \frac{3}{16} \, \rho g H^2$$

where $S_{xx}$ represents the cross-shore component of the radiation stress for waves approaching the beach with their crests parallel to the shoreline. Newton's second law of motion states that changes in momentum flux must be balanced by an opposing force. When this law is applied to the surf zone we obtain the following result

$$\frac{\partial S_{xx}}{\partial x} + \rho g H \frac{\partial \overline{\eta}}{\partial x} = 0$$

where $x$ indicates the cross-shore coordinate, $h$ is the mean water depth and $\overline{\eta}$ is the departure from still water level.

It can now be seen how variations in the wave height and radiation stress affect the water surface gradient in the nearshore. Outside the surf zone, wave height and hence radiation stress increase due to shoaling. The resulting positive gradient in radiation stress ($\partial S_{xx}/\partial x > 0$) must be balanced by a negative water surface gradient ($\partial \overline{\eta}/\partial x < 0$), resulting in a lowering of the water level, known as wave set-down. In the surf zone, wave energy dissipation due to breaking processes results in a decrease of the wave height, and hence a negative gradient in the radiation stress height ($\partial S_{xx}/\partial x < 0$). This gradient is balanced by a positive water surface gradient ($\partial \overline{\eta}/\partial x > 0$), resulting in an elevated water level in the surf zone referred to as wave set-up. Figure 4.22 shows detailed laboratory measurements that confirm the validity of the radiation stress theory, although recent improvements to the theory have been achieved by using more advanced wave theories.

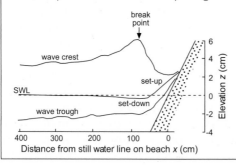

Figure 4.22 – Wave set-up and set-down measured in the laboratory. Measurements were obtained using a wave height of 6.45 cm, a wave period of 1.14 s and a beach gradient of 0.082. SWL refers to still water level. [Modified from Bowen et al., 1968.]

accelerating flow velocities (downrush or backwash). Generally, swash motion is asymmetrical with the magnitude of uprush velocities exceeding that of the backwash. It is generally thought that the onshore asymmetry in the swash motion is responsible for maintaining the beach gradient.

Collectively, wave set-up and swash produce a landward and upward displacement of the shoreline allowing waves to act at higher levels on the beach. The vertical displacement can be measured relative to the still water level and is referred to as **wave runup**. Wave runup can be substantial under energetic wave conditions and significantly increases the potential for shoreline erosion under storm conditions. The two most comprehensive studies of runup statistics on natural beaches are those of Holman (1986) and Nielsen and Hanslow (1991). Both studies obtained the following relationship

$$R_{2\%} = 0.9H_o\xi_0 = 0.36g^{1/2} \tan \beta H_o^{1/2}T \qquad (4.34)$$

where $R_{2\%}$ is the wave runup height exceeded by 2% of the runup events and $H_o$ is the significant wave height measured in deep water. For typical storm conditions ($H_o$ = 2 m and $T$ = 8 s) and an average gradient (tan$\beta$ = 0.1), Equation 4.34 yields a total wave runup of 1.3 m.

## 4.7 SUMMARY

- Natural waves are irregular and statistical techniques need to be employed to describe their characteristics. Spectral analysis is one of the most useful techniques to describe natural waves.
- Waves are generated by wind, and wave properties such as height and period can be predicted using wind speed, wind duration and fetch length. In a developing wave field, wave height and wave period steadily increase until they reach an equilibrium state – this state is referred to as fully arisen sea.
- Linear wave theory is the simplest and most widely used theory to describe the behaviour of waves. A fundamental relationship derived from linear wave theory is the dispersion equation, which expresses the relation between wave length and wave period.
- In deep water, wave motion does not extend to the sea bed and the wave velocity is only dependent on the wave period. Deep water wave velocity increases with wave period and this gives rise to wave dispersion, whereby the longer-period waves in a broad-banded wave field outdistance the shorter-period waves.
- As waves propagate into intermediate and shallow water, the wave motion extends to the sea bed and the waves start to 'feel' the sea bed. Consequently, the wave is slowed down, causing a decrease in the wave length. The shortening of the wave causes an increase in the wave height and this process is known as wave shoaling. During shoaling, wave energy losses may be incurred due to bed friction.

- In intermediate and shallow water, the wave velocity decreases with decreasing water depth. When waves approach the coast at an angle, the wave velocity will vary along the wave crest with the deeper part of the wave propagating at a faster rate than the shallow part of the wave. This results in wave refraction whereby the wave crest rotates with respect to the bottom contours and becomes more aligned with the coastline.

- Waves will break in water that has a depth similar to the wave height. In the surf zone, where most waves are breaking, wave energy is either dissipated by wave breaking or reflected at the shoreline.

- Infragravity wave motion, wave set-up and swash produce a landward and upward displacement of the shoreline allowing waves to act at higher levels on the beach. This displacement is particularly significant during storms and may promote beach erosion.

## 4.8 FURTHER READING

Komar, P.D., 1998. *Beach Processes and Sedimentation* (2nd Edition). Prentice Hall, New Jersey. [For those not deterred by the mathematics inherent in a discussion on wave dynamics, Chapters 5 and 6 of this text provide an excellent and comprehensive treatment of wave processes.]

Open University, 1994. *Waves, Tides and Shallow-Water Processes*. Pergamon Press, Oxford. [Chapter 1 of this text gives a very readable account of the dynamics of ocean waves.]

Pond, S. and Pickard, G.L., 1995. *Introductory Dynamical Oceanography*. 2nd Revised Edition, Butterworth-Heinemann Ltd. [A thorough presentation pitched at students with a junior-intermediate-level physics background.]

Thurman, H.V. and Burton, E.A., 2001. *Introductory Oceanography*. 9th Edition, Prentice Hall. [A useful introductory text including a chapter on waves.]

# 5

# SEDIMENTS, BOUNDARY LAYERS AND TRANSPORT

## 5.1 INTRODUCTION

Sediment refers to both organic and inorganic loose material that is moved from time to time by physical agents including wind, waves, currents and gravity. The sediments found in coastal environments can be either imported from external environments (allochthonous) or locally produced (autochthonous). **Allochthonous sediments** are generally derived from the chemical and mechanical breakdown of continental rocks into grains that consist of either a single mineral or a subset of the minerals contained in the original rock. The two most common allochthonous minerals found in coastal sediments are quartz and clay minerals (*e.g.*, illite, kaolinite, montmorillonite). **Autochthonous sediments** can include material derived from the breakdown of local rocky shorelines, but it is more common for such sediments to consist of grains derived from the body-parts of organisms living in the coastal zone and/or the chemical precipitation of dissolved minerals present in coastal waters. The most common autochthonous minerals are biogenic carbonate and silica. On a global scale, allochthonous sediments account for about 92% of sediment in the modern coastal zone and are delivered by rivers, glaciers, wind, and volcanic eruptions in decreasing order of importance (Open University, 1994).

In Chapter 1 we mentioned that many of the large-scale morphological features found in the coastal zone result from sustained erosion of sediment from one region and its sustained deposition at another. We also demonstrated that it is the existence of persistent spatial gradients in the sediment transport rate that determines these locations of erosion and deposition. The cascade of spatial and temporal scales evident in coastal geomorphology indicates that in order to understand fully large-scale morphodynamic processes (*e.g.*, beach cut and fill, channel switching in deltas) it is necessary to have some appreciation of processes operating at a much smaller scale.

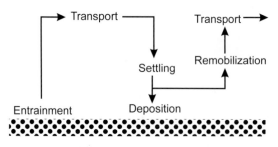

**Figure 5.1** – Schematic representation of the fundamental processes that together constitute sediment dynamics. Most of this chapter is directed towards providing a basic understanding of the cyclic sequence of processes shown here.

The sequence of small-scale processes that control the localized movement of sediment include the following three phases: (1) **entrainment** of sediment up into the flow via fluid-induced stresses and forces acting on the sea bed; (2) **transport** of sediment, as either bedload or suspended load via momentum transfer from the fluid to the sediment; and (3) **settling** and **deposition** of sediment back on the bed via gravity (Figure 5.1). Depending on the speed at which suspended sediment settles to the bed relative to the speed at which flow conditions change, the sequence can re-initialize in one of two ways, namely by re-entrainment of sediment that is at rest on the bed, or by remobilization of a stationary sediment suspension that has not completed its return to the bed. The detailed mechanics of this cycle vary, depending on some key intrinsic properties of the sediment and the fluid.

This chapter begins with a brief description of each of the key sediment and fluid properties, before focusing on fluid and sediment dynamics. The most readily observable outcome of fluid-sediment interaction at the scale discussed in this chapter are the localized, wave-like features on the surface of the sea bed known collectively as bedforms. This chapter concludes with a brief discussion of the development and geometry of the most common bedform types. When reading this chapter it is important to keep in mind that air is as much a fluid as water, and both are important agents for sediment transport in the coastal zone. While many of the concepts presented in this chapter apply to both fluid types, the fluids do differ markedly in their density and viscosity, and hence their ability to move sediment. This chapter focuses on hydrodynamics and sediment transport. Aerodynamics and wind-driven transport are briefly described in Section 8.4.1.

## 5.2 SEDIMENT PROPERTIES

### 5.2.1 Grain size

**Grain size** (as well as sorting and skewness) is the sediment property most widely measured by coastal geomorphologists, and for good reason, since it is a property of first-order importance in a wide range of processes. The simplest measurements that can be made of a grain's size are the lengths of the long (L), intermediate (I) and short (S) axes (Figure 5.2). By convention the I-axis and S-axis are measured at right-angles to the L-axis. The axis

**Figure 5.2** – Definition of the long (L), intermediate (I) and short (S) axis lengths of a sediment grain. [Modified from Allen, 1985.]

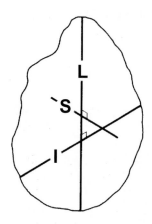

dimensions of large grains can be measured directly with calipers, whereas smaller grains are measured indirectly, usually by either sieving or microscopic inspection. In the case of sieving, it is generally understood that only the I-axis length is obtained. In the case of microscopic inspection, the grain usually sits on the slide in its most stable position (*i.e.*, S-axis normal to the plane of the slide), so only the L- and I-axis lengths can be obtained. A detailed description of the various methods for grain-size analysis can be found in Goudie (1990).

Grains are often classified according to their I-axis length using the **Udden–Wentworth Scheme** (Table 5.1). Notice that the class boundaries are presented both on a millimetre-scale and a phi-scale (φ-scale). The **phi-scale** was developed to account for a particular characteristic of grain-size distributions, which we will discuss further below. For now, it is sufficient to know that the conversion between grain diameter $D$ on the φ-scale to diameter on the millimetre-scale is

$$D = 2^{-\phi} \tag{5.1}$$

and *vice versa*

$$\phi = -\log_2 D \tag{5.2}$$

It should be evident from both Table 5.1 and these two equations that a change from one size class to the next involves a doubling of the axis length.

Whilst it is common to use a single number to represent the grain size of a sediment sample, this number is typically obtained from a statistical analysis of the I-axis length of numerous individual grains. The procedure typically begins with the presentation of the grain size measurements obtained from a sediment sample as a **frequency-histogram** and a **cumulative-frequency curve** (Figure 5.3). There are numerous exceptions, but typically the histogram appears to approximate a log-normal distribution, so when it is plotted on a logarithmic scale such as the φ-scale, the histogram then appears normally distributed. It is for this reason that a log transformation of grain size measurements (performed either graphically by plotting

**Table 5.1** The Udden–Wentworth Scheme of grain size classification.

| mm | φ | Class terms | |
|---|---|---|---|
| | | Boulders | |
| 256 | −8 | | |
| 128 | −7 | Cobbles | |
| 64 | −6 | | |
| 32 | −5 | | |
| 16 | −4 | Pebbles | |
| 8 | −3 | | |
| 4 | −2 | Granules | |
| 2 | −1 | | |
| 1 | 0 | | very coarse |
| 0.5 | 1 | | coarse |
| 0.25 | 2 | Sand | medium |
| 0.125 | 3 | | fine |
| | | | very fine |
| 0.062 | 4 | | |
| 0.031 | 5 | | coarse |
| 0.016 | 6 | Silt | medium |
| 0.008 | 7 | | fine |
| | | | very fine |
| 0.004 | 8 | | |
| | | Clay | |

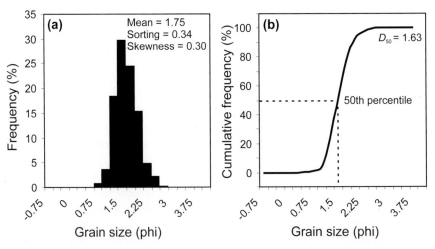

Figure 5.3 – Two methods of presenting grain size data: (a) frequency histogram and (b) cumulative-frequency curve. The example shown is for a sample of medium sand that is very well sorted and fine-skewed (*cf.*, Tables 5.1 and 5.2).

on the phi-scale or numerically by applying Equation 5.2) is performed prior to calculating grain size statistics. It is important to note that the $\phi$-scale on graphs such as those shown in Figure 5.3 is somewhat counter-intuitive if you are not used to working with logarithms. Large positive $\phi$-values indicate finer grain sizes and large negative $\phi$-values indicate coarser grain sizes.

Before the widespread availability of computers, grain size statistics were calculated by graphical means. The most widely used formulae are (Folk and Ward, 1957)

$$Median = \phi_{50} \tag{5.3}$$

$$Mean = \frac{\phi_{16} + \phi_{50} + \phi_{84}}{3} \tag{5.4}$$

$$Sorting = \frac{\phi_{84} - \phi_{16}}{4} + \frac{\phi_{95} - \phi_{5}}{6.6} \tag{5.5}$$

$$Skewness = \frac{\phi_{16} + \phi_{84} - 2\phi_{50}}{2(\phi_{84} - \phi_{16})} + \frac{\phi_{5} + \phi_{95} - 2\phi_{50}}{2(\phi_{95} - \phi_{5})} \tag{5.6}$$

where, for example, $\phi_{50}$ is the 50th percentile of the grain-size distribution plotted on the phi-scale (Figure 5.3b). These formulae are still widely used when the method of sediment analysis does not yield a complete grain-size distribution, because the calculation does not rely on the tails of the distribution (*e.g.*, when dry-sieving is used to analyse most of the sample, but a significant fraction of the material is very fine ($> 4\phi$), and so must be analysed by some other technique). When a complete grain-size distribution is available, the statistical descriptors are calculated using the more accurate and computationally intensive method of moments (Pettijohn *et al.*, 1987)

$$Mean = \bar{x} = \frac{\sum_{i=1}^{n} f_i M_{\phi i}}{100} \tag{5.7}$$

$$Sorting = \sigma = \sqrt{\frac{\sum_{i=1}^{n} f_i (M_{\phi i} - \bar{x})^2}{100}} \tag{5.8}$$

$$Skewness = \sum_{i=1}^{n} \frac{f_i (M_{\phi i} - \bar{x})^3}{100\sigma^3} \tag{5.9}$$

where $f_i$ is the percentage of grains (or percentage of total weight of grains) in each size interval and $M_{\phi i}$ is the midpoint of each size interval in phi units.

The **mean** (1st-moment) is the most common value used to represent the

**Table 5.2** Descriptors for sediment sorting and skewness

| Sorting ($\phi$-scale) | | Skewness | |
|---|---|---|---|
| < 0.35 | Very well sorted | > +0.30 | Strongly fine-skewed |
| 0.35 to 0.50 | Well sorted | +0.30 to +0.10 | Fine-skewed |
| 0.50 to 0.71 | Moderately well sorted | +0.10 to –0.10 | Nearly symmetrical |
| 0.71 to 1.00 | Moderately sorted | –0.10 to –0.30 | Coarse-skewed |
| 1.00 to 2.00 | Poorly sorted | < –0.30 | Strongly coarse-skewed |
| > 2.00 | Very poorly sorted | | |

(from Folk and Ward, 1957)

grain size of sediment, although for some applications particular percentile values are more appropriate. For example, the 65th percentile grain size is often used in determining the roughness of the sea bed. If the distribution is highly skewed then the median or mode may better represent the grain size of the sediment. We have already discussed a classification scheme for the mean grain size (Udden-Wentworth Scheme, Table 5.1). There are classification schemes for the other moment measures as well (Table 5.2). The **sorting** ($2^{nd}$-moment) is controlled in part by the range of sizes present at the sediment source, as well as processes operating during transport and deposition. For example, rapidly deposited sediment from a single transport event is often poorly sorted, whereas frequently reworked and redeposited sediment tends to be well sorted. The **skewness** ($3^{rd}$-moment) is an indicator of the symmetry of the grain size distribution. Negative skewness means that there are more coarse grains than expected in a log-normally distributed sediment and positive skewness means that there are more fine grains than expected. Skewness can arise from the mixing of sediment from two different sources, but can also be indicative of processes operating during transport and deposition. For example, beach sands are typically negatively skewed, because continual agitation by wave action is very effective at resuspending the finer grain sizes, thus leaving an apparent excess of coarser sizes in the bed sediment. There is often a significant shell component in beach sands that also negatively skews the distribution.

## 5.2.2 Grain mass and density

The mass of a grain is an important property that affects its inertia to the forces applied to it by a moving fluid. A grain's **mass** is equal to the product of its volume and density. The volume is clearly related to the grain size. The volume of spherical grains, for example, increases as the grain-diameter cubed. The **density** of a grain is defined as its mass per unit volume, and is largely determined by its mineralogy. The densities of some of the most common minerals found in coastal sediments are listed in Table 5.3. They are separated into 'light' and 'heavy' minerals in the table, because a similar

**Table 5.3** Density (kg m$^{-3}$) of minerals commonly found in coastal sediments.

| Light minerals | Density, kg m$^{-3}$ | Heavy minerals | Density, kg m$^{-3}$ |
|---|---|---|---|
| Aragonite | 2940 | Augite | 2960–3520 |
| Calcite | 2715 | Garnet | 3890–4320 |
| Dolomite | 2860 | Haematite | 5200 |
| Gypsum | 2300–2370 | Hornblende | 3020–3500 |
| Illite | 2600–2900 | Magnetite | 5200 |
| Kaolinite | 2610–2680 | Monazite | 4600–5400 |
| Montmorillonite | 2000–2300 | Pyroxene | 3200–3550 |
| Orthoclase | 2560–2630 | Rutile | 4230–5500 |
| Plagioclase | 2620–2760 | Tourmaline | 2900–3200 |
| Quartz | 2650 | Zircon | 4600–4700 |

(from Gribble and Hall, 1999)

separation often occurs in the environment. For example, the alternating light- and dark-coloured layers of sand present on some beaches are usually layers of concentrated light and heavy mineral types, respectively. The somewhat arbitrary separation between the two mineral types is placed at a density of 2,900 kg m$^{-3}$.

## 5.2.3 Grain shape and roundness

Shape and roundness of grains are sometimes confused with each other, but they are different measures, and have distinctly different interpretations. **Grain shape** is usually defined using some ratio of the axis-lengths. For example, the diagram in Figure 5.4 can be used to classify shape once the L-, I-, and S-axes lengths have been measured. The ratio of the I- and L- axes indicates the degree of grain elongation, and the ratio of the S- and I-axes indicates the degree of grain flattening. Another measure of the grain flatness is the Corey Shape Factor $CSF$ (Corey, 1949)

$$CSF = \frac{S}{\sqrt{LI}} \tag{5.10}$$

where $CSF = 0$ represents a flat disc and $CSF = 1$ represents a perfect sphere. The shape of the original mineral crystal largely determines the shape of a grain, although dissolution during weathering and abrasion during transport have an important modifying effect. The shape of a grain influences both its entrainment and settling. Flat grains tend to be more difficult to entrain and settle more slowly than spherical grains. This is particularly so for coarse sediment such as gravels and pebbles. Sorting on the basis of shape is well known on gravel beaches, for example, with prolate clasts

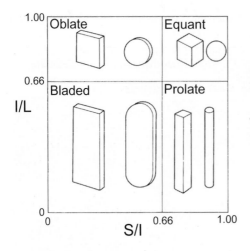

**Figure 5.4** – Zingg's (1935) classification of grain shape. [From Tucker, 1995.] [Copyright © 1995 Blackwell Publishers, reproduced with permission.]

predominantly found at the base of the beach face and oblate clasts found towards the top. The former are easily transportable both up and down the beach, whereas the latter are thrown up the beach during the violent uprush and cannot be removed by the weaker backwash.

**Roundness** on the scale of the grain diameter is a function of shape, for example, a perfect sphere is clearly well-rounded. Usually when we refer to roundness, however, we are referring to the roundness of grain corners and protrusions. In this sense, it is possible to have a spherically-shaped grain that is angular, rather than well-rounded (Figure 5.5). The degree of roundness usually indicates the susceptibility of the grain to chemical weathering and/or the degree of mechanical abrasion the grain has experienced. Angular grains are often indicative of chemically and mechanically resistant minerals or a depositional site that is close to the source. Well-rounded grains tend to indicate chemically unstable minerals, energetic environments where there is constant mechanical abrasion of the grains, or a

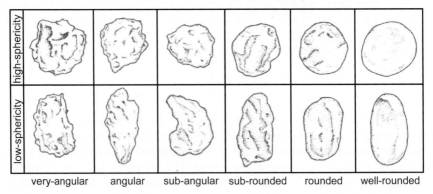

**Figure 5.5** – Power's (1953) classification of grain roundness for grains displaying low sphericity and high sphericity. [From Tucker, 1995.] [Copyright © 1995 Blackwell Publishers, reproduced with permission.]

depositional site that is a significant distance from the source. We will discover shortly that grain roundness influences the friction angle of sediment, which is an important quantity in sediment entrainment and transport.

## 5.2.4 Sediment packing, bulk density and porosity

The arrangement of grains on the sea bed is referred to as sediment packing. Given the wide range of natural grain sizes and shapes that exists there is an almost limitless number of possible packing arrangements. Many sediment grains approximate the shape of a sphere, however, so it is constructive to at least consider some of the packing arrangements possible for spheres. It is important to remember, however, that natural sediments, even if they consist of spherical grains of uniform size, will have packing arrangements that only approximate the arrangements described below. The rate at which grains are deposited on the bed and the direction from which they are deposited have a strong influence that is difficult to account for.

The two basic **packing arrangements** for spheres of uniform size are shown in Figure 5.6 and define the end-members of a spectrum of possible

**(a)** Oblique View

Cubic                                    Rhombohedral

**(b)** Plan View

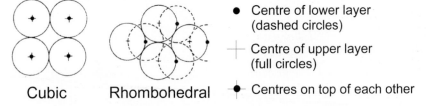

Cubic       Rhombohedral

● Centre of lower layer (dashed circles)

┼ Centre of upper layer (full circles)

┿ Centres on top of each other

**Figure 5.6** – Two basic packing arrangements for spherical grains of uniform size. [Oblique views from Allen, 1985.] [Plan views modified from Dyer, 1986.]

arrangements (see Allen, 1985 for an exhaustive discussion). In the case of the **cubic** arrangement, any two dimensional view of the packing will display a square-grid pattern if lines are drawn to connect the centres of adjacent spheres. The **rhombohedral** arrangement is achieved by translating each layer of a cubic arrangement sideways and backwards (or forwards) a distance of one-radius of the sphere. Any two dimensional view of the packing will then display a rhomboid-grid pattern.

We have already discussed the concept of grain density and its primary dependence on mineralogy (Section 5.2.2). Now that we know something of the way individual grains can pack together we are in a position to define

## Box 5.1 – Grain-Packing and its Effect on Sediment Porosity

In the simplified case of granular spheres, the theoretical packing arrangement that has the smallest bulk density and concentration and the largest porosity is cubic. The rhombohedral arrangement has the largest bulk density and concentration and the smallest porosity (Table 5.4). Natural sediments generally consist of a variety of grain sizes and shapes, which can modify these theoretical values considerably. For example, densely packed natural sediments tend towards the upper end of the theoretical ranges for bulk density and concentration (lower end for porosity), simply because the smaller members of the grain size distribution partly occupy the voids within the packing arrangement of larger members. Some indicative values of the sediment concentration and porosity for natural sands are shown in the bottom half of Table 5.4. While the packing arrangement, degree of sediment sorting and grain shape all have a significant effect on sediment concentration and porosity, the mean grain size generally does not.

**Table 5.4** Measured packing concentrations and porosities of some theoretical and natural materials in water

| Material | Concentration % | | Porosity % | |
|---|---|---|---|---|
| **Theoretical:** | | | | |
| Spheres (cubic packing) | 52.0 | | 48.0 | |
| Spheres (rhombohedral packing) | 74.0 | | 26.0 | |
| **Natural:** | Loose | Dense | Loose | Dense |
| Quartz sands (0.27 mm) | 55.8 | 65.1 | 44.2 | 34.9 |
| Quartz sands (1.04 mm) | 54.1 | 62.8 | 45.9 | 37.2 |
| *Lithothamnium* sands (3.2 mm) | 58.7 | 70.3 | 41.3 | 58.7 |

Note: Loose packing was produced by a simple dumping of sediment onto the bed, whereas dense packing was produced by subsequently agitating the sediment. *Lithothamnium* sands, which are biogenic, have a grain shape that is markedly elongated.

(from Allen, 1985)

the **bulk-sediment density**, which is simply the total sediment mass divided by the total volume of the packed sediment (*i.e.*, grains plus voids). The bulk-sediment density will always be less than the density of constituent grains due to the voids. Directly related to the bulk-sediment density is the **sediment concentration**, which is defined as the total volume of the grains divided by the total volume of the packed sediment (*i.e.*, grains plus voids). The concentration is therefore non-dimensional and can be expressed as a fraction or a percentage. The **sediment porosity** is simply the volume of void spaces contained within the sediment, and in percentage terms it is equal to 100 minus the sediment concentration. In the case of very fine sediments the bulk-sediment density and concentration are controlling factors

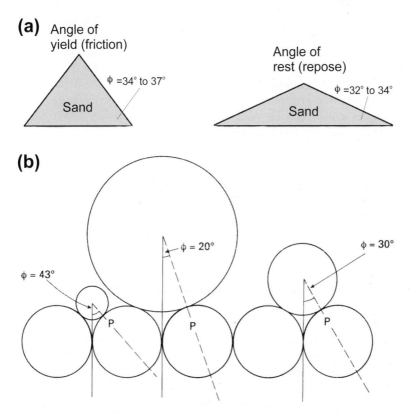

**Figure 5.7** – (a) Schematic showing relationship between angle of yield (friction angle) and angle of rest (repose) for a heap of natural sand grains. (b) Schematic showing the role of friction angle in the initiation of sediment motion. The angle enclosed by the vertical line and the line marked P is the angle the grain must pivot through before it can change its position on the bed. This angle is equivalent to the friction angle. For a perfect sphere the pivot angle is 30° if it is resting on a bed of similar sized grains. If the bed is composed of larger or smaller grains the pivot angle is correspondingly larger and smaller, respectively. [From Pye, 1994.] [Copyright © 1994 Blackwell Publishers, reproduced with permission.]

in their ability to be entrained by waves and currents (Section 5.5.1). Moreover, the sediment porosity is an important factor in beach-groundwater dynamics (Section 8.3.4). Further information on grain packing and its effect on sediment porosity is presented in Box 5.1.

## 5.2.5 Friction angle of sediment

The final sediment property that we will address is the friction angle of sediment, which is also known as the angle of yield. The **friction angle** is the maximum slope angle to which a sediment mass can be tilted, prior to the initiation of grain avalanching down the slope (Figure 5.7a). After avalanching has ceased the new slope angle is referred to as the **angle of rest**, which is always less than the friction angle. On a horizontal bed the friction angle is effectively the angle through which a grain must be rolled in order for it to change its position on the bed, rather than simply falling back to its original position (Figure 5.7b). Experimental data shows that the friction angle depends on grain size, shape, packing arrangement and the surface texture of the grains. Friction angles for natural sand grains are on average 34°–37°, but individual grains can be well outside this range (Pye, 1984). Smaller angles in this range are associated with loosely-packed spherical grains and larger angles with densely-packed angular grains. Friction angles for gravels can reach 45° (Allen, 1985; Trenhaille, 1997). It is not clear why there is a size dependence in the experimental data, since it can be easily demonstrated that the theoretical friction angles for two different sized spheres resting on spheres of similar size are equal. It may simply be that larger grains tend to have larger roughness elements on their surfaces, thus increasing the surface friction.

## 5.3 FLUID PROPERTIES

Our principal interest in fluid properties is to understand how moving fluids such as wind, currents and waves can apply forces and impart momentum to sediments in order to initiate their movement and keep them travelling. So what are the fluid properties that we should consider? If we consider the definition of force, then fluid mass and acceleration are relevant properties. Similarly, considering momentum, mass and velocity are relevant. If we consider any horizontal plane within a body of fluid in which sediment may be present, the total force that is exerted on that plane per unit area is referred to as the total stress, and it consists of a normal stress and a shear stress. The first acts vertical to the plane and the second acts tangential to the plane. We now have a list of the fluid properties directly relevant to sediment dynamics: mass, acceleration, velocity and stress. In most fluid and sediment transport modelling we consider a body of fluid of unit volume, so in order to account for its mass we need only to consider fluid density, which is where we will begin our discussion.

## 5.3.1 Fluid density

**Fluid density** is the mass of the fluid per unit volume. The density of pure fresh water varies with both temperature and pressure, although in shallow coastal waters the latter control can generally be ignored. The density of fresh water is at its maximum at a temperature of 4°C. It is extremely rare for coastal waters to be purely fresh. They generally contain various quantities of dissolved and suspended materials that add proportionately more to the fluid mass than to its volume, and therefore increase its density over that of fresh water. For example, open ocean waters contain dissolved salts that usually have a mass of 35 mg for every kg of water. Coastal waters typically contain less dissolved salts due to dilution of the open ocean water by fresh water running off the land, so the salinity is generally somewhere between 0 and 35 parts per thousand (ppt). Table 5.5 lists some indicative values for the density of coastal waters at 1 atmosphere, as well as the density of air. Notice how much denser water is when compared with air. For a given velocity and acceleration, the force and momentum that water can apply to move sediment is therefore much greater than air is capable of. Nevertheless, water and air achieve comparable sediment transport rates in the coastal zone, due largely to the fact that wind reaches much larger velocities and accelerations than waves and currents.

**Table 5.5** Typical values for fluid density and molecular viscosity

| Fluid type | Fluid density, kg m$^{-3}$ | Fluid viscosity, N s m$^{-2}$ |
|---|---|---|
| **Air:** | | |
| At 10°C | 1.3 | $1.80 \times 10^{-5}$ |
| **Water:** | | |
| Fresh (10°C) | 1000 | $1.06 \times 10^{-3}$ |
| Saline (35 ppt at 10°C) | 1027 | $1.40 \times 10^{-3}$ |

## 5.3.2 Shear stress and viscosity

Most fluids, including air and water, offer some inherent resistance to deformation or flow and are referred to as viscous fluids. This resistance to deformation is provided by molecular forces within the fluid and is termed the **molecular viscosity** of the fluid. The molecular viscosity of water can be measured by a very simple experiment in which a thin body of water is deformed between two smooth plates (Figure 5.8). The bottom plate is kept stationary and the top plate is moved with a velocity $u$. Due to friction between the fluid and the plates we have what is known as the no-slip condition where the layer of fluid in immediate contact with each plate must have the same velocity as the plate. Momentum is passed from one fluid layer to the next, but at a reduced rate due to the internal friction between

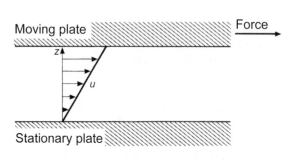

Figure 5.8 – Experimental design used to measure the molecular viscosity of a fluid. A thin film of the fluid is sheared between two plates by moving the top plate horizontally and keeping the bottom plate still. The velocity profile through the fluid is represented by arrows whose length is proportional to the fluid velocity. [Modified from Allen, 1994.

water molecules, so that the fluid velocity decreases with distance away from the moving plate. The result is a linear velocity-gradient within the fluid where the fluid velocity decreases from a maximum within the layer in contact with the top plate down to zero within the layer in contact with the bottom plate. At any level in the fluid the normal stress (vertically-acting force per unit area) is equal to the product of the fluid's mass and the gravitational acceleration. The **shear stress** (tangential-force per unit area) at any level in the fluid $\tau$ can be written as

$$\tau = \mu \, \frac{du}{dz} \qquad\qquad (5.11)$$

where $\mu$ is the molecular viscosity of the fluid. The molecular viscosity thus provides the link between the shear stress and the velocity-gradient at any level in the fluid. For a given fluid volume the molecular viscosity varies with density, thus the viscosity of water is considerably greater than that of air. Moreover, the viscosity of ocean water is greater than that of fresh water (Table 5.5).

## 5.3.3 Flow velocity, acceleration, laminar and turbulent flow

Up to this point we have assumed a basic knowledge of what is meant by the term 'velocity', but it is now necessary to formalise what we mean by this and a number of other terms that describe fluid motion. The flow velocity $u$ is the distance travelled by a fluid parcel per unit of time. Flow acceleration can be either spatial or temporal. **Spatial acceleration** is defined as the change in flow velocity per unit of distance $x$ (i.e., $du/dx$), whereas **temporal acceleration** is the change in flow velocity per unit of time $t$ (i.e., $du/dt$). Figure 5.9 illustrates what we mean by the terms steady/unsteady/oscillatory and uniform/non-uniform flows:

- **Steady** (uniform) **flows** display no temporal (spatial) acceleration so that their velocity is constant through time (space).
- **Unsteady** (non-uniform) **flows** display either an increase or decrease in velocity with time (distance).

**Figure 5.9** – Schematic of current velocity records used to define various flow types. Flows that are constant with respect to time are steady and constant with respect to distance are uniform. Flows that change velocity with respect to time are unsteady or oscillatory, whereas flows that change with respect to distance are non-uniform. [Modified from Leeder, 1999.]

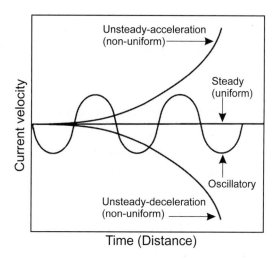

- **Oscillatory flows** display a rhythmic temporal acceleration of the flow as its direction swings back and forth.

In the case of tidal flows, if they are observed over a short enough time period (*c.* 10–15 minutes), they can be approximately steady. If the period of observation is increased beyond one tidal period, however, then tidal flows are seen to be oscillatory.

**Laminar flow** is characterized by fluid motion only in the direction of flow, without any deviation sideways or vertically (Figure 5.10a). Such flows consist of thin layers (lamina) of moving fluid that do not mix with one another. **Turbulent flow** on the other hand is characterized by the movement of fluid parcels mostly in one direction, but with apparently random deviations sideways and vertically. Laminar flow can be represented by a single-component flow vector, whereas turbulent flow representation requires three vector components: one representing movement in the direction parallel to the principal flow direction, the second representing movement orthogonal to the principal flow direction but in the same horizontal plane, and the third representing movement in the vertical direction. Figure 5.10b shows time series of flow velocity measurements from a quasi-steady turbulent flow that has a mean velocity of about 0.45 m s$^{-1}$ in the principal flow direction, but also fluctuates randomly with short-lived periods of temporal flow acceleration and deceleration. The mean flow velocity in the orthogonal and vertical directions are typically zero, but at any one instant in time there is short-lived acceleration/deceleration of the fluid in the positive or negative directions.

In 1883, Osborne Reynolds (1842–1912) performed what is now considered to be a classic experiment on fluid flow. He used novel flow visualization techniques for the time to demonstrate the nature of laminar and turbulent flow. One of the more important outcomes of Reynolds' experiment was the derivation of a parameter, now referred to as the **Reynolds**

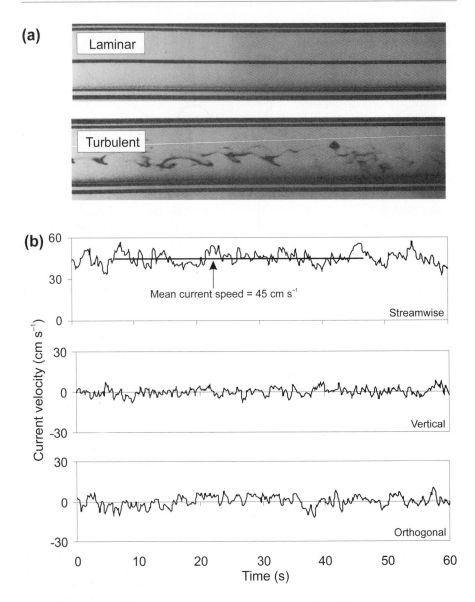

**Figure 5.10** – (a) Laminar (top panel) and turbulent (bottom panel) pipe flow visualized by a horizontal streak of dye along the centreline of flow. [Photographs by N. Johanneson and C. Lowe. Modified from van Dyke, 1982.] [Copyright © 1982 Parabolic Press.] (b) Time series of instantaneous velocity vectors in the streamwise, orthogonal and vertical directions for a turbulent tidal current. The turbulence is indicated by the fluctuations in current velocity about the mean. Time series of current velocity for a steady, laminar flow would show horizontal lines without fluctuations.

**number**, which delimits the conditions associated with either laminar or turbulent flow. The Reynolds number $R_e$ is now used in a variety of applications, but in its most general form it is written as

$$R_e = \frac{\rho u l}{\mu} \qquad (5.12)$$

where $\rho$ is the fluid density, $u$ is the flow velocity, $l$ is a length scale (the pipe diameter in the case of Reynolds's experiments) and $\mu$ is the molecular viscosity of the fluid. For pipe flow, values of $R_e < 500$ are associated with laminar flow and values of $R_e > 2,000$ with turbulent flow. Values of $R_e$ between these extremes are associated with flow that is transitional between laminar and turbulent. The Reynolds number is essentially the ratio between inertial forces and viscous forces acting on the fluid flow. In laminar flow the viscous forces, represented by the molecular viscosity, are sufficient to resist substantial deformation of the fluid and the flow remains ordered. In turbulent flow the inertial forces, represented by the fluid velocity, result in considerable fluid acceleration and flow deformation so that the flow becomes disorganized.

In our discussion of fluid shear stress and viscosity we stated that the tangential force per unit area on any plane within the flow is called the shear stress and it is proportional to the velocity-gradient (Equation 5.11). In laminar flow, such as the experiment shown in Figure 5.8, the proportionality coefficient is the molecular viscosity and it accounts for the shearing occurring at a microscopic scale within the fluid. In the case of turbulent flow there is also shear arising from the friction between adjacent eddies of water with differing momentum. This produces an apparent viscosity that is additional to the molecular viscosity. We call this apparent viscosity due to turbulence the **eddy viscosity** $\xi$. If Equation 5.11 defines the shear stress within laminar flow then the following defines the shear stress in turbulent flow

$$\tau = (\mu + \xi) \frac{du}{dz} \qquad (5.13)$$

Unlike the molecular viscosity, which is constant for a given fluid, the eddy viscosity varies with the flow conditions, increasing as the turbulence intensity increases. Representative values for the eddy viscosity are generally much larger than the molecular viscosity, and indicate the greater mixing and more rapid transfer of momentum that occurs within turbulent flows.

## 5.4 BENTHIC BOUNDARY LAYERS

When a fluid is in relative motion with a boundary such as the sea bed, friction arises between the two. This friction initially affects only the fluid motion in direct contact with the sea bed, but eventually these effects reach to progressively higher elevations in the flow. The region of fluid that is close to the sea bed and influenced by frictional effects is referred to as the **benthic boundary layer**. Modern concepts of boundary layer flow are still firmly rooted in the work of the German hydrodynamicist Ludwig Prandtl.

Numerous experiments have shown that the mean horizontal velocity within the boundary layer increases from zero at the sea bed (the no-slip condition) to a maximum at the top of the boundary layer. A plot of the mean horizontal flow velocity at increasing elevations above the bed defines a curve that we call the **velocity-profile** (Figure 5.11). When a fluid begins to flow over the sea bed the boundary layer is initially laminar and its thickness grows slowly with distance travelled. In most coastal settings laminar flow is short-lived, however, and the boundary layer proceeds through the transitional regime and on to become a turbulent boundary layer. Momentum transfer from the boundary up through the flow is slow in laminar boundary layers because the fluid viscosity is small (only molecular viscosity), whereas it is relatively rapid in turbulent boundary layers because the viscosity is large (molecular plus eddy viscosity). The enhanced rates of momentum exchange that can be achieved by macroscopic mixing in turbulent boundary layers explains why they can grow in thickness at much faster rates and why they have much steeper near-bed velocity gradients than laminar boundary layers (Figure 5.11).

In our earlier discussion of shear stress in a fluid we described it as the tangential component of stress acting on any plane within the fluid. In regard to sediment dynamics it is actually the bed shear stress that we are most interested in. The **bed shear stress** is the tangential component of stress occurring on the fluid plane that is in contact with the bed. Consistent with our earlier discussion, the bed shear stress is also positively related to the velocity-gradient. It should be no surprise that, for a given free-stream velocity and bed roughness, the bed shear stress beneath a turbulent boundary layer is greater than a laminar boundary layer, because of the steeper velocity-gradient.

The preceding introduction to benthic boundary layers is necessarily brief and does not do justice to the enormous literature available. The interested reader should refer to Allen (1985), Allen (1997) and Leeder (1999) for a broader discussion, albeit at a more advanced level. Our goal here is to

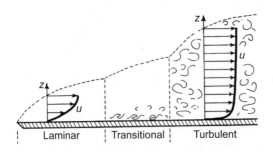

**Figure 5.11** – Schematic illustration of a growing boundary layer and its transformation from laminar to turbulent flow. Notice that the velocity profile in the turbulent boundary layer is steeper, thus larger current velocities impinge on the bed. [From Allen, 1985.]

develop a simple methodology for modelling sediment dynamics in coastal flows, so the following sections on steady and oscillatory boundary layer models is focused primarily on methods for calculating bed shear stress.

## 5.4.1 Boundary layer model for steady flow

Channelized river or tidal flows in deltas and estuaries and wind-driven currents on the inner shelf are all approximately steady flows over short periods of time. In these situations the benthic boundary layer grows with distance travelled by the flow, and if permitted to become fully established, it can occupy the entire water depth. Up to this point we have discussed the benthic boundary layer as a single entity, but in fact it consists of three layers (Figure 5.12a):

- The **bed layer** is generally 1–10 cm thick and can consist of up to two sublayers: the buffer sublayer and viscous sublayer. The presence or absence of the viscous sublayer depends on the hydraulic characteristics of the boundary (see below).

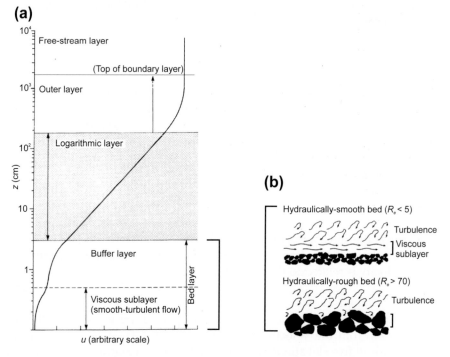

**Figure 5.12** – (a) Schematic current velocity profile plotted on log–linear axes and showing different layers within the boundary layer. The free-stream layer extends from the top of the boundary layer to the water surface. The vertical dimensions shown are indicative of a tidal current some 10 m deep. [Modified from Wright, 1989.] (b) Schematic showing the relationship between flow and bed conditions in the presence and absence of a viscous sublayer. [Modified from Allen, 1994.]

- The **logarithmic layer** is where the velocity increases logarithmically with height above the bed, so that the velocity profile when plotted on a log-linear scale is a straight line. The thickness of the logarithmic layer is generally 1–2 m.
- The **outer layer** is by far the thickest, occupying about 85% of the total boundary layer, which may be tens of metres in suitable water depths.

The stated thickness of each layer above is only indicative, and depends to a large extent on the steadiness of the flow. The more steady the flow, the thicker the layers will become.

The **viscous sublayer**, when present, is a very thin layer of laminar flow (< 1 cm thick) that separates the turbulent buffer sublayer from the bed (Figure 5.12b). It only exists in the presence of hydraulically-smooth beds, which are generally composed of mud (silts and clays) or very fine sand. Hydraulically-rough beds are generally composed of sediment coarser than fine sands. These grain sizes protrude up through the thickness of the viscous sub-layer and destroy it, thus enabling the turbulent flow in the buffer sublayer to impinge directly onto the bed. Hydraulically-smooth and -rough beds can be distinguished according to a version of the Reynolds number termed the boundary Reynolds number $R_e^*$

$$R_e^* = \frac{\rho u_* k'}{\mu}$$   (5.14)

where $u_*$ is the shear velocity and $k'$ is the skin friction roughness length. We will define what we mean by these two terms shortly. Hydraulically-smooth boundaries occur when $R_e^* < 5$, hydraulically-rough boundaries occur when $R_e^* > 70$, and transitional conditions occur between 5 and 70.

The velocity-profile in the logarithmic layer can be described by an equation known as the **Law of the Wall**

$$u = \frac{u_*}{\kappa} \ln\left(\frac{z}{z_0}\right)$$   (5.15)

where $u_*$ is the shear velocity, $\kappa$ is von Karmen's constant (equal to 0.4), $z$ is elevation above the bed and $z_0$ is the hydraulic **bed roughness length**. The shear velocity is related to the all important bed shear stress $\tau_b$ by

$$\tau_b = \rho u_*^2$$   (5.16)

The Law of the Wall provides us with an opportunity to estimate the bed shear stress from the velocity-profile. Consider the flow velocity measurements made at five elevations above the bed shown in Figure 5.13. Notice that we have plotted the velocity profile differently to previous occasions, this time the velocity is on the vertical axis and the height above the bed is on the horizontal axis. The data must be presented in this way, with the dependent variable on the vertical axis, if we are to avoid erroneous estimates of the bed shear stress and bed roughness length. When the velocity-profile is plotted in this way a line of best fit can be drawn through the data of the form $y = ax + b$,

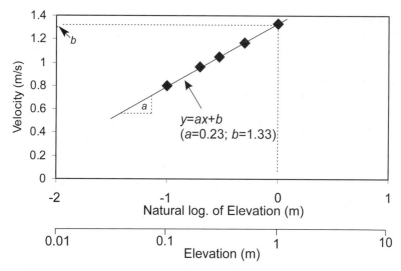

**Figure 5.13** – Illustration of method for estimating bed shear stress and bed rough-ness using the Law of the Wall. The velocity measurements at five elevations above the bed are shown as diamonds and the line of best fit through the data is shown as the solid line. See text for complete explanation.

where $y$ is the velocity, $x$ is the natural logarithm of elevation above the bed, $a$ is the gradient of the line and $b$ is the value of $y$ when $x$ is 0. The bed shear stress $\tau_b$ can be determined from the value of $a$ using

$$a = \frac{u_*}{\kappa} = \frac{1}{\kappa}\sqrt{\frac{\tau_b}{\rho}} \quad \text{hence} \quad \tau_b = \rho\,(\kappa\,a)^2 \qquad (5.17)$$

Moreover, the bed roughness length $z_0$ can be determined from the value of $b$ using

$$b = a \ln z_0 \quad \text{hence} \quad z_0 = e^{-b/a} \qquad (5.18)$$

For the example shown in the figure the regression coefficients $a$ and $b$ are 0.23 and 1.33, which yields a bed shear stress of 8.69 N m$^{-2}$ and a bed rough-ness length of 0.0031 m.

While the method described in Figure 5.13 is a sound means of estimat-ing the bed shear stress and bed roughness in a flow, it requires several measurements of the flow velocity in the logarithmic layer, which is not always possible. Field investigations of flow over numerous sea bed types has yielded an alternative method for estimating the bed shear stress based on a single velocity measurement

$$\tau_b = \rho C_d u_{100}^2 \qquad (5.19)$$

where $C_d$ is the fluid-drag coefficient and $u_{100}$ is the mean horizontal velocity at 100 cm above the bed. Some representative values of the drag coefficient for various sea bed types, together with bed roughness lengths, are listed in Table 5.6.

**Table 5.6** Indicative values for $C_d$ and $z_o$ under steady flow over different bed types

| Bottom type | $z_o$, mm | $C_d$ |
|---|---|---|
| Mud | 0.2 | 0.0022 |
| Mud/sand | 0.7 | 0.0030 |
| Silt/sand | 0.05 | 0.0016 |
| Sand (unrippled) | 0.4 | 0.0026 |
| Sand (rippled) | 6 | 0.0061 |
| Sand/shell | 0.3 | 0.0024 |
| Sand/gravel | 0.3 | 0.0024 |
| Mud/sand/gravel | 0.3 | 0.0024 |
| Gravel | 3 | 0.0047 |

(from Soulsby, 1983)

## 5.4.2 Boundary layer model for oscillatory flow

While the nature of steady and oscillatory boundary layers are generally similar with respect to the basics, there is one major difference that requires us to use a different method for estimating bed shear stress under waves. This difference is the temporal variation in boundary layer structure under

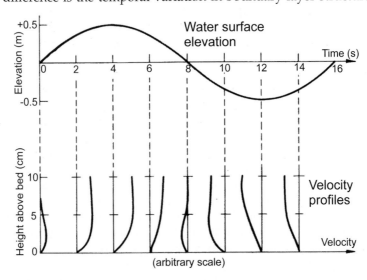

Figure 5.14 – Schematic diagram showing current velocity profiles in oscillatory flow. During the first 8 seconds the flow is in the direction of wave travel. The current speed increases and the boundary layer grows in height in the period 0 to 4 seconds. In the period 4 to 8 seconds the current speed decreases and the boundary layer diminishes. A similar pattern occurs during the period 8 to 16 seconds, however, the flow direction has reversed. The vertical dimensions shown are indicative of a swell wave with a height of 1 m and a period of 16 s. [Modified from Sleath, 1984.]

waves. Figure 5.14 shows the boundary layer velocity profile at several phases of the wave cycle. During each wave cycle the horizontal flow velocity accelerates, then decelerates, crosses zero (changes direction), accelerates and decelerates once more. Consequently, a new boundary layer grows and decays twice during each wave cycle, once on the forward stroke of the wave and once on the backward stroke. Due to the short time periods involved (generally a few seconds), the boundary layer thickness does not reach much more than 10 cm for long-period swell waves and is considerably smaller for short-period wind waves. The relatively thin boundary layer beneath oscillatory flow means that, for a given free-stream velocity and bed roughness, the bed shear stress in oscillatory flow is always larger than beneath steady flow.

The method used to estimate the bed shear stress under oscillatory flow first requires either a measurement or estimate of the maximum horizontal flow velocity at the top of the wave boundary layer $u_o$. If a measurement is not available then linear wave theory can be used to predict this velocity (see Equation 4.14). Next we require an estimate of the bed friction factor under waves $f_w$, which can be obtained from (Nielsen, 1992)

$$f_w = \exp\left[ 5.5 \left( \frac{k_s}{d_o} \right)^{0.2} - 6.3 \right]$$

(5.20)

where $d_o$ is the wave orbital-diameter, also obtainable from linear wave theory (see Equation 4.13) and $k_s$ is known as the **Nikuradse roughness length**, which is analogous to the bed roughness length described for steady flows (Section 5.4.1). An empirical recipe is available to estimate the Nikuradse roughness length, but it is not straightforward. The first step is to understand that there are at least two contributors to the total roughness. The first is the grain (skin friction) contribution $k'$ and is related to the roughness of individual grains making up the bed. The second is the bedform contribution $k''$ and is related to the roughness provided by ripples and dunes that may be present on the bed. Under conditions of extreme sediment transport, for example sheet flow, there is an additional apparent roughness, which we will not consider here. In most cases the total Nikuradse roughness length will simply be the sum of the grain and form roughness contributions

$$k_s = k' + k''$$

(5.21)

Experimental data suggest the following formulae for estimating each roughness contribution (Nielsen, 1992)

$$k' = D$$

(5.22)

and

$$k'' = 8 \frac{\eta^2}{\lambda}$$

(5.23)

where $D$ is the mean grain diameter of the bed sediment, $\eta$ is the bedform height and $\lambda$ is the bedform spacing. Now that we are able to obtain an estimate for the maximum flow velocity at the top of the boundary layer and the wave friction factor we can calculate the maximum bed shear stress under waves $\tau_w$ using

$$\tau_w = \frac{1}{2} \rho f_w u_o^2 \qquad (5.24)$$

As an example let us consider a shallow-water wave with a height of 0.5 m and a period of 15 s in a water depth of 5 m. You should be able to calculate the wave semi-excursion using Equation 4.15, which is 1.67 m, and the maximum horizontal flow velocity at the top of the boundary layer using Equation 4.16, which is 0.35 m s$^{-1}$. Now assume that the grain diameter of the bed sediment is 0.5 mm, and the bed is covered with ripples 3 cm in height and 20 cm apart. You should be able to calculate the Nikuradse roughness length using Equations 5.21–5.23 to be 0.037 and the wave friction factor using Equation 5.20 to be 0.024. Finally, the bed shear stress using Equation 5.24 is 1.51 N m$^{-2}$. For comparative purposes, now take the maximum oscillatory velocity and substitute it into Equation 5.19 with a representative drag coefficient for rippled sands from Table 5.6. You should get a shear stress of 0.77 N m$^{-2}$, which confirms that for a similar free-stream velocity the bed shear stress is much larger under waves than currents. Now that we have the tools to calculate the bed shear stress under either steady or oscillatory boundary layers we can now move onto the details of sediment dynamics.

## 5.5 SEDIMENT DYNAMICS

The dynamic behaviour of sediment in a moving fluid is strongly determined by the grain size of the sediment. For grain sizes greater than 63 $\mu$m the grains are free to behave individually and single-grain properties are most important (*e.g.*, grain size). For grain sizes less than 63 $\mu$m the sediment is cohesive due to electro-static forces. The dynamic behaviour of cohesive sediment depends less on single-grain properties and more on bulk-sediment properties (*e.g.*, floc size and water content). This distinction in sediment response to a moving fluid is most apparent when considering sediment entrainment and sediment deposition.

### 5.5.1 Sediment entrainment and resuspension

A cohesionless grain at rest on a bed of similar grains experiences no acceleration so all of the forces acting on the grain must be in equilibrium. The forces involved are the lift, drag and weight forces. The lift force arises due to the slightly faster flow across the top of the grain compared to the lower part of the grain, which results in a pressure differential. The idea is similar to the simple experiment of blowing over the top of a piece of paper, which causes a velocity and pressure differential between the top and bottom sides of the paper that causes it to lift. The drag force arises due to the friction between the fluid and the particle, and it therefore acts horizontally in the direction of mean flow. The weight force arises from the mass of the grain being acted upon by gravity. In order for the grain to move it must pivot over the adjacent grain through an angle equal to the friction angle (Figure 5.7b). This can only occur if the lift and drag forces overcome the weight force acting on the grain. In principle, mathematical expressions can be derived to describe the nature of these forces, which can then be manipulated to solve for the critical flow conditions required to initiate grain motion (*e.g.*, Allen, 1997). In practice, however, the drag and particularly the lift forces are difficult to quantify. For this reason the flow conditions necessary to initiate sediment movement are usually predicted on the basis of data from numerous laboratory experiments (*i.e.*, empirical data).

Shields (1936) was the first to conduct exhaustive experiments in this area and proposed that, when grains begin to move in a steady current, a relationship exists between the bed shear stress and the grain size. More recently, Soulsby (1997) has compiled all of the subsequent steady flow data as well as the oscillatory flow data and presented them on the one diagram (Figure 5.15). In order to make the data performed under a vast range of experimental conditions comparable, the **critical bed shear stress** required to initiate grain motion is non-dimensionalized in a parameter known as the **Shields parameter**

$$\theta_c = \frac{\tau_c}{gD(\rho_s - \rho)} \tag{5.25}$$

The non-dimensional grain diameter $D_*$ that appears on the horizontal axis in Figure 5.15 is given by

$$D_* = D\left[\frac{\rho^2 g(s-1)}{\mu^2}\right]^{1/3} \tag{5.26}$$

For non-dimensional grain diameters less than *c.* 10 there is an inverse relationship between the grain diameter and the critical Shields parameter. Or put more simply, as the grain diameter decreases the bed shear stress required to initiate sediment motion increases. This is due to a combination of factors. Grains in the silt and clay size range are readily compacted to the point that individual grains do not protrude through the viscous sublayer and are therefore not exposed to turbulent flow (hydraulically smooth conditions; Figure 5.12b). Moreover, grains in this size range are cohesive and therefore experience an electrostatic attraction to the bed. Both of these factors increase in effect as the grains become smaller, thus making them difficult to entrain despite their small mass. For non-dimensional grain diameters greater than 10 the complications arising from grain cohesion disappear and the grain sizes are sufficiently large to break up the viscous

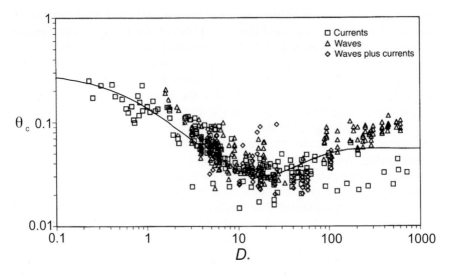

**Figure 5.15** – Modified Shields diagram showing empirical data and best fit function (Equation 5.27) for predicting the critical Shields parameter necessary to initiate sediment motion. Equation 5.25 can be used to determine the equivalent critical bed shear stress. For the case of waves this approach yields the critical maximum bed shear stress for the wave cycle. [Modified from Soulsby, 1997.]

sublayer, thus turbulent flow impinges directly on the bed (hydraulically rough conditions, Figure 5.12b). In these circumstances the weight force is of overriding importance so there is a positive relationship between the critical bed shear stress and the grain size.

Based on the data shown in Figure 5.15 the following expression has been proposed to predict the critical Shields Parameter (and bed shear stress) required to initiate grain motion

$$\theta_c = \frac{0.30}{1 + 1.2D_*} + 0.055[1 - \exp(-0.02D_*)] \tag{5.27}$$

It is important to note that the data used to produce Figure 5.15 is largely based on experiments involving sediments composed of uniform grains. Poorly sorted sediments may lead to a situation where surface grains are resting on a bed of larger or smaller grains. In the former case the pivot angle through which the grain must move is larger than the case for well sorted sediment (Figure 5.7b). In the latter case the pivot angle is smaller. There is a corresponding effect on the critical bed shear stress required to initiate motion, thus Equation 5.27 only applies to well sorted sediments.

Although the modified Shields diagram (Figure 5.15) predicts the critical bed shear stress required to initiate motion of cohesive sediments, it oversimplifies the process for this sediment type – it really only predicts the critical conditions to initiate movement of single grains of silt from a compacted bed. In a natural environment such as an estuary, however, silty and clayey sediments display a wide range of compactions. Recently deposited sediment is poorly compacted and has a high porosity and water content (small bulk density), whereas 'aged' sediment is well compacted and has a low porosity and water content (large bulk density). These two situations produce markedly different responses to an erosive tidal current (Section 7.4.2). We know that poorly compacted sediments are entrained far more easily than the aged sediments represented in Figure 5.15, but there has been little quantitative work to date and no reliable predictive equation is available.

## 5.5.2 Modes of sediment transport

Once in motion, the transport path (or mode of transport) that a sediment grain takes is largely determined by the mass of the grain and the speed of the current. Sediment transport is traditionally categorized into two modes (Figure 5.16):

- **Bedload** – grains transported as bedload are supported by either continuous contact (**traction**) or intermittent contact (**saltation**) with the bed. In the case of traction the grains slide or roll along, maintaining contact with the bed at all times. This is a relatively slow form of transport and is typical when weak currents are transporting sands or strong currents are transporting pebbles and boulders. In the case of saltation the grains take short hops along the bed. Saltation is typical when moderate currents are transporting sand or strong currents are transporting gravel and pebbles.
- **Suspended load** – grains transported as suspended load are supported by the turbulence in the fluid. The grains may make intermittent contact with the bed, but on average they spend most of their time in suspension. The grain paths of suspended load are distinguishable from saltation, due to their irregularity, which arises from the grains being buffeted by turbulent eddies in the current. Suspension transport is typical when moderate currents are transporting silts or strong currents are transporting sands. Grains transported as **wash load** are permanently in suspension, and typically consist of clays and dissolved material.

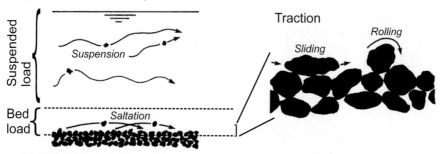

Figure 5.16 – Schematic representation of sediment transport modes showing grain paths. Note that bedload includes both saltation and traction. [Modified from Allen, 1994.]

## BOX 5.2 – RELATIVE TRANSPORT SPEEDS OF LARGE CLAST-SIZES

Sandy sediments are transported across the entire range of transport modes, and in general the speed at which sand is transported is closely related to the transport stage $\psi$ (Equation 5.28) – relatively slow transport speeds occur during bedload and relatively fast transport speeds during suspended load. Gravels and pebbles, on the other hand, move almost entirely as bedload and the transport speed of individual clasts depends not only on the transport stage, but also on their relationship to the background sediment mass. Based on the ratio of the size of an individual clast to the modal size of the surrounding sediment mass, there are three possibilities (Figure 5.17; Orford *et al.*, 1991):

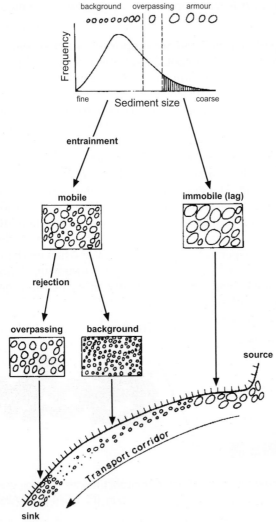

**Figure 5.17** – Schematic diagram demonstrating the importance of the clast size distribution to the relative speeds that individual clasts are transported. Paradoxically, larger clasts can sometimes travel faster than smaller clasts. See text for further explanation. [Modified from Orford *et al.*, 1991.]

- If the ratio is less than half, then the grain movement will be impeded as the clast will lose its surface position and be subsumed into the background. In other words, smaller clasts will be located within the interstices of larger clasts and will not participate in sediment transport.
- If the ratio is greater than about two, then the clast will have a propensity to move faster than the modal value for the sediment mass. This occurs because the clast will project further into the boundary layer, and will therefore experience larger fluid drag forces. The preferential transport of such larger clasts is termed **overpassing**.
- Clasts that are much bigger than the modal clast size and so have very large ratios may sink into the sediment (due to their large mass), become immobile, and form a **lag** deposit.

The net result of differential transport speeds is that an initial, poorly-sorted sediment will become a better-sorted deposit through constant rejection of the larger grains. In the context of gravel beaches, these processes may lead to segregation of the gravel deposit and the development of graded shorelines (Bird, 1996). A classic example of such a graded beach is Chesil Beach, south England, where over a distance of 30 km the clast size decreases from c. 10 cm to 1 cm at an approximately linear rate.

Francis (1973) performed a series of flume experiments to investigate the relationship between the sediment transport mode and the intensity of the transporting current. His results can be summarized by introducing the concept of transport stage $\Psi$

$$\Psi = \frac{u_*}{w_s} \tag{5.28}$$

which is the ratio of the shear velocity $u_*$ to the grain settling velocity $w_s$. The former is a surrogate for the forces driving the transport of sediment and the latter is a surrogate for the forces driving the deposition of sediment. For low $\Psi$, only bedload is transported. There is no clear threshold-value for $\Psi$ that marks a boundary between bedload and suspended load transport. As $\Psi$ increases, the amount of material transported as bedload declines and the proportion travelling as suspended load correspondingly rises. Moreover, the speed at which the grains are being transported increases. For high $\Psi$, almost all transport is as suspended load moving at a speed that is close to the speed of the fluid. The preceding discussion is based on experimental data for sand- and granule-sized material. A discussion of the factors controlling the speed at which coarser material is transported can be found in Box 5.2.

## 5.5.3 Models for calculating the transport rate

In general terms the **sediment transport rate** $q$ can be defined as the mass of sediment transported per unit cross-sectional area of flow per unit time.

Consider a unit width of fluid that extends from the stationary bed level to the height of the highest saltating grains being transported as bedload. The bedload transport rate $q_b$ is then the product of the mass of sediment in that cross-sectional area of fluid and the grain velocity, which is equivalent to saying

$$q_b = \int_0^{z=a} u_g(z)C(z)dz \qquad (5.29)$$

where $u_g$ is the grain velocity as a function of height above the bed $z$, $C$ is the concentration of grains as a function of height above the bed, and $a$ is the height of the bedload layer. While Equation 5.29 is theoretically valid it is rather impractical. The bedload layer is typically a couple of centimetres high at most, so it is difficult to measure grain velocities and concentrations directly. An alternative approach that uses easily determined parameters is required. From our preceding discussion of transport stage we know that as the shear velocity (and bed shear stress) of a current increases, the faster sediment is transported. It seems reasonable to expect then, that the sediment transport rate is proportional to the bed shear stress exerted by the flow, or more precisely the excess bed shear stress above the critical amount required to initiate grain motion. Numerous field and laboratory experiments indicate that the bedload transport rate can be predicted using an equation of the form (e.g., Meyer-Peter and Muller, 1948)

$$q_b = A(\tau_b - \tau_c)^{1.5} \qquad (5.30)$$

where $\tau_b$ is the bed shear stress, $\tau_c$ is the critical bed shear stress and $A$ is a proportionality coefficient that depends on the properties of the sediment.

The suspended load transport rate can also be determined using Equation 5.29, except that integration occurs over the water column height from the top of the bedload layer to the water surface. In many cases it is possible to make suitable measurements to utilize this approach, however, it requires a significant amount of expensive instrumentation. Typically the only data available to predict the suspended sediment transport rate is the mean current velocity.

An alternative approach to the sediment transport problem was proposed by Bagnold (1963, 1966), who equated a transporting current to a machine and the sediment transport to work done by that machine. In this approach, which is referred to as the **energetics approach**, the work done (sediment transport rate) is proportional to the power of the machine (transporting current). For bedload transport Bagnold proposed

$$q_b = \frac{\tau_b u e_b}{\tan\phi - \tan\beta} \qquad (5.31)$$

where $\tau_b u$ is the current power, $e_b$ is the bedload efficiency factor, $\phi$ is the friction angle of the sediment and $\beta$ is the slope-angle of the bed. The

**Table 5.7** Indicative values for bedload and suspended load efficiency factors under steady and oscillatory flow

| Flow type | Bedload efficiency, $e_b$ | Suspended load efficiency, $e_s$ |
|-----------|---------------------------|----------------------------------|
| Steady | 0.15 tan $\phi$ | 0.01 |
| Oscillatory | 0.10 | 0.02 |

(from Bagnold, 1966; Bailard, 1982)

bedload efficiency is smaller than one, because the current is not 100% efficient at transporting sediment – some of the power is lost due to frictional dissipation as heat. For suspended load transport Bagnold proposed

$$q_s = \frac{\tau_0 u e_s}{(w_s/u) - \tan \beta} \tag{5.32}$$

where $e_s$ is the suspended load efficiency factor. The transport efficiency factors depend on both flow conditions and sediment properties (Table 5.7). For equivalent sediment properties, waves are generally more efficient at transporting sediment than steady currents.

In order to apply any of the sediment transport models just presented to a practical situation that involves steady currents or waves, all that is required are measurements of the bed slope, grain size, grain settling velocity and current speed. The bed shear stress is estimated using the appropriate method described in Section 5.4. Situations that involve combined waves and steady currents, however, are far more complex. Nonlinear interaction between the waves and the current produces a combined boundary layer that is not well described by either of the individual boundary layer models that we have presented. Combined wave-current boundary layer models exist, but they are beyond the level of an introductory text. We conclude our discussion of boundary layer sediment transport by simply pointing out that in combined flows waves are mostly responsible for suspending sediment and the steady current transports it away.

## 5.5.4 Sediment deposition

Sediment deposition involves the settling of grains towards the bed, from either the bedload or suspended load. In the case of the former, when the bed shear stress and fluid turbulence are insufficient to keep the sediment moving, the grains simply cease their horizontal movement and rapidly come to rest. In the case of suspended load the grains must settle a considerable distance through the fluid before coming to rest. Sediment will begin to settle when the upward acting fluid forces are insufficient to overcome the downward acting weight force on the grain.

When a single cohesionless grain is suspended in a stationary fluid and is then released, it will initially experience a period of acceleration towards the bed. This acceleration is short-lived, however, because the relevant forces quickly equilibrate so that the grain's acceleration drops off to zero at the same time as it reaches its **terminal** (maximum) **fall velocity**. It should be no surprise that the grain is falling through the fluid due to the downward acting weight force, but what is the balancing force when the grain is travelling at its terminal fall velocity? The answer is a combined buoyancy and fluid-drag force. Consider the simple case of a perfectly spherical grain in a stationary body of water. The balance of forces can be expressed by

$$\frac{4}{3}\pi r^3 \rho_s g \quad = \quad \frac{4}{3}\pi r^3 \rho g \quad + \quad \frac{1}{2}C_d\rho\pi r^2 w_s^2 \qquad (5.33)$$

| downward-acting weight force | = | upward-acting buoyancy force | + | upward-acting fluid-drag force |

The weight force is the product of the sphere's volume, its density and the gravitational acceleration. The buoyancy force is the product of the sphere's volume, the water density and the gravitational acceleration. The fluid-drag force is the product of the drag coefficient, the water density, the sphere's surface area and its velocity squared. The half at the front is there because the drag force is only acting over the leading hemisphere of the grain as it falls through the water.

The balance of forces in Equation 5.33 can be re-arranged to yield an expression for the terminal fall velocity of the sphere

$$w_s = \sqrt{\frac{8gr}{3C_d}\frac{(\rho_s - \rho)}{\rho}} \qquad (5.34)$$

So if we have a spherical grain with a known radius (or diameter) and density, settling in a fluid with a known density, then we can use Equation 5.34 to calculate the fall velocity provided we also know the value of the drag coefficient. For small grain Reynolds numbers the flow around the grain as it falls through the fluid is laminar and the drag coefficient can be estimated using

$$C_d = \frac{24}{R_G} \qquad (5.35)$$

where $R_G$ is the grain Reynolds number in which the grain settling velocity and the grain diameter are the appropriate velocity and length scales used in Equation 5.12. For larger Reynolds numbers, which may relate to larger settling velocities, or larger grains in a less viscous fluid, the prediction of the drag coefficient is less straightforward. Equations 5.34 and 5.35 constitute **Stokes Law of settling** and are strictly valid only for spheres settling in a fluid with a grain Reynolds number less than 20. For grains settling in water this restricts Stokes Law to grain sizes less than 0.15 mm diameter. Or to put it another way, it can only be used in the case of very fine sand, silt or clay. Stokes Law is not valid at all in air, due to the low viscosity of air and the large density difference between sediment grains and air.

In the case of grains coarser than very fine sand, the flow separation around the grains as they settle through the fluid complicates the drag force and leads to unpredictable behaviour. Moreover, the grain shape and roundness modify the drag and therefore the fall velocity, the effects being larger for larger grain sizes. For this reason the fall velocity is usually measured directly (see Lewis and McConchie, 1994 for description of method), or calculated from empirical equations. Soulsby (1997) has analysed most of the available data for sands of typical shapes and has proposed the following formula to estimate the grain settling velocity

$$w_s = \frac{\mu}{\rho D} \left[ \sqrt{(10.36^2 + 1.049\, D_*^3)} - 10.36 \right] \tag{5.36}$$

where $D_*$ is the dimensionless grain size given by Equation 5.26. The settling velocity of sand grains increases directly with grain size and grain density, and increases inversely with fluid density and fluid viscosity.

We now turn our attention to the settling of cohesive sediment (*i.e.*, grain diameters > 4$\phi$). Stokes Law is appropriate for calculating the settling velocity of fine particles if they remain dispersed, but in coastal waters they generally combine to form flocs, which are several particles held together to produce a composite particle. Moreover, several flocs may combine to produce an aggregate (Figure 5.18). The process of **flocculation** results from the electrical charges on individual particles, but is also strongly influenced by organic coatings. The face of a clay platelet has a slight negative charge and the edge a slight positive charge. When two platelets come into close proximity, the face of one particle and the edge of the other are electrostatically attracted. The probability of the particles coming together in freshwater is low, however, because the negatively charged faces of the two particles, which have a much larger surface area than the edges, will tend to repel the particles from one another. In the case of seawater the probability of effective attraction improves considerably. Seawater is a strong electrolyte which helps to neutralize the negatively-charged faces, thus facilitating electrostatic attraction between the particles. In order for flocculation in seawater to be significant, individual particles need to be brought within very close proximity to one other. This is achieved by either Brownian motion within a relatively concentrated suspension, differential settling of variously sized particles, and/or fluid turbulence. A detailed discussion of flocculation can be found in Eisma (1993).

Figure 5.18 — Indicative dimensions of a clay particle, a floc (group of clay particles) and an aggregate (group of flocs). [Modified from Eisma, 1993.]

*1 μm*

Individual clay particle

*10 to 20 μm*

Individual floc

*50 to 200 μm*

Individual floc group (aggregate)

The effect of flocculation is to considerably increase the fall velocity of clay-sized particles. Flocs have the combined mass of their component particles, and therefore settle more rapidly than the component particles would as individuals. Since collisions of individual particles are a necessary precursor to flocculation, it should also be apparent that the more concentrated the suspension of particles the more likelihood there is for flocculation. The end result is that there is often a clear relationship between the settling velocity of cohesive sediment and the suspension concentration, although, the relationship is not consistent from one site to another. Local factors such as water biochemistry and the presence/absence of organic films on grains can also play an important role in the flocculation process. These factors are difficult to account for in any general sense and have hindered attempts to develop a predictive equation for the settling velocity of cohesive sediment.

## 5.6 BEDFORMS

**Bedforms** are quasi-regular patterns on the sea floor that develop during bedload and, to a lesser extent, suspended load transport. Flow conditions control the bedform morphology (crest pattern and dimension), and in turn the bedforms locally modify both flow conditions and sediment transport modes. Bedforms are also one of a variety of bed roughness elements that increase the friction factor and therefore reduce the energy of waves and currents.

### 5.6.1 Dimensions and relation to flow regime

The nomenclature and the pattern of flow and sediment transport over a typical bedform shape is shown in Figure 5.19a. The flow displays a spatial acceleration across the stoss slope, and transports sediment towards the crest (brink point). As the flow reaches the crest it separates from the bed and drives an eddy in the lee of the bedform. The sediment transported to the crest is piled to a slope-angle greater than its yield angle and it consequently avalanches downslope. This cycle of sediment erosion from the stoss slope and deposition on the lee slope results in the bedform migrating in the direction of flow. The relationship between the bedform and the flow is intimate, with small changes in either causing a corresponding change in the other.

The example just described is for unidirectional flow. The pattern for oscillatory flow over asymmetric wave ripples can be seen in Figure 5.20. In this case the relatively strong onshore stroke of the wave forms a vortex behind the wave ripple. Provided the onshore flow persists, the vortex remains trapped in the lee of the ripple, but when the flow reverses, the vortex is thrown upward off the bottom and a small cloud of suspended sediment is ejected into the water column. The sediment cloud is moved seaward by the offshore stroke of the wave. Because the offshore stroke of

**(a)**

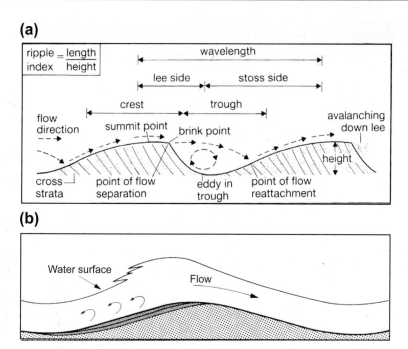

**(b)**

Figure 5.19 – Schematic diagram showing nomenclature, flow and sediment transport pattern over bedforms in unidirectional flow: (a) ripple or dune [From Tucker, 1995.] [Copyright © 1995 Blackwell Publishers, reproduced with permission.]; and (b) antidune. [From Reineck and Singh, 1980.] [Copyright © 1980 Springer-Verlag, reproduced with permission]

the wave is relatively weak, no vortex forms in the lee of the ripple. The net suspended sediment transport during one wave cycle is therefore in the seaward direction, which produces the asymmetry in the ripple shape. If the orbital velocity-magnitude is symmetrical, then the wave ripples are symmetrical and a vortex (with associated sediment cloud) occurs on both sides of the ripple crest each wave cycle. Refer to Box 5.3 for further discussion.

Figure 5.20 – Schematic diagram showing suspended sediment transport by onshore asymmetric wave motion over sharp-crested ripples. See text for further explanation.

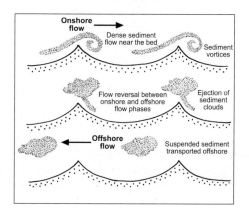

## Box 5.3 – Acoustic-Visualisation of Sediment Suspension Over Wave Ripples

Recent technology has considerably improved our ability to investigate near-bed sediment suspensions beneath waves and currents. One of the most useful instruments available is the acoustic back-scatter sensor (ABS), which consists of a sound source and a receiver or hydrophone. When the acoustic signal (sound) that is emitted by the ABS encounters a sediment suspension, the signal scatters in all directions. The ABS measures the intensity of the signal that is scattered back towards the instrument. The acoustic backscatter strength is then calibrated against known suspended sediment concentrations to yield a suspension concentration 'map' like the one shown in Figure 5.21.

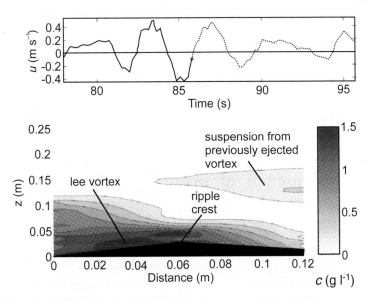

**Figure 5.21** – Top panel shows the time series of oscillatory flow velocity above the ripple crest. The flow velocity history that is responsible for the suspension pattern shown in the lower panel is marked in bold. The bottom panel shows a snapshot in time of the suspension concentrations over the ripple crest. See text for further explanation. [From Villard and Osborne, 2002.] [Copyright © 2002 International Association of Sedimentologists, reproduced with permission.]

This particular map shows a slice through the suspension concentration that is parallel to the direction of wave travel, so that the wave ripple is seen in cross-section. In this example, waves were travelling from left to right, and at this particular time the seaward stroke of the wave was decelerating to zero (top panel). Some of the key features of suspension dynamics described in the earlier schematic diagram (Figure 5.20) are clearly evident in this data. A vortex containing large concentrations (c. 1.5 g l$^{-1}$) of suspended sediment is clearly evident in the lee of the ripple crest. Moreover, well above the ripple crest is another vortex of suspended sediment that was ejected from the ripple crest during an earlier wave cycle. The suspension concentration in the latter is smaller (c. 0.25 g l$^{-1}$) due to some deposition since ejection. Although only a few experiments have produced this type of acoustic imagery to date, the ABS has already advanced our understanding of suspension dynamics considerably. The future widespread use of ABS in field experiments is expected to produce a wealth of new research opportunities.

Most bedforms fall into the categories of either ripples, dunes or antidunes. These bedform types can be distinguished largely on the basis of dimension (Table 5.8). **Ripples** and **dunes** can be further described as either 2-dimensional or 3-dimensional based on the plan view of the crest line. Straight-crested bedforms are 2-dimensional, and sinuous through to linguoid bedforms are 3-dimensional (Figure 5.22). **Antidunes** are low relief bedforms that have heights indicative of ripples, but lengths indicative of dunes. They are restricted to rapid, shallow flows that have a wavy water surface. This type of flow is called super-critical flow. Antidunes are so named because they have the potential to migrate in the opposite direction to the flow, hence the steepest slope is on the upstream side of the dune and the gentlest slope is on the downstream side. Migration is achieved by scour on the upstream side by the breaking surface wave as it slowly migrates upstream (Figure 5.19b). The distinction between flow regimes can be determined using the **Froude Number** $F_r$

$$F_r = \frac{u}{\sqrt{gh}} \tag{5.37}$$

**Table 5.8** Dimensions of bedforms and associated flow conditions

| | Unidirectional flow | | Oscillatory flow | | |
|---|---|---|---|---|---|
| | **Ripples** | **Dunes** | **Antidunes** | **Rolling-grain ripples** | **Vortex ripples** |
| Length (spacing) | 0.1–0.2 m | 0.6–30 m | 0.1–1 m | 0.02–1 m | 0.02–1 m |
| Height | < 0.06 m | 0.06–1.5 m | 0.01–0.1 m | A few mm | A few cm |
| Ripple index | 8–15 | > 15 | Not applicable | > 10 | 4–10 |
| Typical flow velocity | Low | Moderate | High | Low and high | Moderate |
| Typical flow depth | > a few cm | A few dm | A few cm to dm | Up to half the wavelength | Up to half the wavelength |
| Typical grain size | 0.03–0.6 mm | > 0.3 mm | All sand and gravel | All sand | All sand |

(after Reineck and Singh, 1980; Sleath, 1984; Nielsen, 1992)

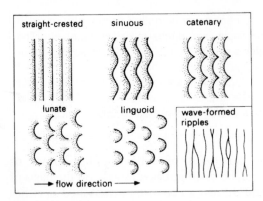

**Figure 5.22** – Schematic diagram showing plan view of crest lines for 2-dimensional and 3-dimensional bedforms. [From Tucker, 1995.] [Copyright © 1995 Blackwell Publishers, reproduced with permission.]

Sub-critical flow occurs when $F_r<1$ and supercritical flow occurs when $F_r>1$. Ripples and dunes are the products of sub-critical flow and antidunes are the product of super-critical flow.

While ripples and dunes can be distinguished solely on their dimensions, they can also be distinguished by several other factors that relate to their behaviour. The sizes of both ripples and dunes are controlled by the flow velocity (and bed shear stress), but only dune size is further controlled by flow depth (Table 5.8). Specifically, the size of both ripples and dunes increases with flow velocity and the size of dunes also increases with water depth. For a given flow velocity and water depth, larger grain sizes yield larger ripples and dunes. This last statement is true for ripples in sediment up to a grain size of 0.6 mm. Beyond this grain size ripples do not exist. This is understood to be related to the fact that beyond this grain size the bed is hydraulically rough for all flow velocities capable of transporting sediment, thus precluding the existence of a viscous sublayer at the base of the boundary layer. The precise details are yet to be established, but the implication is that ripples are somehow scaled to the bed layer thickness whereas dunes are scaled to the entire flow depth.

## 5.6.2 Bedform stability fields

Flow conditions change more quickly than the bed configuration. This is particularly true for dunes since a substantial volume of sediment must be moved to change their size and shape. It is not uncommon, therefore, for bedforms to be in disequilibrium with the flow conditions, particularly in unsteady tidal flows. Nevertheless, it is instructive to know for what flow conditions each bedform type is stable. This information can be presented in bedform stability diagrams such as those shown in Figure 5.23. Such diagrams indicate a 'field' of combined grain sizes and flow velocities for which each bedform type is stable. An analogous bedform stability diagram exists for oscillatory flow over wave ripples (Figure 5.23b). Only two stability fields exist in this case: one field for vortex ripples and one field for rolling grain (low-steepness) ripples. The former are dominated by

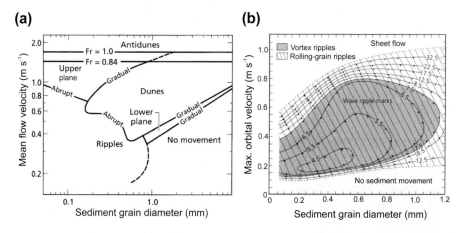

**Figure 5.23** – Bedform stability diagram for (a) unidirectional flow and (b) oscillatory flow beneath wind waves and swell. [Unidirectional flow modified from Southard and Boguchwal, 1990.] [Copyright © 1990 Society for Sedimentary Petrology, reproduced with permission.] [Oscillatory flow modified from P.A. Allen, 1997 after J.R.L. Allen, 1985.] [Copyright © 1997 Blackwell Publishers, reproduced with permission.]

suspended sediment transport (Figure 5.20) and the latter by bedload transport.

## 5.7 SUMMARY

- Sediments are typically classified according to statistical descriptors of their grain size distribution namely the mean (1st moment), sorting (2nd moment) and skewness (3rd moment). In order to calculate these statistical descriptors the size data are first log-transformed onto a phi-scale. All three descriptors indicate something of the history of the sediment, as do the grain shape and roundness.
- Turbulent boundary layers consist of several sublayers that can be recognized by the shape of their velocity profile. These layers are the bed layer (including viscous sublayer and buffer sublayer), the logarithmic layer and the outer layer.
- The bed shear stress is a fundamental parameter in sediment transport calculations. It can be estimated for steady flow using the Law of the Wall and at least two velocity measurements in the logarithmic part of the boundary layer. It can be estimated for oscillatory flow using an empirical recipe based on estimates of the bed roughness and flow velocity at the top of the wave boundary layer.
- The critical bed shear stress required to entrain a sediment grain from a well-sorted bed can be predicted from a Modified Shields Diagram. For sediments in the silt and clay size classes the critical bed shear stress

increases with decreasing grain size, due to the fact that these sediments form hydraulically smooth beds and the grains are cohesive. For coarser sediments the critical bed shear stress increases with grain size, reflecting the overriding importance of the weight force. The situation is more complicated in the case of poorly sorted sediments or cohesive sediments with a high water content.

■ Sediment is transported as either bedload or suspended load. In the case of the former the grains are supported by continuous (traction) or intermittent (saltation) contact with the bed. In the case of the latter the grains are supported by upward-directed fluid momentum related to turbulence. The sediment transport rate can be calculated using a variety of equations. Most of them involve either the excess bed shear stress or the fluid power, which is the product of the bed shear stress and flow velocity.

■ Settling of cohesionless sediment is governed by grain properties such as the size (mass) and shape, whereas settling of cohesive sediment is governed by the suspension concentration and the development of flocs.

■ Bedforms can be classified according to their dimensions and related flow regimes. Current and wave ripples are small-scale bedforms and dunes and antidunes are large-scale bedforms. Ripples and dunes are the products of subcritical flow and antidunes are the product of supercritical flow. Ripples seem to occur in relation to hydraulically smooth bed conditions whereas dunes seem to occur in relation to hydraulically rough bed conditions.

## 5.8 FURTHER READING

Allen, J.R.L., 1984. *Sedimentary Structures: Their Character and Physical Basis.* Elsevier. [The most comprehensive work available on the variety of bedforms known and the physical processes responsible for them.]

Allen, J.R.L., 1985. *Principles of Physical Sedimentology.* George Allen and Unwin. [A detailed physical account of all aspects of sediment dynamics by one of the most active and respected researchers in this field.]

Allen, P.A., 1997. *Earth Surface Processes.* Blackwell Science. [A valuable advanced text that includes chapters on sediment dynamics.]

Leeder, M., 1999. *Sedimentology and Sedimentary Basins: From Turbulence to Tectonics.* Blackwell Scientific Publications. [Another valuable advanced text that includes chapters on sediment dynamics.]

Pye, K., 1994. Properties of sediment particles. In: K. Pye (editor), *Sediment Transport and Depositional Processes.* Blackwell Scientific Publications, 1–24. [Provides a detailed overview of sediment properties.]

# 6

# FLUVIAL-DOMINATED COASTAL ENVIRONMENTS – DELTAS

## 6.1 INTRODUCTION

Coastal **deltas** are accumulations of allochthonous (terrigenous) sediment deposited where rivers enter into the sea. River sediments may also accumulate at the head of coastal embayments if the coastline is drowned (bayhead delta), but these deposits are generally controlled by estuarine processes and will not be discussed here (see Section 7.2.1). This chapter focuses on situations where the delta is sufficiently large to cause the adjacent coastline to prograde. This implies that sediments must be delivered by the river faster than they are dispersed by waves, tides and ocean currents. For a river to deliver sufficient sediment to cause coastal progradation, the drainage basin usually needs to be large, although high rates of precipitation and catchment denudation are also sufficient conditions. Large drainage basins are mostly restricted to the tectonically passive trailing-edge of continents where the drainage divide is well inland from the coast. High rates of precipitation and catchment denudation mostly occur in mid to low latitudes. Inman and Nordstrom's (1971) census of major deltas placed 57% along trailing-edge coastlines (*e.g.*, the Amazon), and 34.5% along coasts fronting marginal seas that are often protected from ocean waves by island arcs (*e.g.*, the Klang). Many in this latter group are located in the mid to low latitudes (*e.g.*, the Fly).

There is an enormous volume of literature on deltaic environments. Most research either addresses case studies of individual deltas or struggles to formulate an all-encompassing classification of deltas in order to make some sort of broader sense of the case studies. There are several major controlling factors, all of which can vary between wide limits, so the possible combinations seem endless (Figure 6.1). Rather than focusing on delta classification schemes, this chapter will concentrate on the morphodynamic processes occurring in deltaic environments. This approach still requires us to adopt a

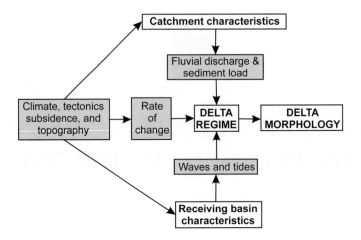

**Figure 6.1** – Interaction of factors responsible for the observed global variation in delta morphology. [Modified from Elliot, 1986.]

classification scheme in order to structure and scope our discussion. We will use the scheme proposed by Galloway (1975), because it has received widespread acceptance, it is reasonably straightforward, and it offers a suitable context for focusing on processes. Other delta classification schemes are reviewed in Reading and Collinson (1996), Boggs (1995) and Postma (1990).

## 6.2 GALLOWAY'S CLASSIFICATION OF DELTAS

Galloway's scheme is a tripartite classification with three end-members (Figure 6.2):

■ **Fluvial-dominated deltas** are characterized by large catchments, river discharge into protected seas with minimal nearshore wave energy, and a small tidal prism.
■ **Wave-dominated deltas** are characterized by exposure to open-ocean swell.
■ **Tide-dominated deltas** are characterized by a tidal prism larger than the fluvial discharge.

Relative rather than absolute magnitude of river, wave and tide power is the important controlling factor.

Fluvial-dominated deltas are the only delta-type that is indisputably a delta. Reading and Collinson (1996), and others, have noted that many deltas classified as wave-dominated may be better classified as beach-ridge strand plains, since most of their sediment is supplied by littoral drift rather than river processes. Similarly, many so-called tide-dominated deltas should probably be considered estuaries since they are transgressive systems. Nevertheless, there are major deltas that meet the criteria of coastal pro-

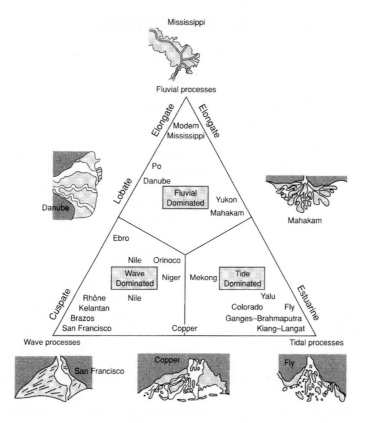

**Figure 6.2** – Galloway's (1975) classification of fine-grained deltas. The end-member types are fluvial-, tide- and wave-dominated. [From Briggs *et al.* 1997.] [Copyright © 1997 Taylor & Francis Books, reproduced with permission.]

gradation and yet are modified substantially by wave or tidal processes (*e.g.*, the Niger and the Ganges-Brahmaputra). We therefore adopt Galloway's scheme and acknowledge its shortcomings. The next three sections focus on the morphodynamics of fluvial-dominated deltas, with a brief description of how waves and tides modify delta morphology in particular settings.

It is important to understand that Galloway's scheme seeks only to classify medium to fine-grained deltas, which include almost all the major deltas of the world. A different type of coastal delta occurs on the tectonically-active leading-edge of continents. These settings are characterized by proximal drainage divides, steep hinterland slopes, high rates of denudation, and narrow-deep continental shelves. This combination of factors produces coarse-grained, steeply-sloped, fan- and braid-type deltas. While not as large as their finer-grained counterparts, these deltas can be locally important in terms of coastal progradation. We conclude this chapter with a brief summary of coarse-grained delta morphology, and highlight the contrasts with their larger, finer-grained counterparts.

# 6.3 GENERIC DELTA MORPHOLOGY

Delta morphology is a result of factors common to both the river catchment and the receiving basin (coastal waters), as well as factors unique to each (Figure 6.1). The combination of all these factors, together with their rate of change through time, determines the delta regime (*i.e.,* river-, wave-, or tide-dominated), which in turn controls the delta morphology. There are numerous possible permutations, each leading to a unique delta morphology. Often only one factor needs to be different to produce a completely different style of delta. For example, two deltas that have exactly the same catchment characteristics and exactly the same receiving basin characteristics, except one has a larger basin subsidence rate, will evolve into two markedly different delta morphologies. The slowly subsiding delta will prograde furthest seaward with a relatively thin sediment wedge, whereas the rapidly subsiding delta will have the thickest sediment wedge, due to the constantly increasing accommodation space at the river entrance. The two deltas will also develop different entrance conditions. Because the delta with the slow subsidence rate progrades the furthest, it will be the first to become exposed to open ocean waves and tides, whilst the other remains fluvially-dominated. This is just one example of how varying a single factor can produce markedly different deltas.

Although the detailed morphology of deltas varies from one example to the next, depending on the delta regime, there are three morphological units common to almost all deltas (Figure 6.3): the delta plain, delta front and prodelta. The **delta plain** is the sedimentary platform that mantles recent coastal progradation. The sedimentary deposits of the delta plain are called topset beds and they lie unconformably over the sediments of past delta fronts, which are called foreset beds. Each foreset bed marks a previous location of the delta front during the delta's progradation. Each foreset bed onlaps the bottom-set beds, which are the product of prodelta sedimenta-

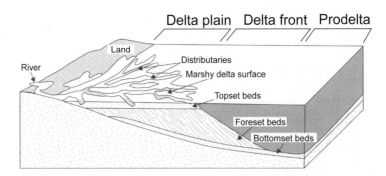

**Figure 6.3** – The three generic morpho-stratigraphic units found in all fine-grained deltas: delta plain (topset beds), delta front (foreset beds) and prodelta (bottomset beds). [Modified from Haslett, 2000.]

tion. In general terms we perceive a situation where the river continuously delivers sediment to the coastline and causes the **delta front** to prograde seaward through deposition of the relatively coarse sediments closest to the coast and the finer sediments further seaward. The finest sediments are deposited on the **prodelta**. As the delta front progrades horizontally the delta plain aggrades vertically. The resulting sediment distribution through the delta wedge is then a general fining in the seaward direction and a general coarsening upwards in the vertical direction. Reading and Collinson (1996) review much of the geological and sedimentological literature on deltas. The following paragraphs describing the delta plain, delta front and prodelta provide a synthesis of their expansive review.

## 6.3.1 Delta plain

The delta plain can be traversed by either a single or multiple channels, whose role it is to distribute water and sediment from the river valley out to the presently active coastline. The number of these **distributary channels** active on the plain at any one time is mostly related to the slope of the plain and the grain size of the sediment load. Rivers carrying a coarse-sediment load across a steep delta plain have several simultaneously active distributary channels, whereas rivers carrying a fine-sediment load across a gently-sloped delta plain have relatively few (although there may be numerous abandoned channels). Distributary channels convey predominantly unidirectional flow on the upper delta plain and behave much like fluvial channels, but they may convey reversing tidal flows on the lower plain and behave more like estuarine channels. There is a corresponding zonation of morphological features within the interdistributary areas. The **upper delta plain** includes fluvial features such as levees, floodplains, lakes and swamps between the active channels, all of which are fed sediment by overbank flow during floods. On the other hand the **lower delta plain** includes estuarine features such as brackish lagoons, tidal flats, shallow bays, marshes and relict beach ridges.

The distributary channels are relatively high-energy environments, so they tend to contain only the coarsest fractions of the sediment load. The interdistributary areas, however, are low-energy environments dominated by deposition of fine-sediment from suspension. Due to the continual initiation, growth and abandonment of distributary channels associated with delta switching (Section 6.5), the distribution of sediment on the delta plain can be complex. It is not uncommon, for example, for relatively coarse sediments from an active or recently abandoned distributary channel to be juxtaposed with relatively fine sediment deposited under quiescent conditions in an interdistributary lagoon.

The surface slope of the delta plain is generally horizontal to very gently seaward dipping, and this slope is always maintained, even when the delta front is prograding. Thus the topset beds of the delta plain are also horizontal to very gently dipping. For topset beds to aggrade the basin must subside, otherwise the distributary channels will continually erode the delta plain. The thickest topset beds are associated with the largest rates of basin subsidence. The main causes of basin subsidence are isostatic compensation for sediment loading and sediment compaction.

## 6.3.2 Delta front

The delta front is the most active part of the delta and the most interesting from a morphodynamic viewpoint. It is also where the uniqueness of each delta is most apparent. Most of the sediment issuing from the presently active distributary mouths is deposited on the delta front. As the river water exits the distributary mouth it expands and decelerates, thus losing its competency to transport sediment. The coarsest sediment is deposited first and produces a **distributary mouth bar**. The morphology of the bar depends on the hydrodynamic behaviour of the river effluent as it leaves the distributary mouth, plus any modification caused by waves and/or tides. The details are discussed more fully in Section 6.4. In summary, using the terminology of Coleman (1982), the three types of effluent and related bar morphology are: (1) an axial jet, which deposits a lunate bar that is relatively narrow and with subdued relief (Figure 6.4a); (2) a planar jet, which deposits a broad radial bar or a broad middle ground bar with high relief and bifurcating channel (Figure 6.4b); and (3) a buoyant plume, which deposits long, straight, subaqueous levees and a distal bar (Figure 6.4c). The distributary mouth bars associated with jets are situated closer to the distributary entrance than those associated with buoyant plumes. Tides tend to rework sediments at the distributary mouth into linear shoals that are oriented perpendicular to the coastline (*i.e*, parallel to the direction of tidal flow; Figure 6.4d). Waves tend to rework the sediments into swash bars and beach ridges that are oriented parallel to the coastline (*i.e.,* parallel to the refracted wave crests; Figure 6.4e).

## 6.3.3 Prodelta

The large distance between the prodelta and the distributary mouth means that only the finest sediments reach this zone, generally very fine silts and clays. These sediments settle from suspensions carried by either jets or buoyant plumes. The lack of bedload transport and a generally horizontal to gently-sloping bed means that the prodelta aggrades vertically to produce relatively flat-lying bottomset beds (Figure 6.3). Not all deltas include a prodelta – if present, they are located in water depths beyond the regional wave base and are also largely unaffected by tides. The only potentially active process on the prodelta is **mass wasting**, which refers to the movement of bed sediment downslope under the action of gravity. If mass wasting is unimportant then the prodelta surface will generally be smooth and featureless. If it is important then a variety of slumps, depressions and erosive channels and gullies can disrupt the surface (Figure 6.5).

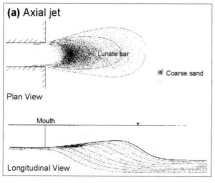

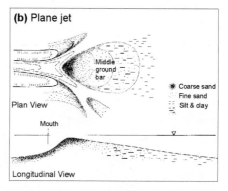

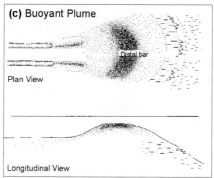

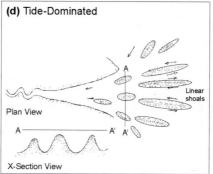

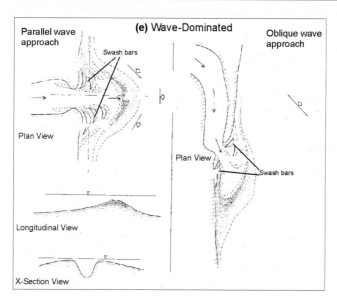

**Figure 6.4** – Morphology of distributary mouths dominated by: (a) axial jets, (b) plane jets, (c) buoyant plumes, (d) strong tides and (e) strong waves. [From Wright, 1977.] [Copyright © 1977 Geological Society of America, reproduced with permission.]

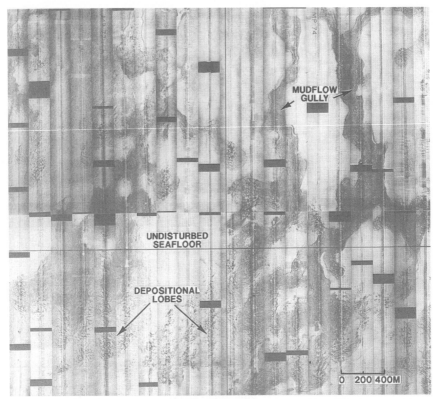

**Figure 6.5** – Sidescan sonar mosaic showing mass wasting features in the Mississippi River delta, USA. [From Coleman et al., 1998.] [Copyright © 1998 Coastal Education and Research Foundation, reproduced with permission.]

Reading and Collinson (1996) identified the following three conditions necessary for mass wasting: (1) a very high sedimentation rate, which leads to retention of pore waters and ultimately high pore water pressure following burial, resulting in a low shear strength for the sediment; (2) biodegradation of organic detritus and the production of methane gas, which further weakens the sediment; and (3) a shock mechanism (*e.g.*, storm waves or tectonic activity), providing the trigger for mass wasting. Mass wasting events can be initiated in the prodelta zone, for example, as collapse depressions. Reading and Collinson describe these as shallow (< 3 m) bowl-shaped depressions about 100 m wide that result from liquefaction of the sediment, typically by storm waves. It is more likely, however, for mass wasting to be initiated on the more steeply-sloping delta front and their effects extended to the prodelta. For example, a rotational slump or submarine slide that develops into a turbidity current and erodes a channel down the delta front and across the prodelta (Box 6.1). These channels are slightly sinuous, up to several kilometres long, up to 20 m deep and up to 1500 m wide (Reading and Collinson, 1996).

# BOX 6.1 – GRAVITY-DRIVEN TURBIDITY CURRENTS

In Chapter 5 we discussed at some length the sediment transport mechanics based on fluid drag over the top of a sedimentary bed. There are a variety of additional transport modes whose mechanics are based on the downslope component of the sediment weight force (product of mass and gravity). These gravity-driven transport modes include soil creep, slides, debris flows and turbidity currents. We briefly discuss the mechanics of turbidity currents here as an example.

**Turbidity currents** are negatively buoyant, highly-concentrated, near-bed suspensions (typically composed of sand, silt and clay) that flow downslope under their own weight (Figure 6.6). There is no motion of the overlying water necessary to provide drag – gravity does all of the work. Turbidity currents characteristically consist of a head and a body. The body typically travels 16% faster than the head and so is constantly feeding the head region with fluid and sediment. The velocity of the turbidity current increases with the thickness of the body and with the density difference between the suspension and the ambient fluid.

Body                              Head

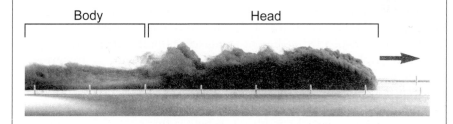

**Figure 6.6** – Turbidity current produced in the laboratory displaying the characteristic head and body regions. [From Allen, 1994.] [Copyright © 1994 Blackwell Publishers, reproduced with permission.]

As the turbidity current flows downslope, the Kelvin-Helmholtz waves that develop at the head (Figure 6.6 and Box 6.3) serve to entrain relatively clear fluid, thus reducing the bulk density and negative buoyancy of the suspension. Sediment is also deposited at the base of the turbidity current. In order for a turbidity current to be sustained, these mechanisms that reduce the bulk density of the suspension must be balanced by the body region entraining sediment and continually feeding it to the head. It is this sediment entrainment by turbidity currents that is responsible for eroding many of the channels observed on the lower delta front and prodelta. They are able to carve these often enormous channels during a relatively small number of transport episodes, because their large bulk density and hence their large mass generates more momentum than relatively clear currents of similar discharge. Detailed discussions of turbidity currents, and other gravity-driven transport mechanisms, can be found in Allen (1985), Allen (1997) and Leeder (1999).

## 6.4 DELTA-FRONT MORPHODYNAMICS

In a classic paper, Bates (1953) introduced the notion that delta-front morphology could be understood in terms of hydrodynamic jets. In a subsequent paper that is now also a classic, Wright (1977) built on Bates' foundation to provide a complete synthesis of river mouth morphodynamics. This section is a summary of these two papers.

The hydrodynamic behaviour of the river effluent as it enters the receiving basin is controlled by the current speed at the river mouth, the slope of the sea bed seaward of the river mouth and the vertical density distribution through the water column. Depending on how these factors combine, the effluent can behave as either a jet or a plume (Table 6.1). The principal determining factor is the density pattern, examples of which are illustrated in Figure 6.7. When the water is well mixed vertically the **isopycnals** (lines of equal density) are vertical and the water density at the surface and the bed are similar (*i.e.*, **homopycnal** conditions). When the water is stratified the isopycnals are nearly horizontal, with the least dense water mass floating on the most dense water mass. If the river effluent is the least dense, because it is fresher than the basin water, then it travels seaward on the surface (*i.e.*, **hypopycnal** conditions). If the river effluent is the most dense, because it has sufficient suspended sediment to makes its bulk density greater than the coastal water, then it travels seaward along the bed (*i.e.*, **hyperpycnal** conditions).

Given that the water exiting a river mouth is usually fresh and loaded with sediment and the coastal water is saline, there is always the potential for hypo- or hyperpycnal conditions. Whether this potential is realized or

**Table 6.1** Morphodynamic summary of distributary mouth processes

|  | Jet | | Plume | | Tides | Waves |
|---|---|---|---|---|---|---|
|  | Axial | Planar | Surface | Near-bed |  |  |
| River current velocity | Large | Large | Moderate | Moderate | Small | Small |
| Offshore slope | Steep | Gentle | Relatively steep | Relatively steep | Relatively gentle | Relatively steep |
| Density conditions | Homopycnal | Homopycnal | Hypopycnal | Hyperpycnal | Homopycnal | Homopycnal |
| Morphology | Narrow, lunate bar | Broad, radial bar or middle-ground bar and bifurcating channel | Subaqueous levees and distal bar | Gullies and channels from mudslides and turbidity currents | Linear channels and shoals aligned perpendicular to the coast | Swash bars and beach ridges aligned parallel to the coast |

Figure 6.7 – Schematic diagram showing three water
density patterns: (a) homopycnal conditions, (b) hypopycnal
conditions and (c) hyperpycnal
conditions. The lines are isopycnals labelled with indicative
water densities in kg m$^{-3}$
Homopycnal conditions are
associated with river effluents
behaving as jets. Hypopycnal
and hyperpycnal conditions are
associated with river effluents
behaving as positively- and
negatively-buoyant plumes,
respectively.

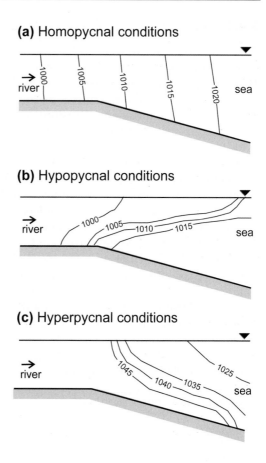

**(a)** Homopycnal conditions

**(b)** Hypopycnal conditions

**(c)** Hyperpycnal conditions

whether mixing and homopycnal conditions prevail depends on the
strength of the river current (which will tend to promote mixing) and the
difference in density between the river and coastal waters (which will tend
to promote stratification). The prevalence of one over the other is represented in the **densimetric Froude Number $F'_r$**

$$F'_r = \frac{u}{\sqrt{gh'\left(1 - \frac{\rho_R}{\rho_C}\right)}} \tag{6.1}$$

where $u$ is the river current velocity (or the plume velocity in hypopycnal
conditions), $g$ is the gravitational acceleration, $h'$ is the depth of flow (or the
thickness of the plume in hypopycnal conditions), and $\rho_R$ and $\rho_C$ are the
densities of the river and coastal water, respectively. For supercritical values
of $F'_r$ greater than unity the effluent behaves as either an axial or plane jet
and for subcritical values less than unity it behaves as a buoyant plume.
Subcritical plumes may become critical due to an increasing river discharge
(and velocity), decreasing water depth, or decreasing density ratio.

## 6.4.1 Jets

**Axial jets** occur where river water exits a distributary mouth at high speed into a deep basin under homopycnal conditions (Table 6.1; Figure 6.8a). There is no interaction between the jet and the sea bed, hence the energy of the jet is dissipated almost entirely by turbulent mixing between the jet and the coastal water (turbulent diffusion). The jet consists of a core region that extends from the river mouth a short distance seaward. The mean current velocity is constant throughout the core region, and the turbulence intensity is large. Beyond the core region is the zone of established flow. In this zone turbulent eddies cause the jet to exchange water and momentum with the basin water. This results in the expansion of the jet, a deceleration of its velocity and a reduction in the turbulence intensity with distance from the distributary mouth. The expansion rate, indicated by the spreading angle, is constant at about 12°, so the corresponding deceleration of the cross-section-ally averaged velocity of the jet is also constant (Box 6.2). Beyond the core region, the jet loses its competency to transport sediment and it deposits sediment in the form of a lunate bar (Figure 6.4a). In cross-section, this bar rises gently from the distributary mouth to the bar crest and then slopes steeply seaward. The lateral extent of the bar is relatively narrow and consistent with the width of the jet. The coarsest sediment is deposited around the margins of the core region and the sediment progressively fines in the seaward direction. Axial jets and lunate bars may occur at newly created

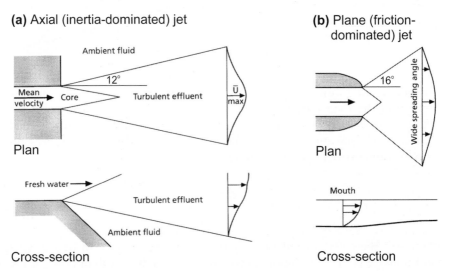

**(a)** Axial (inertia-dominated) jet

**(b)** Plane (friction-dominated) jet

**Figure 6.8** – Schematic illustration of (a) an axial (inertia-dominated) and (b) a plane (friction-dominated) jet, showing both plan and cross-section views. Note that axial jets do not interact with solid boundaries after exiting the mouth and are free to expand in all directions. Plane jets are strongly influenced by the sea bed. [From Leeder, 1999; modified from Wright, 1977.] [Copyright © 1994 Blackwell Publishers, reproduced with permission.]

## Box 6.2 – Flow Velocity in an Axial Jet

Axial jets are simple to produce in the laboratory and are relatively well understood, because the complications that arise due to interaction with a solid boundary are absent. The behaviour of the jet with distance from the orifice can be described completely if the current velocity at the orifice and the width of the orifice are known (e.g., Allen, 1985; Allen, 1997). Figure 6.9a shows that the jet spreads laterally at a roughly constant rate, the angle between the longitudinal axis of the jet and the margin being about 12°. The spreading is due to turbulent eddies along the margin of the jet causing an exchange of fluid and momentum between the jet and ambient waters. Since the total discharge and momentum flux through any cross-section in the jet must be equal (*i.e.*, conservation of mass and momentum), the cross-sectionally averaged velocity in the jet must decrease with distance

**(a)**   **(b)**

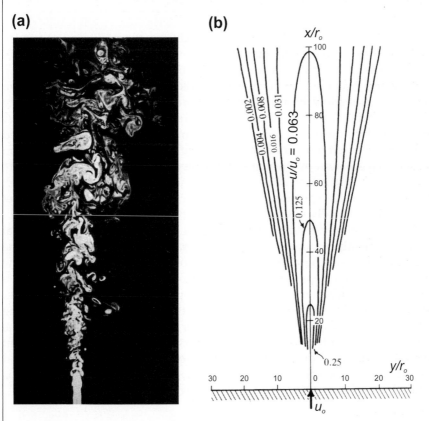

**Figure 6.9** – (a) Laboratory visualization of an axial jet showing an approximately constant rate of expansion from the orifice. [From van Dyke, 1982.] [Copyright © 1982 The Parabolic Press.] (b) Velocity contours in an axial jet. The distance in the *x* and *y* directions is normalized by the radius *r* of the jet and the velocity *u* is normalized by the velocity at the orifice $u_o$. [From Allen, 1985.]

from the orifice. If the angle of spread is constant it can be shown that the spatially-averaged velocity across the section, $\bar{u}$, decreases with distance at a constant rate given by

$$\bar{u} = \frac{w}{x\ tan\theta}\ u_0$$

where $w$ is the width of the orifice, $x$ is the distance from the orifice, $\theta$ is the angle of spread and $u_0$ is the velocity at the orifice. In reality, the velocity in the river entrance is not constant across the section, but is a maximum at the centre and a minimum at the margins. This leads to a velocity distribution across the jet similar to that shown in Figure 6.9b. As expected, there is a decrease in velocity with distance from the orifice, but there is also a decrease in velocity between the centreline and the adjacent margins.

distributary mouths, but the necessary conditions generally do not persist. As deltas develop the receiving basin becomes shallower so the jet is eventually affected by the sea bed, and therefore becomes a plane jet.

**Plane jets** occur where river water exits a distributary mouth at high speed into a shallow basin under homopycnal conditions (Table 6.1; Fig 6.8b). The vertical expansion of the jet is restricted by the shallow water depth, and in order to conserve mass and momentum this is compensated for by an increased horizontal spreading angle that is about 16°. In addition to energy dissipation by turbulent diffusion along the sides of the jet, there is also energy dissipation at the base of the jet by friction with the sea bed. Due to the greater spreading angle and bed friction, velocities in the jet decrease more rapidly with distance so neither the core region or the established flow region extend as far seaward as they do in the case of an axial jet. Consequently sediment deposition is rapid and occurs closer to the entrance. The resulting radial bar slopes rapidly up from the mouth to the bar crest and then slopes gently seaward. Continued deposition results in a positive feedback situation. As the water depth over the bar gets shallower through deposition, frictional dissipation of the jet increases, thus enhancing sedimentation and further shallowing over the bar. In order to continue delivering sediment to the coast the jet must expand laterally to the point where it bifurcates, forming two channels either side of a middle-ground bar. When the flow splits, the vertical and lateral mixing in each channel is reduced so each can continue further seaward.

## 6.4.2 Buoyant plumes

When neither the river water exiting a distributary mouth or the coastal water are turbulent enough to induce thorough mixing then the waters over the delta front will be stratified. In this case the river water proceeds seaward as a plume that is either positively (hypopycnal conditions) or negatively (hyperpycnal conditions) buoyant. **Positively-buoyant plumes** travel across the surface and **negatively-buoyant plumes** travel along the bed. The former situation is the more common, and is best developed when the river water exits the entrance with low to moderate velocity into relatively deep water (Table 6.1). Moreover, the river discharge should exceed the tidal prism, but should be insufficient to keep out the tidal intrusion of salt water into the lower reaches of the distributary channel (similar to a salt-wedge estuary; Section 7.3.1). These conditions produce stable stratification with the fresh river water floating on the denser, saline coastal water. The situation of negative buoyancy arises if the river water is carrying so much sediment in suspension that the bulk density of the fresh river water is larger than the density of the saline coastal water. This usually requires suspended sediment concentrations of the order of 100 g l$^{-1}$, which is rare. One example is the Huanghe River delta, which delivers silt-sized sediment from the Mongolian plateau in the interior of China to the Gulf of Bohai (Wright *et al.*, 1986). Here we will restrict our discussion to the more common, positively-buoyant plume.

The margins of buoyant plumes are marked by sharp boundaries known as **fronts**. These fronts can at times be recognized simply by an obvious colour difference between the plume and the coastal water. They may also be marked by lines of flotsam composed of foam, algae, or debris carried by the river. The lines of flotsam that mark fronts provide a clue to the nature of flow within the plume (Figure 6.10). The plume is flowing seaward, being driven through the river entrance by the sloping water surface and related pressure-gradient force (see Box 3.1). Because the plume is positively buoyant and elevated slightly above the coastal water, there is also a gentle slope from the centre of the plume out towards the margins (*i.e.*, the front). This sloping water surface is also associated with a pressure gradient force and therefore a secondary, lateral surface flow. The diverging surface flow at the centre of the plume draws water vertically upwards and the convergent flow at the margins of the plume drives water vertically downwards to complete the circulation. It is the surface flow convergence at the plume front that accumulates the flotsam. The convergent flow at the base of the plume means that as sediment is transported seaward its lateral dispersion near the bed is limited, so it tends to deposit in linear, subaqueous levees.

Figure 6.10 – Schematic cross-section of a buoyant plume showing circulation in the plane transverse to the flow. The principal flow direction is either into or out of the page. [From Wright, 1977.] [Copyright © 1977 Geological Society of America, reproduced with permission.]

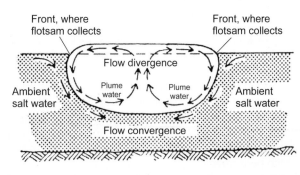

## BOX 6.3 – MIXING AND SEDIMENTATION IN A 'CRITICAL' PLUME

As a buoyant plume expands laterally with distance from the river mouth it becomes progressively thinner in order to conserve mass. If at the same time the plume velocity remains roughly constant, then the densimetric Froude number may eventually reach a value of one (Equation 6.1). If this occurs, then the weight force of the plume is insufficient to counteract the inertial forces driving the plume, so the shear along the density interface between the base of the plume and the coastal water will distort the interface into a series of internal waves (Figure 6.11a). Wright (1977) proposed that these waves promote mixing between the basal plume water and the coastal water, thus causing the plume to lose mass, momentum, and competency to transport sediment. This results in the coarser sediment being deposited to form the distal bar. The plume's deceleration eventually causes the Froude number to drop below unity again, thus suppressing mixing and enabling the plume to continue travelling seaward with the fine sediment load (Figure 6.11b). The fine sediment is progressively deposited seaward beyond the distal bar.

**(a)**

**(b)**

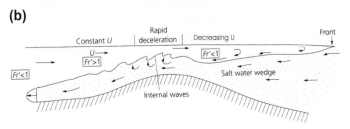

**Figure 6.11** – (a) Laboratory visualized internal waves produced on the interface between two liquids in relative motion to each other. In this case the waves are Kelvin-Helmholtz waves. [From van Dyke, 1982.] [Copyright © 1982 The Parabolic Press.] (b) Schematic longitudinal view of a buoyant plume showing the changes in densimetric Froude number as the plume thins in the seaward direction. When the plume is at its thinnest the Froude number reaches supercritical values and internal waves develop along the base of the plume. [From Allen, 1997; modified from Wright, 1977.] [Copyright © 1997 Blackwell Publishers, reproduced with permission.]

**Figure 6.12** – Satellite image showing part of the Mississippi River delta, USA, an archetypal fluvial-dominated delta displaying the classic 'birds-foot' morphology. [Modified image from the NASA/Science Photo Library, USA.]

15 km

These levees grow seawards and steadily extend the length of the distributary channel. In addition to long, subaqueous levees, deposition from positively-buoyant plumes is also responsible for a distal bar. The mechanics of distal sedimentation from the plume are complex, but they are controlled, at least in part, by the changing value of the densimetric Froude number as the plume thins with distance travelled from the distributary entrance (Box 6.3).

Positively-buoyant plumes and long, linear levees produce a delta front morphology that often has a 'birds-foot' appearance. This is the hallmark of fluvial-dominated deltas; including the Mississippi River delta, USA (Figure 6.12). The fresh water discharge rate of this river varies between 8,400 and 28,000 $m^3\ s^{-1}$, whereas the mean tide range is only 0.2 m. This predominance of fresh water flow, together with an annual sediment load of $2.1 \times 10^8$ tonnes (mostly consisting of fine sediment), results in the regular occurrence of sediment-laden positively-buoyant plumes, which are actively extending the delta front region (Wright, 1978; Milliman and Meade, 1983).

## 6.4.3 Tide- and wave-effects

Along meso- and macro-tidal coastlines the primary effect of the tide on delta front processes is to cause strong mixing and the maintenance of homopycnal conditions. The river effluent typically behaves as a plane jet that is interacting with strong tidal currents flowing roughly parallel with it (*i.e.*, perpendicular to the coastline). The result is elongated distributary mouth bars separated by tidal channels (Figure 6.4d). In order for the system to be a delta there must be a net transport of sediment seaward to achieve coastal progradation. The situation is made complex, however, by the fact that flooding tidal currents oppose the river effluent and ebbing currents flow with it. This results in **mutually-evasive sediment transport zones** in which one side of the channel conveys sediment mostly in the

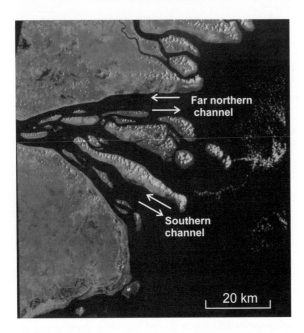

Figure 6.13 — Satellite image showing the Fly River delta, Papua New Guinea, an archetypal tide-dominated delta displaying the classic 'funnel-shape' morphology. Also note the linear sand ridges and tidal channels aligned perpendicular to the coast. Arrows indicate direction of net beload transport at the distributary entrance. [Modified image from Geoscience Australia, reproduced with permission.]

seaward direction while the other side conveys it mostly in the landward direction.

Mutually-evasive transport is well developed in the tide-dominated Fly River delta, Papua New Guinea (Figure 6.13). The fresh water discharge rate of this river is relatively large with an annual average of 7,000 $m^3$ $s^{-1}$. The tidal discharge rate, however, is about 18 times the river discharge rate. The annual sediment load delivered to the delta is $8.5 \times 10^7$ tonnes. The dominant tidal discharge results in homopycnal, reversing tidal currents controlling sedimentation. The net bedload transport (by combined river and tidal flow) at the entrance of the far northern channel is directed landward on the north-side and seaward on the south-side (Figure 6.13). There is a similar mutually-evasive net bedload transport pattern in the southern channel. Despite some sediment being transported landward at specific sites in this system, on the whole, most sediment is being transported seaward to feed an actively prograding delta front (Harris *et al.*, 2002).

When delta systems are connected to a steep shoreface the delta front may be exposed to high-energy waves. Waves also enhance mixing, hence any buoyancy is destroyed and the river effluent again behaves as a modified plane jet. Waves encountering the jet experience a shortening of their length and a corresponding increase in their steepness. As a result the waves break further offshore in a water depth greater than they otherwise would. This widens the surf zone on the delta front, which further enhances mixing and causes wave setup (Section 4.6.5). Wave setup has the effect of reducing the water surface slope through the distributary mouth and therefore reducing the river current velocity, thus the bar tends to be deposited closer to the mouth than it otherwise would be in the absence of waves

Figure 6.14 – Satellite image showing the Shoalhaven River delta, Australia, an archetypal wave-dominated delta displaying the classic 'arcuate-shape', swash-aligned shoreline. At the time of photography the Shoalhaven entrance was closed. [Modified image from Geoscience Australia's Oz Estuaries database.]

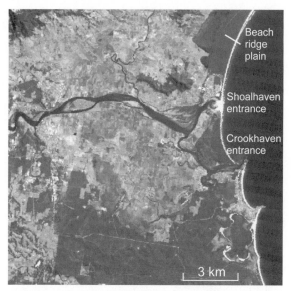

(Figure 6.4e). The seaward extent of the delta front is further limited by the reworking of delta front sediments by waves in the form of shoreward-migrating swash bars. As these bars weld to the beach, the delta plain progrades as a series of beach ridges and swales. When waves approach with their crest parallel to the coastline they are refracted symmetrically around the distributary mouth bar. When they approach obliquely, however, they cause longshore transport and spit growth, with the river entrance constantly migrating downdrift.

The Shoalhaven River delta, Australia is an archetypal wave-dominated delta (Figure 6.14). The river discharge is relatively modest, but episodic; the mean annual discharge rate is 57 m$^3$ s$^{-1}$, but is punctuated by flood events with a discharge rate of 5,000 m$^3$ s$^{-1}$. Sediment delivery to the delta front occurs almost entirely during flood events, since the entrance is typically closed for long periods between events. Closure is achieved by persistent wave energy pushing delta sediments landward. The significant wave height exceeds 1.5 m for 50% of the time and exceeds 3 m for 5% of the time (Wright, 1976). The dominant wave energy has shaped the long, arcuate beach to the north of the entrance, which is backed by a wide beach ridge plain that is also composed of delta sediment.

## 6.5 DELTA SWITCHING

Presently active deltas are relatively young features, having developed in the last 7,000-10,000 years since sea level reached its present high-stand position. The oldest deltas, fronting coastal valleys that were rapidly infilled, have undergone several cycles of what is termed **delta switching** during the Holocene. This is where the active region of coastal progradation switches

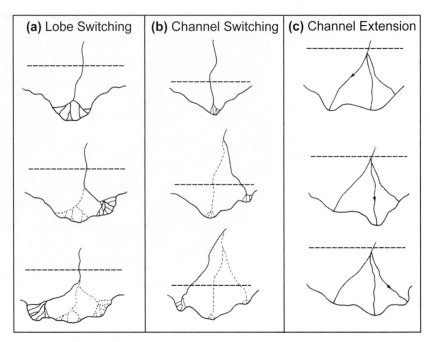

**Figure 6.15** – Schematic plan view of the three styles of delta switching: (a) lobe switching; (b) channel switching; and (c) channel extension. The border between the fluvial valley and the delta plain is indicated by the dashed, horizontal line. [Modified from Coleman, 1982].

from one location on the delta to another. The interval between switching varies between a few hundred to a few thousand years, depending on the size of the delta. Coleman (1982) identifies three types of switching: lobe switching, channel switching and channel extension (Figure 6.15).

**Lobe switching** occurs when the active area of delta progradation consists of a network of numerous distributary channels. After a period of time the active lobe becomes shallow, over-extended and inefficient at conveying sediment all the way across the delta plain. The entire network is then abandoned in favour of another more competent channel cut through a shorter, steeper section of the delta plain (Figure 6.15a). When this happens, the initial lobe becomes completely inactive with respect to fluvial processes and may undergo subsequent erosion by waves and tides. New lobes typically develop in coastal bays between previous lobes, or they extend over the top of previously subsided lobes. Lobe switching is favoured where there are low offshore slopes, and low wave- and tide-energy. **Channel switching** occurs well upstream, somewhere on the fluvial plain (Figure 6.15b). A new course is cut for the river and a new delta develops. This style of switching produces coastal progradation over a wide area, although it may be patchy. It seems to be favoured where there are intermediate offshore slopes, and high wave- and tide-energy. **Channel extension** involves the fluvial channel branching on the upper delta plain into two or more distributary channels that continue to the coast generally with no further branching (Figure 6.15c). At any one time a single distributary channel will carry the bulk of the river flow and sediment load, thus undergoing rapid progradation while the other channels are largely dominated by transgressive estuarine processes. Both channel switching and channel extension are initiated for the same reasons as lobe switching; *i.e.*, overextension of the presently active distributaries, which reduces their competency to convey sediment across the delta plain.

## 6.6 COARSE-GRAINED DELTAS

Up to this point we have concentrated on deltas deposited in relatively shallow coastal waters that consist predominantly of mixed sand and mud delivered to the coast by channelized flows as both bedload and suspended

### (a) Fan delta

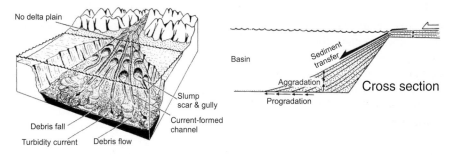

### (b) Braid delta

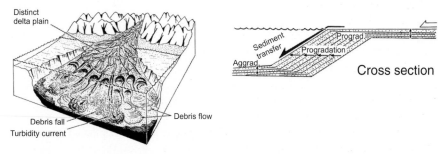

Figure 6.16 – Schematic showing both oblique and cross-section views of two styles of coarse-grained deltas. (a) Fan deltas are a subaqueous extension of subaerial alluvial fans, thus there is no delta plain and no subaerial progradation. (b) Braid deltas display a distinct delta plain and undergo subaerial progradation. [From Nemec, 1990.] [Copyright © 1990 International Association of Sedimentologists, reproduced with permission.]

load. This is the delta style characteristic of tectonically-passive, trailing-edge coastlines. Along tectonically-active, leading-edge coastlines deltas are relatively rare, but they can be locally important areas of coastal progradation. These deltas are characteristically coarse-grained, due to their proximity to the sediment source, and consist mostly of gravels, cobbles and boulders. Coarse-grained deltas also tend to have much steeper slopes than their fine-grained counterparts, due to the narrow and deep continental shelf that characteristically fronts leading-edge coastlines. Coarse-grained deltas can be classified as either the fan- or braid-type (Figure 6.16).

**Fan deltas** are composed of the coarsest sediment (largely cobbles and boulders) and occur where a subaerial fan feeds directly into coastal waters or a bedload-dominated river discharges into very deep coastal waters. Slopes on the underwater part of the fan generally exceed 27° and may reach 35–40° for angular material (Nemec, 1990). Almost all progradation and aggradation occurs beneath the water level of the receiving basin, hence the delta plain is virtually absent from this style of delta (Figure 6.16a). **Braid deltas** are composed of a slightly finer and better sorted sediment mix (largely gravels and cobbles) that is delivered by braided rivers that cross a narrow coastal plain before reaching the coastline. Offshore slopes are generally not as steep, thus permitting a morphological style that is similar to fine-grained deltas. The three generic morphological units – delta plain, delta front and prodelta – can all be recognized (Figure 6.16b).

An important difference between coarse- and fine-grained deltas is the

**Table 6.2** Summary of sediment transport mechanisms for coarse- and fine-grained deltas

| Morphological unit | Fan deltas | Braid deltas | Fine-grained deltas |
|---|---|---|---|
| Delta plain | Land-derived slope failure and avalanches; bedload transport by unconfined river flow | Land-derived slope failure and avalanches; bedload transport by poor to well-confined river flow | Bedload and suspended load transport in well-confined channels |
| Delta front | Land- and delta front-derived, unconfined avalanches and debris flows | Bedload transport by friction dominated jets; delta front-derived, poor to well confined turbidity currents; land-derived avalanches | Bedload and suspended load transport by waning jets; suspended load transport by plumes; delta front-derived slumps and channelized turbidity currents |
| Prodelta | Land- and delta front-derived, unconfined avalanches and debris flows | Delta front-derived debris flows and poorly channelized turbidity currents | Suspended load transport by plumes; delta-front derived turbidity currents |

mode of sediment transfer across the various zones of the delta (Table 6.2). The mechanism for sediment delivery to the delta front of fans is predominantly gravity-driven transport, with some poorly confined river flow occurring during flash-flooding. The river effluent is most likely to behave as a weak axial jet, but it has little impact on the delta front morphology. Episodic avalanches and debris flows that initiate at the top of the slope and continue all the way to the base in one episode are responsible for building the long, steeply sloping delta front (Figure 6.16a). In contrast with fine-grained deltas, the coarsest and finest materials are generally well mixed across the delta front, even out to the deepest water depths. Horizontal progradation and vertical aggradation is restricted to the seaward margin of the delta front, thus there is little change in the coastline position unless the basin becomes sufficiently shallow to permit the development of a braid delta.

On braid deltas a mixture of gravity-driven and stream-driven transport occurs across the delta plain. The upper delta plain receives material via avalanching and debris flows, usually triggered by heavy precipitation. This material makes its way across the lower delta plain to the delta front as bedload carried by numerous braided streams. At the delta front these streams debouch into the ocean as plane jets, immediately depositing their bedload at the top of the subaqueous slope. When the top of the slope becomes over-steepened material then travels down the subaqueous delta front by gravity-driven processes. The gravity-driven transport of coarse material occurs as unconfined debris flows while finer material is transported as poorly confined debris flows and turbidity currents.

## 6.7 SUMMARY

- The largest deltas commonly occur along tectonically-passive trailing edge coastlines. A large catchment area, a high annual rainfall, a high catchment denudation rate, and low wave- and tide-energy at the coast all have a positive impact on the areal size of the delta. A rapid subsidence rate will produce the thickest deltaic wedge and a slow subsidence rate will produce the most expansive deltaic wedge.
- Fine-grained deltas consist of three distinct morpho-stratigraphic units: the delta plain with topset beds, the delta front with foreset beds and the prodelta with bottomset beds. The delta front is the region of active coastal progradation and displays the greatest morphodynamic variability.
- Delta front morphology includes subaerial and subaqueous levees and a variety of distributary mouth bars. The precise nature of these features depends on the nature of the river effluent and any modifications caused by tides and waves.
- When homopycnal conditions occur and the densimetric Froude number is much greater than unity the river effluent behaves as a jet. When

hypopycnal conditions occur and the densimetric Froude number is less than unity the river effluent behaves as a positively-buoyant plume. In special circumstances hyperpycnal conditions occur and the river effluent behaves as a negatively-buoyant plume.

- Jets create deltaic bars situated proximal to the distributary mouth and plumes create deltaic bars that are distal. Tides cause deltaic bars to become elongated in the seaward direction. Waves cause deltaic bars to migrate shorewards to become beach ridges.

- Coarse-grained deltas occur on tectonically-active leading-edge coastlines. They differ from their fine-grained counterparts in many respects. They are composed predominantly of gravels, cobbles and boulders, rather than sands, silts and clays. They also occupy steeper and deeper receiving basins. The most striking contrast, however, is that sediment transport mechanics in coarse-grained deltas are dominated by gravity-driven processes whereas in fine-grained deltas they are dominated by fluid dynamic processes.

## 6.8 FURTHER READING

Colella, A. and Prior, D.B., 1990. *Coarse-grained Deltas*. Special Publication Number 10 of the International Association of Sedimentologists, Blackwell Science, Oxford. [A useful compilation of papers on the relatively under-studied coarse-grained deltas.]

Reading, H.G. and Collinson, J.D., 1996. Clastic coasts. In: H.G. Reading (editor), *Sedimentary Environments: Processes, Facies and Stratigraphy*, Blackwell Science, Oxford, 154–231. [A valuable synthesis of the extensive literature on delta morphology, sedimentology and stratigraphy].

Stone, G.W. and Donley, J.C. (editors), 1998. The world deltas symposium: A tribute to James Plummer Morgan (1919-1995). Special Thematic Section: *Journal of Coastal Research*, **14**, 695–916. [A broad collection of recent papers covering many aspect of delta geomorphology.]

Wright, L.D., 1977. Sediment transport and deposition at river mouths: A synthesis. *Geological Society of America Bulletin*, **88**, 857–868. [Seminal paper describing the relationship between effluent dynamics and morphology of distributary-mouth bars.]

# 7

# TIDE-DOMINATED COASTAL ENVIRONMENTS – ESTUARIES

## 7.1 INTRODUCTION

Dalrymple *et al.* (1992) define an **estuary** as '… the seaward portion of a drowned valley system which receives sediment from both fluvial and marine sources and which contains facies influenced by tide, wave and fluvial processes.' Modern estuaries first developed when coastal river valleys were flooded at the end of the last postglacial marine transgression. Sea level stabilized around 6,000 years before present, and since that time estuaries have been infilling. Estuaries that received a large sediment influx relative to their accommodation space during the Holocene have completely infilled and are now prograding as coastal deltas (Chapter 6). Estuaries that received a more modest sediment supply still have accommodation space and are continuing to infill. Sediment enters an estuary from both the land (river) and the sea (waves and tides). The fundamental difference between deltas and estuaries is that, averaged over many years, the net sediment transport in deltas is seaward whereas in estuaries it is landward. Deltas are progradational systems that are presently extending the coastline, whereas estuaries occupy coastal embayments that are presently infilling.

The various possible combinations of coastal processes (river, waves and tide), together with some additional processes that are uniquely estuarine, result in a rich variety of estuary types. As estuaries fill their valleys, an evolving morphodynamic regime further increases this variety. Not surprisingly then, estuaries are amongst the most widely researched coastal environments. There are many aspects of estuaries that could not be covered in this chapter and we highly recommend consulting the extra readings. The purpose of this chapter is to summarize what is known about the most generic aspects of estuary morphology as well as the uniquely estuarine processes that have not been dealt with elsewhere in this book.

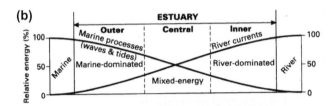

Figure 7.1 – (a) Plan view of an estuary showing facies and hydraulic boundaries. (b) Chart showing the change in energy regime along the estuary axis, i.e., the changing mix of wave, tide and river processes. [From Dalrymple et al., 1992.] [Copyright © 1992 Society for Sedimentary Geology, reproduced with permission.]

## 7.2 ESTUARY MORPHOLOGY

Dalrymple et al. (1992) proposed that most estuaries can be divided into three zones, which are unique with respect to their energy regime, sediment type and morphology. These zones are the inner zone, central zone and outer zone (Figure 7.1). The importance of river processes is greatest at the estuary head (landward end) and diminishes towards the estuary entrance (seaward end), whereas the importance of marine processes is greatest at the entrance and diminishes towards the head. Considering these processes in concert, the energy regime in the **inner zone** is river-dominated, in the **outer zone** it is marine-dominated (waves and tides) and in the **central zone** it is mixed (tide and river processes). The effect of waves is mostly restricted to the coast or seaward margin of the outer zone. The inner and outer zones are the most energetic and are predominantly sediment transfer zones. Dalrymple et al. (1992) argue that for the system to be infilling, the net bedload transport direction (averaged over several years) must be landward in the outer zone. This combined with the net seaward-directed transport in the inner zone results in sediment convergence in the central zone, which is a sediment sink. In keeping with the relative energy levels, the inner and outer zone generally contain the coarsest sediment and the central zone contains the finest sediment.

### 7.2.1 Estuary classification

The appearance of estuaries varies dramatically, depending on the palaeo-valley configuration, entrance conditions and degree of infilling. The first is determined by geological inheritance, the second by the predominance of waves or tides at the coastline and the third by the amount of available sediment since all estuaries are roughly the same age (*i.e.*, Holocene). There are many estuary classification schemes, but the one we have adopted was proposed by Dalrymple *et al.* (1992). We have chosen their scheme because of its focus on both morphology and processes. The two end-members of the scheme are wave- and tide-dominated estuaries, which each have characteristic energy regimes, morphology, and stratigraphy. The terms wave- and tide-dominated are based on an assessment of the relative energy levels of waves and tides in the outer zone of the estuary. While classification schemes are a useful framework for synthesizing information, it should always be remembered that a particular estuary may only display elements of the archetypes discussed here. Palaeo-valley configuration and sediment supply may be of overriding importance in some situations.

The outer zone of a **wave-dominated estuary** consists of a **barrier system** that may be subject to washover during storms and a **tidal inlet** (Figure 7.2). Deltas associated with the jet-like tidal flow through the narrow inlet develop both outside (**ebb-tide delta**) and inside (**flood-tide delta**) the estuary. This environment is dominated by wave processes, but their effects decline rapidly with distance from the coast, due to wave breaking over both the barrier and tidal deltas. There is also a rapid decline in tidal energy with distance from the coast, due to the restricted inlet cross-section, but some tidal energy persists into the estuary to shape the flood-tide delta. The outer zone is often composed of medium-size marine sands. A distinguishing characteristic of wave-dominated estuaries is the very low energy level in

**Figure 7.2** – (a) Chart showing the change in energy regime along the axis of a wave-dominated estuary. (b) Plan view of the estuary showing positions of principal morphological features. (c) Section view along the estuary axis showing stratigraphy. [From Dalrymple *et al.*, 1992.] [Copyright © 1992 Society for Sedimentary Geology, reproduced with permission.]

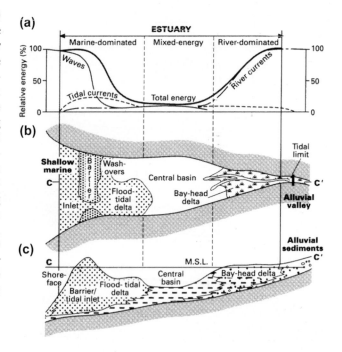

the central zone. If the estuary is relatively young with plenty of accommodation space, a deep central **mud basin** accumulates the finest sediments,
analogous to a prodelta environment. If the estuary is mature, then the central zone is infilled and dominated by salt marshes or mangrove flats composed of predominantly muddy sediments. The inner zone consists of a
**bay-head delta** deposited by the river flow exiting the alluvial valley, typically as a plane jet. The morphological features previously described for the
plain and front environments of fluvial-dominated deltas (Section 6.3) can
also be recognized in bay-head deltas, albeit at a generally smaller-scale.

The outer zone of a **tide-dominated estuary** consists of **linear sand bars**
that delineate usually **multiple tidal channels** (Figure 7.3). The appearance
is similar to a tide-dominated delta, but importantly, these bars do not
extend beyond the drowned valley, since the system is infilling a coastal
embayment rather than building out the coastline. This environment is
dominated by tidal processes and tidal energy often increases towards the
landward end of the outer zone, due to tidal shoaling (Section 7.3.2).
Although waves are subordinate, their influence can sometimes extend further into a tide-dominated estuary, simply because the entrances of these
estuaries are more open than wave-dominated estuaries. Due to the high-

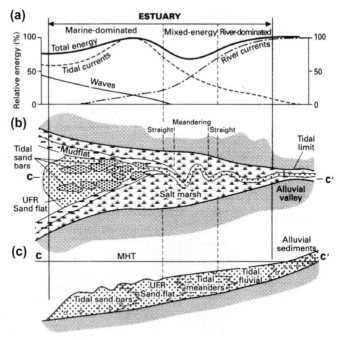

**Figure 7.3** – (a) Chart showing the change in energy regime along the axis of a tide-
dominated estuary. (b) Plan view of the estuary showing positions of principal
morphological features. (c) Section view along the estuary axis showing stratigraphy.
[From Dalrymple et al., 1992.] [Copyright © 1992 Society for Sedimentary Geology,
reproduced with permission.]

energy tidal currents in the outer zone, the sediment is predominantly sand. Wave energy in the central basin is insignificant, whereas tidal energy is still relatively high, hence the central zone of tide-dominated estuaries is more energetic than that in wave-dominated estuaries. The relatively straight multiple channels of the outer zone reduce to a single **meandering tidal channel** in the central zone. This channel is influenced by river processes during times of high discharge, but it is tide-dominated for most of the time and includes extensive intertidal morphology. The central zone is again a sink for fine sediment, however, sediment accumulates in the central zone not because of a rapid decline in energy, but due to the control of gravitational circulation and tidal distortion on cohesive sediment dynamics (Section 7.4). Due to the persistent tidal energy through the central and into the inner zones, fluvial sediments become progressively mixed with estuarine sediments in the inner zone and there is no discrete fluvial delta. The entire pattern of straight channels in the outer zone, a meandering channel in the central zone and a return to straight channel in the inner zone is characteristic of tide-dominated estuaries.

## 7.2.2 Tidal channels and the intertidal zone

Up to this point we have only considered energy gradients and their control on morphology along the estuary axis. At any position along the estuary axis, there is also an energy gradient towards the estuary margin. This is mostly due to the relationship between bed elevations and tidal water levels. To demonstrate this point, consider the channel cross-section shown in Figure 7.4, and note that maximum tidal current velocities generally occur at mid tide, while the current is slack around high and low tide. That part of the channel situated below mean low water is the **subtidal zone**, which is almost always submerged and conveys water at all stages of the tide. The subtidal channel is the most energetic zone, because it conveys both maximum flood and ebb currents. That part of the channel situated between mean low water and mean high water is the **intertidal zone**, which is alternately submerged and exposed during every tide cycle. The lower intertidal zone is submerged during maximum tidal currents, but the water is shallower and the current velocities are correspondingly smaller so its energy level is moderate compared to the subtidal zone. The upper intertidal zone is not submerged for large periods of time and it is never exposed to the maximum tidal currents, due to its elevation above mid tide, so its energy level is low. That part of the channel situated above mean high water is the **supratidal zone**, which is only submerged during spring high tides and is therefore mostly exposed. The supratidal zone is a very low energy environment.

While tidal currents are the most important energy source in the estuary on a large scale, small wind waves breaking in shallow water along channel margins can cause considerable erosion. At Gentlemans Halt, for example, the water level is most often located just above mean low water and just

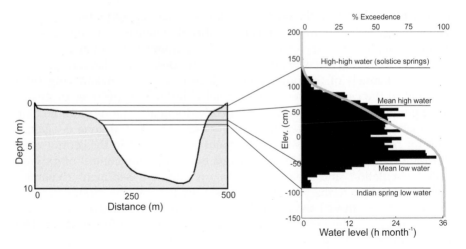

**Figure 7.4** – The left panel is a cross-section of the estuary channel at Gentlemans Halt, Hawkesbury River, Australia, showing tidal water levels. The right panel is a chart showing a histogram (black bars) and percentage-exceedence curve (grey line). The histogram represents the number of hours per month that the water level is located at each elevation in the intertidal zone. The percentage-exceedence curve shows the percentage of time in the month that each elevation in the intertidal zone is submerged.

below mean high water (Figure 7.4). These two locations on the intertidal profile therefore experience the effects of wind waves most often, and are marked by local increases in the profile gradient. Similarly, the lower intertidal zone is the one most often exposed to scour by the maximum tidal currents and is therefore steeper than the upper intertidal zone.

In most estuaries the lower intertidal zone is devoid of vegetation, due to excessive bed shear stress preventing seedlings taking anchor in the sediment. The upper intertidal zone is less energetic, however, and in tropical and subtropical environments it is frequently colonized by **mangroves** (Figure 7.5 and 7.6a). Common species present in Australian estuaries include *Avicennia marina* and *Rhyzophora stylosa* (Lear and Turner, 1977). The convoluted interference to flow caused by the trunk and root systems of these trees enhances sediment deposition. In addition, there is a steady supply of organic detritus (*e.g.*, leaves, branches). It is common for *Avicennia* to grow on the exposed channel margin, performing the role of the pioneer species, with *Rhyzophora* growing behind. In temperate environments the mangroves are replaced by salt-tolerant grasses and reeds known collectively as **salt marsh** (Figure 7.6b). A common species succession in the United Kingdom, from most exposed to least is: *Salicornia*; to *Halimione portulacoides* or *Aster maritima*; and finally to *Puccenelia maritima*, *Limonium vulgare* or *Armeria maritama* (Pethick, 1984). Salt marshes are generally restricted to the supratidal zone, so that the upper intertidal zone in temperate environments remains unvegetated. If there is sufficient accommodation space, salt

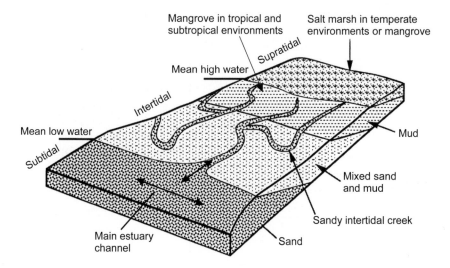

**Figure 7.5** – Diagram showing typical morphology and sediment distribution for the intertidal margin of an estuary channel. [Modified from Boggs, 1995]

marsh may also occur in the supratidal zone behind subtropical mangrove forests, which is where the greatest salinities occur, because of infrequent inundation and high evaporation rates. If localized salt concentrations become too high for even the most salt tolerant mangrove or marsh species to survive, then bare **salt pans** develop.

If the upper intertidal or supratidal zones are extensive, then they may be drained by small **intertidal creeks** that feed into the main estuary channel (Figure 7.5). Due to their elevation these creeks are dry for a significant part of the tidal cycle. At high water, when the creeks are full (or overflowing on spring tides), the tidal current is usually slack. The strongest currents conveyed by the creeks occur late in the falling tide. This occurs because the water level in the main channel drops more quickly than the water can

**Figure 7.6** – (a) In sub-tropical and tropical environments, the upper intertidal and supratidal zone may be dominated by mangrove species like these in the Hawkesbury River, Australia. [Photo M.G. Hughes.] (b) In temperate environments, the supratidal zone may consist of salt marsh like this example from Gibraltar Point, England, where salt marsh environments are present between dune ridges. [Photo G. Masselink.]

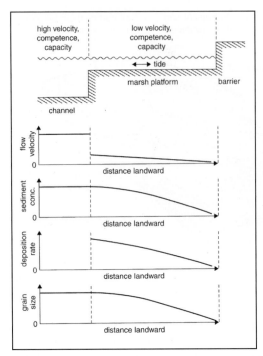

**Figure 7. 7** – Conceptual model that summarizes the sediment dynamics of the intertidal zone. The model was developed for salt marsh environments, but also applies generally to mangrove environments. Axes on all graphs are arbitrary scale. [From Allen, 1996.] [Copyright 1996 © Geologist's Association, reproduced with permission.]

drain from the tidal flat, resulting in the development of strong hydraulic gradients, which can drive currents with peak velocities approaching 1 m s$^{-1}$ at times (French *et al.*, 1993). Current speeds across the tidal flat and marsh, outside of the channels, rarely exceed a few cm s$^{-1}$. The sudden change in flow conditions as the water level moves from the channel confines out onto the intertidal platform results in a general pattern of decreasing suspended sediment concentration, deposition rate and grain size with distance from the channel (Figure 7.7). Sediment grain size typically grades from mixed sand and mud to primarily mud furthest from the channel (Figure 7.5).While the sedimentological behaviour shown in Figure 7.7 is generally correct, it ignores the importance of local topography and roughness elements affecting the flow across the intertidal zone. Localized morphology can have a significant effect on the rate of vertical accretion in mangroves and salt marshes (Box 7.1).

## BOX 7.1 – ACCRETION RATES IN AN ENGLISH SALT MARSH

The potential accelerated rate of sea-level rise in the near future makes salt marshes particularly vulnerable environments. The expected response of these environments to sea level rise depends primarily on the rate of sediment supply. If it matches the rate of sea level rise then the vertical accretion rate of the salt marsh will also match sea level rise and the marsh area will be maintained. If the rate of sediment supply is insufficient then the marsh will erode and become drowned. The actual response will of course be much more complex, and reflect the heterogeneity of the intertidal morphology (Figure 7.5). The wide range of vertical accretion rates within a macrotidal, back-barrier salt marsh (Hut Marsh) is evident in Figure 7.8.

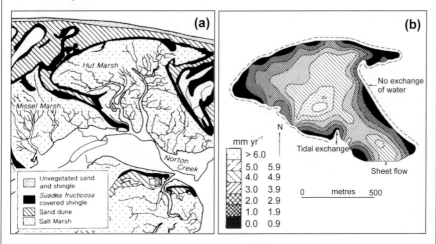

Figure 7.8 – (a) Map showing geomorphological setting of Hut Marsh on Scolt Head Island, North Norfolk, United Kingdom. (b) Map showing five-year mean vertical-accretion rates measured on Hut Marsh (1986–1991). [From French et al., 1995] [Copyright © 1995 Coastal Education and Research Foundation, reproduced with permission.]

French et al. (1995) found that on a spatial scale comparable to the size of the salt marsh there is an inverse relationship between the accretion rate and marsh elevation. The north, east and west margins of Hut Marsh are the most elevated, and have accretion rates of only 0–3 mm yr$^{-1}$. The widest range in accretion rates occurred around the mid tide level, because accumulation rates consistently higher than the average were measured along the margins of intertidal creeks. The centre and southeast corner of Hut Marsh display the densest networks of intertidal creeks, and marsh accretion rates in these areas reach 6 mm yr$^{-1}$. These larger accretion rates are thought to result from high suspended sediment concentrations in the channelized flows (c. 1,000 mg l$^{-1}$), which drift overbank on spring high tides and settle adjacent to the channel margin.

The response of salt marsh to sea level rise is likely to be complex and, if the rate of sediment supply is close to critical, some sections of marsh may disappear while other sections grow, depending on the local intertidal morphology.

## 7.2.3 Estuary evolution

As estuaries infill, their morphodynamic behaviour evolves towards that of a coastal delta. Dalrymple et al. (1992) believe that the key indicator for the switch from an estuary to a delta is the meandering tidal channel that is characteristic of the central zone of both tide- and mature wave-dominated estuaries. They argue that the presence of the tight meanders indicates that net bedload transport is landward in the channel seaward of the meanders, *i.e.*, the system is behaving as an estuary. The absence of the meandering zone indicates that the net bedload transport is seawards throughout the system, *i.e.*, the system is behaving as a delta. The evolutionary path towards a delta differs between the two estuary archetypes.

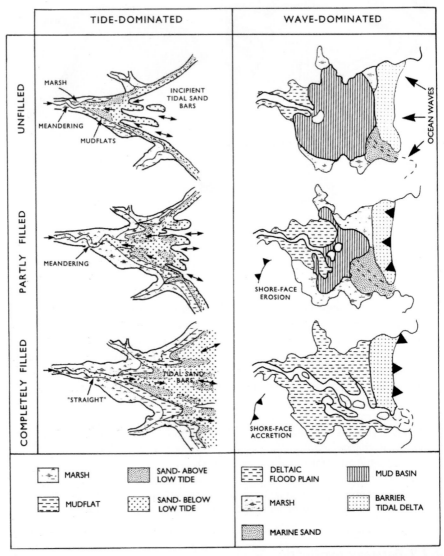

**Figure 7.9** – From top to bottom, stages in the filling of a wave- and tide-dominated estuary [From Trenhaile, 1997. Originally modified from Roy *et al.*, 1980 and Dalrymple *et al.*, 1992.] [Copyright © 1997 Oxford University Press, reproduced with permission.]

Wave-dominated estuaries infill through seaward progradation of the bay-head delta and landward extension of the flood-tide delta (Figure 7.9). Eventually, fluvial and marine sands bury the central basin muds. As the central basin shrinks, the tidal channel in the outer zone links up with the fluvial channel in the inner zone to produce a meandering tidal channel. At this point in time, from its seaward to landward end, the wave-dominated estuary has the straight-meandering-straight tidal channel that characterizes tide-dominated estuaries. From this time on, the evolutionary path of the two estuary types is similar (Roy, 1984; Dalrymple *et al.*, 1992). Tide-dominated estuaries begin by rapidly infilling their outer zone through energetic tidal currents redistributing sand into broadening and seaward extending sand ridges. As this process continues the three estuary zones steadily migrate seaward along the drowned valley, progressively filling it as they go. The position of the meandering tidal channel migrates seaward as well. Ultimately it reaches the present coastline and the straightened channel upstream can then competently convey sediments directly to the coastline and begin building a delta (Dalrymple *et al.*, 1992; Harris, 1988).

## 7.3 ESTUARY HYDRODYNAMICS

A key element of estuary hydrodynamics is the mixing of salt and fresh water masses that are delivered to the estuary by tide and river flows, respectively. We therefore begin our discussion with an explanation of the factors controlling estuarine mixing and, most importantly, demonstrate how partial mixing can produce estuarine currents that are additional to tide and river flows. In Chapter 3 we discussed how global tides are generated and described the behaviour of tidal amphidromes in ocean basins. Although this background gives us considerable insight into the nature of coastal tides, it provides little insight into the complexities of tide behaviour inside estuaries. We therefore conclude our discussion of estuary hydrodynamics by describing the role that channel morphology plays in controlling tide behaviour.

### 7.3.1 Stratification, mixing and gravitational circulation

Whether fresh river water and saline coastal water remain segregated or combine in an estuary is determined by the effectiveness of molecular diffusion and turbulent mixing. **Molecular diffusion** refers to the movement of salt molecules from areas of high concentration in the salt water body to areas of low concentration in the fresh water body. The length scale of this type of mixing is very small. The far more effective **turbulent mixing** involves the movement of parcels of fresh water into the salt water body and *vice versa* by eddies in the current. When these processes are ineffective, the two water masses remain segregated and the estuary is stratified. As these two processes become more effective the estuary becomes partially- to well-mixed.

**(a)** Stratified estuary

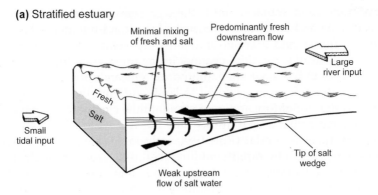

**(b)** Partially mixed estuary

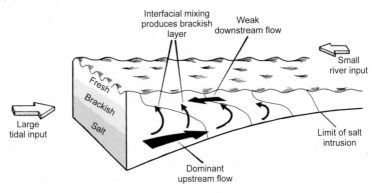

**(c)** Well mixed estuary

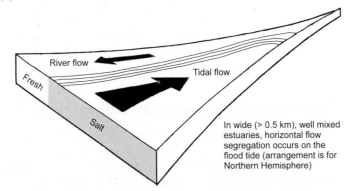

**Figure 7.10** – Diagram illustrating three main types of estuaries based on density stratification: (a) stratified, (b) partially-mixed and (c) well-mixed estuary. Water masses are indicated by shading on the front face of each block. Salinity contours (arbitrary scale) are indicated by thin lines on the side face of each block in (a) and (b) and on the top face of the block in (c). In (a) and (b) vertical mixing is indicated by thin arrows and non-tidal currents are indicated by thick arrows. [Modified from Pethick, 1984.]

**Stratified estuaries** commonly occur along microtidal coasts where there is sufficient river discharge to develop a fresh surface water mass, but the discharge rate is insufficient to completely expel the lower saline water mass from the estuary or generate sufficient turbulence to cause effective mixing (Figure 7.10a). Salinity contours are horizontal and close together, thus forming a **halocline**. A **salt wedge**, defined on the upper surface by the halocline and on the lower surface by the estuary bed, tapers in thickness towards the landward end of the estuary and is generally stationary, although it may move to a limited extent with variations in the fluvial discharge or the tide. If it becomes too dynamic, however, turbulence is generated and the stratification breaks down. In stratified estuaries, mixing occurs along the interface between the fresh and salt water mostly by molecular diffusion, and is therefore rather weak, which helps to maintain the stability of the stratification.

**Partially-mixed estuaries** develop if the tidal energy is sufficient to cause increased shear along the halocline and the development of internal waves (Figure 6.11a). These waves, and more importantly the turbulence caused by the shear between river/tidal flow and the estuary channel, results in considerable mixing (Figure 7.10b). Salinity contours dip steeply near the surface and near the bed, marking layers of predominantly fresh and salt water. In the central part of the water column, however, the salinity contours are gently sloping and indicate mixing between the fresh and salt to produce a brackish layer. The salinity decreases towards the landward end of the estuary at both the surface and at depth.

**Well-mixed estuaries** occur when mixing is so effective that the salinity gradient in the vertical direction vanishes entirely. If well-mixed estuaries are sufficiently broad for the Coriolis force to be effective, the river and sea water may become horizontally segregated (Figure 7.10c). This segregation is best developed on the flooding tide when the fresh- and salt-water masses are flowing in opposite directions. In the Southern Hemisphere, the river flow is steered by the Coriolis force towards the left bank (looking downstream) and the landward-directed flooding tide is deflected left towards the opposite bank. During the ebb both the tide and river flow are deflected towards the same bank. The corresponding arrangement for the Northern Hemisphere is illustrated in Figure 7.10c.

In some estuaries there is a circulation of water that is relatively weak compared to tidal currents, but is sufficient in many cases to influence the net transport of fine suspended sediment. This so called **gravitational circulation** consists of a near-bed current that is directed landward and a surface current that is directed seaward (Figure 7.10b). The circulation is best developed in partially-mixed estuaries, because they have a strong horizontal salinity gradient along the estuary axis. The pressure gradient force that we described in Box 3.1 was caused by a gradient in the water surface elevation. In partially-mixed estuaries the pressure-gradient force is due to a gradient in the water density, which drives water from the high pressure (most saline) region at the estuary entrance to the low pressure (least saline)

## BOX 7.2 – SPENCER GULF, AUSTRALIA: AN INVERSE ESTUARY

The small river basins draining into the Spencer Gulf, South Australia means that saline coastal waters entering the gulf are not significantly diluted. Rather the salt in the waters occupying the shallow head of the estuary becomes concentrated, due to the high evaporation rates in this arid, subtropical environment. This results in a strong salinity gradient along the estuary axis, with maximum salinities at the head of the estuary (Figure 7.11).

Given the direct relationship between water salinity and water density, a strong salinity gradient corresponds to a strong density gradient. Following our derivation of the pressure-gradient force due to a sloping water surface in Box 3.1, we can determine the hydrostatic pressure $P$ at any given depth as

$$P = \rho gh$$

**Figure 7.11** – Map of Spencer Gulf, South Australia, showing depth-averaged salinity contours in practical salinity units (PSU), as observed in March, 1984. Ocean water is typically about 35 PSU. Note the increase in salinity northwards towards Port Augusta, which is at the head of this estuary. [Figure provided by M. Tomczak, personal communication.]

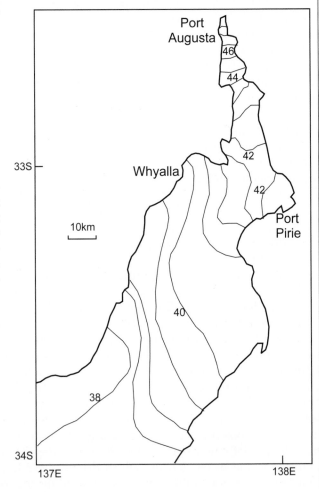

where $\rho$ is the water density, $g$ is the gravitational acceleration and $h$ is the height of water above the depth of interest. Now consider an estuary with a horizontal water surface, but a difference in water density between the entrance and the head of the estuary. The hydrostatic pressure at the entrance $P_E$ and head $P_H$ is, respectively

$$P_E = \rho gh \text{ and } P_H = (\rho + \Delta\rho) gh$$

where $\rho$ is the water density at the estuary entrance and $\Delta\rho$ is the difference in density between the estuary entrance and the estuary head. The pressure gradient along the estuary is then

$$\frac{\Delta P}{\Delta x} = \frac{P_H - P_E}{\Delta x} = gh \frac{\Delta\rho}{\Delta x}$$

So analogous to the pressure-gradient force produced by a gradient in water surface elevation (Box 3.1), along any horizontal plane in the water column there is a pressure-gradient force produced by the density gradient, which drives a current from the area of high pressure to the area of low pressure. In normal estuaries where the salinity decreases towards the head of the estuary it will be a landward-directed current. In inverse estuaries where the salinity increases towards the head of the estuary, such as the Spencer Gulf, it will be a seaward-directed current.

region at the estuary head. This landward flow is generally only apparent near the bed, because the seaward river flow near the surface usually overwhelms it. In arid environments where fresh water discharge into an estuary may be absent for long periods of time and there is a high evaporation rate, the water salinity can actually increase towards the head of the estuary. This situation is referred to as an **inverse estuary**, and the gravitational circulation is directed seawards (Box 7.2).

## 7.3.2 Tidal dynamics in estuary channels

In Chapter 3, we discussed that the tide is a shallow-water wave and that its speed $C$ is given by

$$C = \sqrt{gh} = \frac{L}{T} \tag{7.1}$$

thus

$$L = T\sqrt{gh} \tag{7.2}$$

where $g$ is the gravitational acceleration, $h$ is the water depth, $L$ is the tidal wavelength and $T$ is the tidal period. For a semi-diurnal tide with a period of 12.42 hours the tidal wavelengths for various water depths are listed in Table 7.1. The Amazon River, Brazil, has a tide-affected channel length of 850 km and can accommodate several tidal wavelengths (Defant, 1961). High tide therefore occurs at several locations along the estuary at one time.

**Table 7.1** Tidal speed and wavelength based on linear theory for shallow water waves

| Depth, m | Tidal speed, m s$^{-1}$ | Tidal wavelength, km |
|---|---|---|
| 2 | 4.4 | 198 |
| 4 | 6.3 | 280 |
| 6 | 7.7 | 343 |
| 8 | 8.9 | 396 |
| 10 | 9.9 | 443 |
| 15 | 12.1 | 542 |
| 20 | 14.0 | 626 |
| 25 | 15.7 | 700 |
| 30 | 17.2 | 767 |

Calculations made using equations (7.1) and (7.2)

The Amazon represents an extreme example, however, most estuaries only accommodate a single or a fraction of a tidal wavelength. For example, the much shorter Hawkesbury River, Australia, has a tidal length of 145 km and an average depth of 5 m, so it can accommodate about one-half a tidal wavelength. This means that the time of high tide at the estuary head corresponds with low tide at the estuary mouth.

The behaviour of the tide in long estuary channels (*i.e.*, longer than about one-quarter of a tidal wavelength) is quite different to that in short estuaries. In long estuaries, the tide displays characteristics of progressive and/or standing wave behaviour, whereas in short estuaries a time-varying hydraulic gradient between the tidal water level outside the estuary and the water level inside is what produces periodic water level oscillations and currents. In this section we begin by discussing the tidal dynamics of long estuaries and conclude with the dynamics of short estuaries.

The movement of the tide wave in an estuary controls the vertical change in water level and the horizontal motion of the water column. The vertical rise in water level from low to high tide results from the passage of the wave front and the fall from high to low results from the passage of the rear of the wave. For roughly half a tidal period, a horizontal current flows in the landward direction and is called the **flood current**. Similarly, for the other half of the tidal period an **ebb current** flows in the seaward direction. Between the flood and ebb currents, when the flow direction reverses, the current is **slack**. The timing of the water level rise and fall relative to the ebb and flood currents depends on whether the tide behaves as a progressive or a standing wave.

If we consider a tidal channel that is straight and infinitely long, so no reflection occurs, then the tide behaves as a **progressive tide** (Figure 7.12a). At successive positions along the estuary the time of high and low tide occurs increasingly later, *i.e.*, there is a **tidal lag**. At any particular point in

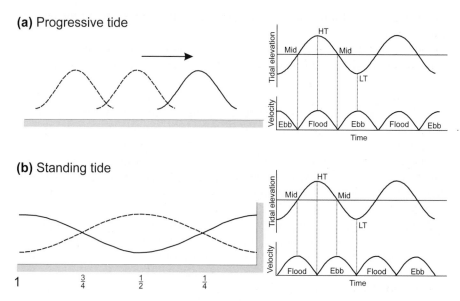

**(a)** Progressive tide

**(b)** Standing tide

**Figure 7.12** – Diagram showing the nature of (a) progressive and (b) standing tides. The left-hand panels show the dynamic behaviour of the water level along the estuary. The fractions shown beneath the standing tide indicate the location of the estuary entrance if the estuary length is an integer-multiple of one-quarter the tidal wavelength and therefore resonant. The right-hand panels are graphs showing the relationship between tidal water level and current velocity. See text for further discussion.

the estuary, the maximum flood velocity occurs at high tide, the maximum ebb velocity occurs at low tide and slack water occurs at mid tide. You might wonder how is a tide ever like a progressive wave in an estuary, since estuaries are not infinitely long. In fact the estuary need only be sufficiently long for friction to steadily diminish the height of the tide wave to zero before it reaches a significantly reflective boundary.

As tides travel up river, they experience reflection from channel margins, channel bends and a rising bed. In particular situations the reflected tide and the incoming tide interact to form a **standing tide**. If we consider a straight channel that has a vertical boundary at the estuary head and an incoming tide that is perfectly reflected at this boundary, the resulting standing wave pattern is shown in Figure 7.12b. There is generally an anti-node at the estuary head. At the estuary entrance a node will occur if the estuary length is an odd multiple and an antinode will occur if it is an even multiple of one-quarter the tidal wavelength. Maximum tide range occurs at the antinodes and minimum tide range occurs at the nodes. In the case of a perfect standing tide, the tidal range at the antinodes inside the estuary will be twice the range of the coastal tide outside the estuary. Moreover, the maximum flood velocity occurs at mid tide on the rising tide, the maximum ebb velocity occurs at mid tide on the falling tide, and slack water occurs at high and low tide (Figure 7.12b). If the estuary length is one-quarter of the wavelength, then high water occurs simultaneously through out the estuary, as does low water.

Wright *et al.* (1973) showed that a standing tide is responsible for producing the funnel shape that is characteristic of many macrotidal estuaries. In order for a channel shape to be stable the system must maximize entropy, or in other words, eliminate any gradients in the sediment transport rate along the channel. It turns out that the funnel shape, characterized by a decrease in width towards the estuary head, maximizes entropy for a situation of increasing tide range towards the estuary head. This increasing tide range is precisely the behaviour that a standing tide displays in an estuary that is one-quarter of the tidal wavelength (Figure 7.12b). It should be no surprise then, that the length of the estuary funnel is usually one-quarter the tidal wavelength.

Although there are some estuary configurations that favour either progressive or standing tide behaviour, in most cases the tide in the estuary displays a combination of both. Slack water commonly occurs 1 to 2 hours after high or low tide and maximum current velocity occurs shortly after mid tide. Often the tide behaviour is more progressive near the estuary mouth and more standing near the estuary head, where most reflection occurs.

Estuary channels tend to get narrower and shallower with increasing distance from the coast. The reduction in water depth causes a reduction in the speed and hence the wavelength of the tide (Equation 7.1). This has the same effect as it does on shoaling wind waves (Section 4.5.2). In order to conserve the energy flux, the energy density must increase so that the wave height, or in this case the tide range, must increase. Opposing this shoaling

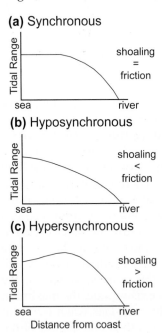

**(a)** Synchronous

**(b)** Hyposynchronous

**(c)** Hypersynchronous

**Figure 7.13** – Graphs showing tidal range as a function of distance from the coast, which illustrate the definition of: (a) synchronous, (b) hyposynchronous and (c) hypersynchronous estuaries. The relative strength of shoaling versus friction effects is also indicated in each case. The scale on the axes is arbitrary.

and convergence effect is friction between the tide and the estuary channel, which dissipates energy and thus reduces the tide range. These two competing processes are rarely balanced and three scenarios are possible (Figure 7.13):

- **Synchronous estuaries** are those where the shoaling and friction effects are balanced in the lower estuary so that the tide range is constant with distance from the coast. Eventually, however, friction becomes overwhelming and the tide range decreases to zero at the tidal limit (Figure 7.13a).
- **Hyposynchronous estuaries** are those where friction is of overriding importance everywhere and the tide range decreases throughout the estuary (Figure 7.13b).
- **Hypersynchronous estuaries** are those where the shoaling effect is dominant in the lower estuary and there is an increase in tidal range with distance from the coast. Eventually, however, friction overwhelms the shoaling effect and the tide range decreases to zero at the tidal limit (Figure 7.13c).

Long estuaries tend to be hypersynchronous and short estuaries tend to be hyposynchronous, but the synchronicity can be strongly influenced by the entrance conditions (Box. 7.3).

A common feature of long estuaries is **tidal distortion**, which is usually expressed as a short rising tide and a longer falling tide. Tidal distortion is a manifestation of a progressive steepening of the wave front as it travels along the estuary (Figure 7.15a). Off the coast, the tide range is a small proportion of the water depth so there is little difference in the speeds of the crest and trough. Inside the estuary, however, the tide is a significant proportion of the water depth and the deeper water under the crest travels faster than the shallower water under the trough. This can be demonstrated by re-writing Equation 7.1 in an equivalent, but slightly different form

$$c = \sqrt{g(\bar{h} \pm a)} \qquad (7.3)$$

where $\bar{h}$ is the local tidally-averaged water depth in the estuary and $a$ is the tidal amplitude ($-a$ for low tide and $+a$ for high tide). Figure 7.15b shows the speeds of the tide crest and trough and the speed differential as a function of water depth. For water depths of 20 m and more the two speeds are similar and the speed differential tends towards zero, so that the tide is symmetric. In shallower water depths, particularly < 5 m depth, the speed differential increases rapidly with a corresponding increase in the tidal distortion. In particular circumstances the distortion becomes so pronounced that the front of the tide is vertical, much like the front of a breaking wave in the surf zone. Indeed, it is possible to surf the tidal bore in the Severn River, United Kingdom. Because the discharge volume through the channel on the flooding tide closely matches the discharge volume on the ebbing tide, the inequality between the flood and ebb durations must produce a **velocity-magnitude asymmetry** between the tidal currents (Figure 7.15c). In order

## Box 7.3 – Synchronicity in New South Wales Estuaries

**(a)** Long, ria-type estuary

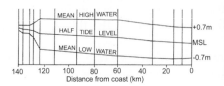

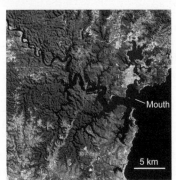

Mouth

5 km

**(b)** Short, barrier-type estuary

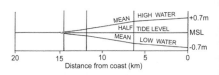

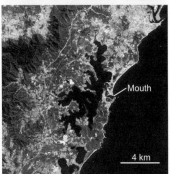

Mouth

4 km

**(c)** Long, river- type estuary

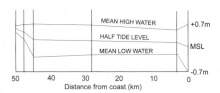

Mouth

2 km

**Figure 7.14** – Left-hand panels show the elevations of mean high water, half tide level and mean low water as a function of distance from the coast for three different estuaries in New South Wales, Australia. Tidal range, and thus synchronicity, is indicated by the change in vertical distance between mean high and low water along the estuary. Aerial photographs of each estuary, highlighting the differing entrance conditions, are shown in the right-hand panels. The estuaries are: (a) Hawkesbury River, (b) Lake Macquarie and (c) Manning River. [Tidal elevations are modified from New South Wales Government, 1992.] [Satellite images are modified from Geoscience Australia's OzEstuaries data base.]

The coastline of New South Wales, Australia, is a wave-dominated environment. The modal deepwater significant wave height is 1.5 m, and it exceeds 4 m 1% of the time (Lawson and Abernathy, 1975). The narrow continental shelf and steep shoreface means that 96% of the offshore wave energy reaches the coastline (Wright, 1976). The coastal tide has a mean range of 1.6 m, thus it is a microtidal environment. All of the estuaries along this coast are wave-dominated estuaries (Roy, 1984). Nevertheless, the variation in hinterland relief and sediment supply along the coast has yielded a variety of estuary entrance conditions.

The flooding of deeply incised river valleys and a low sediment supply have produced long ria-type estuaries with open, hydraulically efficient entrances. These estuaries are hypersynchronous (Figure 7.14a). The flooding of broad shallow river valleys and a low sediment supply have produced short barrier-type estuaries with constricted, hydraulically inefficient entrance conditions. Frictional attenuation is of overwhelming importance over the entire short estuary length, and they are hyposynchronous (Figure 7.14b). The flooding of moderately incised river valleys with a large sediment supply has produced long river-type estuaries that also have constricted entrances. These estuaries display mixed synchronicity. Frictional attenuation is significant through the short entrance channel, but once inside the estuary the tidal channel is sufficiently long to permit shoaling to occur (Figure 7.14c).

for the discharge volume to be conserved, the shorter duration of the flood current requires it be of larger magnitude than the ebb current.

Wave-dominated estuaries are often short (less than one-quarter the tidal wavelength) and their entrance channel is restricted, so that the tide does not display wave-like behaviour inside the estuary. The periodic rise and fall of the water level in these estuaries results from the periodic filling up and draining of the estuary due to different water levels inside and outside the entrance. When the coastal tide outside the estuary is rising it eventually becomes higher than the water inside the estuary, thus the water surface slopes into the estuary and drives the flood current. The total volume of water entering the estuary on the flooding tide is called the estuary's **tidal prism**, and it raises the water level in the estuary by an amount roughly corresponding to the volume of water divided by the surface area of the estuary. In a similar fashion, when the tide outside the estuary is falling it eventually becomes lower than the water inside the estuary, thus the water surface slopes out of the estuary and drives the ebb current. The volume of water exiting the estuary then reduces the water level in the estuary by a corresponding amount.

While the coastal tide is clearly the driver of water levels and currents inside short estuaries, it should be evident that it is not the same situation as the tide wave propagating through the estuary. Nevertheless, tidal distortion and velocity-magnitude asymmetry occur in short estuaries just as they do in long estuaries, although the mechanism is different. Friedrichs and Aubrey (1988) demonstrated that tidal distortion could develop due to

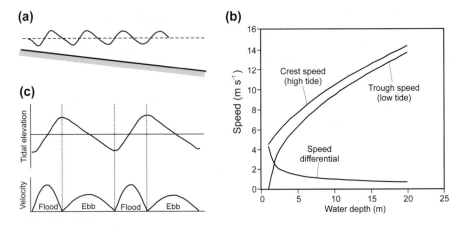

**Figure 7.15** – (a) Diagram showing the increasing distortion of the tide wave as it progresses along the estuary. (b) Graph showing the velocity of the crest (high tide) and trough (low tide) of a shallow water wave as a function of water depth. Also shown is the velocity differential between the two, which increases markedly in very shallow water depths. Calculations were made using Equation 7.3. (c) Graphs showing the relationship between tidal water level and current velocity. Note the velocity-magnitude asymmetry between the dominant flood current and the subordinate ebb current. Scales on the axes are arbitrary.

channel shape. Consider a channel that has a large cross-sectional area in the intertidal zone relative to the subtidal zone (Figure 7.16a). Between mid and high tide when the maximum flood current velocity is expected to occur, there is inefficient water exchange through the estuary. This hydraulic inefficiency is due to large roughness elements in the form of mangroves, salt marsh and creek channels in the upper intertidal zone (Section 7.2.2). As a result the flooding tide is slow to turn and therefore has a longer duration than the ebbing tide (Figure 7.16b). Between mid and low tide, when the maximum ebb current velocity is expected to occur, almost all of the flow volume is conveyed in the subtidal part of the channel, which is hydraulically efficient. As a result the ebbing tide is quick to turn and therefore has a shorter duration than the flooding tide. If the ebb tide is shorter than the flood tide then conservation of discharge volume demands that it must have the largest velocity-magnitude (Figure 7.16b).

The sense of velocity-magnitude asymmetry can be consistent throughout an estuary or it can vary along the estuary channel. Estuaries or channel reaches that display a flooding tide that is larger in velocity-magnitude and shorter in duration than the ebbing tide are said to be **flood-dominant**, whereas those that display an ebbing tide that is largest in magnitude and shortest in duration are said to be **ebb-dominant**. Flood- or ebb-dominance often translates directly to net landward or seaward sediment transport, respectively. It turns out that the velocity-magnitude asymmetry need not be large to produce a dominant sediment transport direction. In Section

Figure 7.16 – (a) Channel cross-section characteristic of short, wave-dominated estuaries in which the volume of water in the intertidal channel is a substantial proportion of the total channel volume. (b) Graphs showing the relationship between tidal water level and current velocity. Note the velocity-magnitude asymmetry between the dominant ebb current and the subordinate flood current. Scales on the axes are arbitrary.

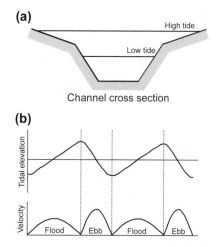

(a)

High tide

Low tide

Channel cross section

(b)

Tidal elevation

Velocity

Flood   Ebb   Flood   Ebb

5.5.3 we explained that the sediment transport rate is related to the stream power, or product of the flow velocity and bed shear stress (Equation 5.31). Since the bed shear stress is related to the flow velocity squared (*e.g.,* Equation 5.19), the sediment transport rate is actually related to the velocity cubed. Thus a small difference in the velocity magnitude between the flood and ebb tide can lead to a large difference in the total amount of sediment transported by each, and therefore a net sediment transport. In general, **flood-dominant estuaries** tend to infill their entrance channels by continually pushing coastal sediment landward and as a result are often intermittently closed, whereas **ebb-dominant estuaries** tend to flush sediment seawards from their entrance channels and as a result are often stable (Friedrichs and Aubrey, 1988). This relationship between flood- or ebb-dominance and the net sediment transport direction is really only true for sandy sediments, for which there is a direct and consistent relationship between the sediment transport rate and the bed shear stress. In the case of finer, cohesive sediments there are many factors that complicate the prediction of net sediment transport based on velocity-magnitude asymmetry (Section 7.4.2).

## 7.4 THE TURBIDITY MAXIMUM ZONE: COHESIVE SEDIMENT DYNAMICS

In some estuaries there is a well-defined zone in which the suspended sediment concentration (*i.e.,* water turbidity) is on average higher than the waters further seaward or landward (Figure 7.17). This **turbidity maximum zone** (TMZ) is characterized by suspended sediment concentrations of the order of 100 mg $l^{-1}$ in microtidal estuaries such as the Hawkesbury River, Australia (Hughes *et al.*, 1998). In macrotidal estuaries such as the Severn River, United Kingdom, the concentrations can reach 20,000 mg $l^{-1}$ (Kirby,

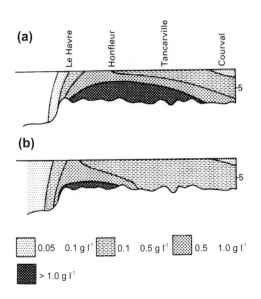

Figure 7.17 – Section view along the axis of the Seine Estuary, France, showing the TMZ during two different river discharges: (a) 200 m³ s⁻¹ and (b) 800 m³ s⁻¹. [From Dyer, 1994; modified from Avoine, 1981.] [Copyright © 1994 Blackwell Publishers, reproduced with permission.]

1988). The position of the TMZ varies with tidal stage and river discharge (Figure 7.17). For example, the TMZ in the Seine River, France, moves seawards when the river discharge increases and oscillates back and forth during the ebb and flood stages of each spring tide (Avoine, 1981). The magnitude of the peak suspended sediment concentration observed in the TMZ is determined by the amount of fine-sediment stored in the zone and the strength of the tidal currents available to regularly stir the sediment into suspension.

## 7.4.1 Generation mechanisms

In partially-mixed estuaries, gravitational circulation results in a weak seaward-directed flow in the surface waters and a weak landward-directed flow in the bottom waters (Figure 7.18). If the superimposed tidal currents are approximately symmetrical, this gravitational circulation is sufficient to control the net transport of suspended sediment. Sediment brought downstream by the river will begin to settle upon reaching the estuary. Because the material is fine, it may not settle until it is well down the estuary. Nevertheless, once the sediment settles to the bottom waters it will experience a net transport landward over many tide cycles. There is a **null point** in the estuary where the near-bed river current and the near-bed part of the gravitational circulation are roughly equal in their competency to transport sediment. Since the currents are equal in competency and opposite in direction, this null point is where the sediment carried landward by the gravitational circulation will accumulate. In reality tidal currents smear the null point along the estuary axis into a null zone, or TMZ, that is roughly the dimension of the **tidal excursion length** (horizontal distance a parcel of water is moved over a tide cycle). The TMZ is located near the landward

**Figure 7.18** – Section view of a partially-mixed estuary showing the pattern of non-tidal currents (bold arrows), sediment settling (thin arrows) and the position of the null point for suspended sediment transport. The concentration of suspended sediment is indicated by the concentration of dots. [Modified from Dyer, 1986.]

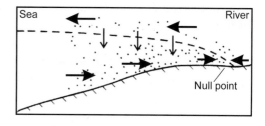

limit of salt intrusion, usually where water salinities are in the range 1 to 5 ppt, and thus insufficient to competently drive the gravitational circulation any further landward. An example of a TMZ resulting from gravitational circulation is the TMZ in the microtidal Rappahannock Estuary, USA (Nichols and Poor, 1967).

In macrotidal estuaries the role of tidal distortion in generating a TMZ becomes paramount. In long estuaries, the tidal distortion is such that flood currents display the largest velocities, although they are of shorter duration than the ebb currents. The further landward the tide travels, the more enhanced this velocity-magnitude asymmetry becomes (Figure 7.19). Since the sediment transport rate is generally related to the current velocity cubed, the asymmetry in the sediment flux is even stronger. Averaged over many tide cycles, there is a residual transport of sediment landward along estuary. Although the tidal distortion ensures that the flood current is more competent to transport sediment than the ebb current, there will be a point in the estuary where the competency of the flood current is matched by the competency of the river-assisted ebb current. This is the null point and focus for fine sediment accumulation. An example of a TMZ resulting from tidal

**Figure 7.19** – Section view of a well-mixed macrotidal estuary showing the change in tidal distortion, velocity-magnitude asymmetry, and net sediment transport direction along the length of the estuary. [Modified from Allen et al., 1980.]

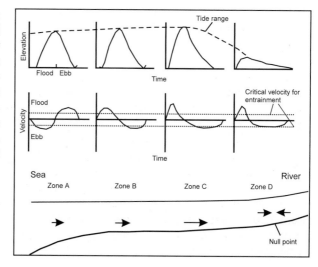

distortion is the TMZ in the macrotidal Gironde Estuary, France (Allen *et al.*, 1980).

## 7.4.2 Entrainment and velocity-suspension lags

The turbidity maximum zone is always most active and suspended sediment concentrations are always largest during spring tides. During neap tides, the suspended sediment concentrations usually drop off dramatically, and in some cases the turbidity maximum may disappear all together. Although the link between the magnitude of tidal current velocities and suspended sediment concentration in the TMZ is obvious, it is not straightforward by any means. For example, over a spring–neap tide cycle in the Tamar Estuary, UK, the maximum and minimum suspension concentrations in the TMZ lag behind the time of maximum and minimum tide range by up to several days (Uncles *et al.*, 1994). Moreover, within a single tide cycle, the maximum and minimum concentrations lag behind the time of maximum and minimum current velocity by up to several hours. These **velocity–suspension lags** suggest a complexity in the sediment entrainment process that goes beyond our discussion in Section 5.5.1. This complexity arises from the cohesive nature of the fine-sediments that dominate the TMZ.

Figure 7.20 shows three possible scenarios for the relationship between bed shear stress and suspended sediment concentration over a tidal cycle in the TMZ. In Figure 7.20a there is a critical threshold stress required for erosion of the bed, below which there is no sediment in suspension. Once the critical stress is exceeded, there is an increase in suspension concentration with increasing shear stress. Note also that the same bed shear stress produces the same suspension concentration whether the velocity is increasing or decreasing. This scenario is characteristic of coarse-sediment behaviour

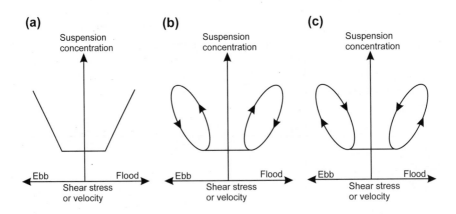

**Figure 7.20** – Three scenarios for the response of the suspended sediment concentration to changes in bed shear stress (or velocity) over a tide cycle. See text for further discussion.

and does not lead to velocity-suspension lags. Figure 7.20b also shows a situation where there is a critical shear stress for erosion. Once the critical stress is exceeded, this scenario displays a **hysteresis loop** in which the response of the suspension concentration depends on the history of bed shear stress. In this case the loop is clockwise (anticlockwise) on the flood (ebb) tide so that the largest concentrations are associated with increasing bed shear stress or the accelerating current. Figure 7.20c shows a similar situation, but with an anticlockwise (clockwise) loop on the flood (ebb) tide so that the largest concentrations are associated with decreasing bed shear stress or the decelerating current. It is these latter two scenarios that produce velocity-suspension lags. These are idealized scenarios and in real estuaries the sediment dynamics of the TMZ can display a combination of these scenarios over the flood and ebb stages of the tide. Moreover, the critical entrainment stress and the size of the hysteresis loop can be different between the flood and ebb tide.

Dyer (1994) describes several factors unique to cohesive sediments that can lead to velocity-suspension lags. For example sediment compaction increases and hence porosity and water content decrease with depth in a muddy substrate. We discussed in Section 5.5.1 how the entrainment of cohesive sediment depends on these bulk properties. The sediment at the surface is least compacted and has the highest water content so it is easily entrained, but can become rapidly exhausted. As the bed is eroded the more compacted and more difficult to erode deeper sediments become exposed, thus requiring increasingly larger shear stresses to sustain the same suspension concentration. This can lead to a situation where sediment that is already in suspension can be maintained there by shear stresses that are actually smaller than those required to erode the bed. Another important factor is the small settling velocity of silts and clays. The fact that it may take longer for a suspension to settle than it takes the tide to accelerate, decelerate and reverse direction can lead to complex relationships between the bed shear stress and the suspension concentration. In summary, the combined effect of cohesive sediment behaviour and unsteady tidal currents means that the suspension concentrations in the TMZ are rarely in equilibrium with the hydrodynamic forcing. This is particularly true of muddy, macrotidal estuaries that have highly-concentrated suspensions.

## 7.4.3 Settling of highly-concentrated suspensions: Fluid muds

When suspended sediment is settling during slack water, near bed concentrations in the TMZ of some estuaries reach levels greater than 10 g l$^{-1}$. These highly-concentrated near-bed suspensions are termed **fluid mud,** and they have a different settling behaviour to the more common dilute suspensions. When the tidal current goes slack, the suspension goes through a sequence of settling, deposition and consolidation (Dyer, 1986). The suspension may start out being relatively uniform with depth, but eventually an

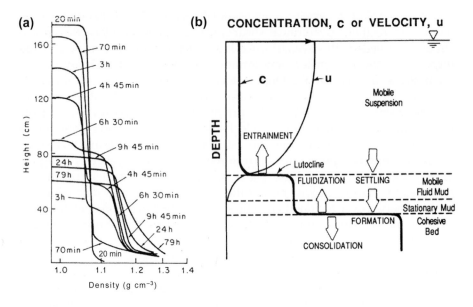

**Figure 7.21** – (a) Density profiles through time measured in an experimental suspension. [From Been and Sills, 1981.] [Copyright © 1981 Thomas Telford, reproduced with permission.] (b) Diagram showing a near-bed suspension concentration profile and relationship between the current velocity profile and the fluid mud layer. [From Mehta, 1989.] [Copyright © 1989 American Geophysical Union, reproduced with permission.]

interface develops between relatively clear water above and the suspension below (Figure 7.21a). This interface is called the **lutocline** and it marks a sharp vertical change in the suspended sediment concentration and the bulk density of the fluid. Above the lutocline, sediment is free to settle at its single particle or aggregate fall velocity. Below the lutocline, the sediment experiences **hindered settling**, because the concentration of particles is so high that the paths for settling towards the bed are narrow and convoluted, and the upward drag of escaping fluid is significant. The fluid is escaping vertically because the lutocline is steadily falling over time so that more sediment is being accommodated into less volume. This increases the pore water pressure causing water to be expelled upwards. At some point the bulk density at the bottom of the suspension is so high that no further settling can take place and this marks the surface of the depositing bed. As deposition continues from the bottom of the suspension the bed rises up towards the falling lutocline. When the two meet deposition is complete and the process of consolidation continues through the slow expulsion of pore water.

The time scale for fluid mud to completely settle is long compared with the semi-diurnal tide, so that the current goes from slack back to maximum velocity before settling is complete. If the settling of the fluid mud is taking place as the tide is going into neaps, then the velocity profile during subsequent tides may not penetrate the lutocline and the fluid mud beneath will continue to settle as a stationary suspension. If the settling is occurring during slack water on springs, however, then the velocity profile on subsequent tides will almost certainly extend down below the lutocline and a mobile fluid mud layer develops (Figure 7.21b). Because of the extreme sediment concentrations involved, mobile fluid mud layers can result in enormous tonnages of sediment movement over a single tide cycle.

## 7.5 SUMMARY

- Estuaries are systems that involve net landward transport of sediment when averaged over several years. Estuaries are also ephemeral systems over geological time scales – once an estuary is infilled it becomes a prograding coastal delta or barrier system.
- The large-scale morphology of an estuary is controlled by palaeo-valley configuration, entrance conditions and degree of infilling. Based on their entrance conditions estuaries can be classified as either wave- or tide-dominated.
- Wave-dominated estuaries are fronted by barrier, tidal inlet and tidal delta morphology. The inner zone consists of a bay-head delta. These estuaries are characterized by a high-energy outer and inner zone, and a low-energy central zone. The coarsest sediments are found in the inner and outer zones, and the finest sediments are found in the deep central mud basin.
- Tide-dominated estuaries are fronted by linear sand ridges and multiple straight channels. The central zone consists of a single meandering channel and broad intertidal mud flats, salt marshes or mangroves. The inner zone consists of a straight channel and limited intertidal morphology. All zones in the estuary are relatively high energy compared with wave-dominated deltas. The coarsest sediments are found in the outer and inner zones, and the finest sediments in the central zone.
- Intertidal areas can be vegetated by tree-species such as mangroves in tropical to subtropical environments. In temperate environments they may be vegetated by reed- and grass-species known collectively as salt marsh.
- In long estuaries (greater than one-quarter the tidal wavelength), the tide displays wave-like behaviour. Particular channel dimensions favour progressive wave behaviour and others favour standing wave behaviour. In most estuaries, however, the tide displays elements of both. A tidal lag between the time of high (and low) tide at the coast and at the head of the estuary is a common element of progressive wave behaviour, and the occurrence of maximum current velocity shortly after mid tide and slack water shortly after high and low tide are common elements of standing wave behaviour.

- In short estuaries, a time-varying hydraulic gradient between the tidal water level outside the estuary and the water level inside is what produces periodic water level oscillations and currents.
- Flood-dominant estuaries develop due to distortion of the tide wave as it progresses along the estuary. Ebb-dominant estuaries develop due to the difference in the hydraulic efficiency of the intertidal and subtidal channel.
- The turbidity maximum zone (TMZ) in an estuary is where the largest suspended sediment concentrations occur. It is best developed in the central zone of long, muddy estuaries. The location of the TMZ corresponds to the null point in suspended sediment transport driven by either gravitational circulation or velocity-magnitude asymmetry. The former is most important in partially-mixed micro/mesotidal estuaries, and the latter is most important in well-mixed macrotidal estuaries.

# 7.6 FURTHER READING

Dalrymple, R.W., Zaitlin, B.A., and Boyd, R., 1992. Estuarine facies models: Conceptual basis and stratigraphic implications. *Journal of Sedimentary Petrology*, **62**, 1130–1146. [Provides a broad summary of the fundamental geological and sedimentological aspects of estuaries. It also contains the estuary classification scheme used in this chapter.]

Dyer, K.R., 1986. *Coastal and Estuarine Sediment Dynamics*. Wiley, Chichester. [An advanced text on estuaries that includes chapters on estuarine sedimentation and cohesive sediment dynamics.]

Dyer, K.R., 1998. *Estuaries: A Physical Introduction*. Wiley, Chichester. [A comprehensive introduction to tidal dynamics, stratification and mixing in estuaries.]

Pethick, J., 1984. *An Introduction to Coastal Geomorphology*. Edward Arnold, London. [Chapters 8 and 9 of this undergraduate text provide a good introduction to tide and sediment dynamics in the intertidal zone.]

Roy, P.S., 1984. New South Wales estuaries: Their origin and evolution. In: B.G. Thom (Editor), *Coastal Geomorphology in Australia*. Academic Press, Sydney. [A definitive paper on wave-dominated estuaries.]

# 8

# WAVE-DOMINATED COASTAL ENVIRONMENTS – BARRIERS

## 8.1 INTRODUCTION

In Chapter 1 we discussed an evolutionary classification of depositional coastal environments based on the dominance of either fluvial, tide or wave processes (Figure 1.6). Chapters 6 and 7 dealt with environments mainly controlled by river and tide processes, and in this chapter we will discuss wave-dominated coastal environments. To most people, the principal landform of wave-dominated coasts is the beach. However, beaches are but one component of wave-dominated coasts. The shoreface, which is the underwater slope that lies seaward of the subaerial beach, is also dominated by wave-processes. Additionally, the coastal dunes behind beaches can also be considered ubiquitous elements of wave-dominated coastal environments. The coastal dune, beach and shoreface are strongly linked by sediment transport pathways and morphodynamic feedbacks. Collectively, they make up coastal barriers and these landforms can be considered the basic depositional elements of wave-dominated coasts (Roy *et al.*, 1994). The following three sections of this chapter will discuss shoreface, beach and coastal dune morphology and processes. The final section will look at the bigger picture and deals with barrier morphology, processes and evolution.

## 8.2 SHOREFACE

The **shoreface** can be defined as the upper part of the continental shelf that is affected by contemporary wave processes and extends from the limit of wave runup to the depth limit for wave-driven sediment transport (Cowell *et al.*, 1999). Most shoreface profiles are characterized by a concave-upward shape, but the degree of concavity and general steepness of the profile is highly variable (Figure 8.1). Nearshore bars may be present on the upper part of the shoreface.

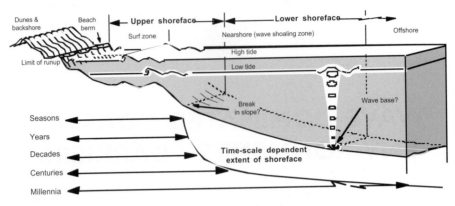

Figure 8.1 – Schematic showing shoreface profile. The seaward limit of the shoreface is dependent on the time scale of interest. [From Cowell et al., 1999.] [Copyright © 1999 John Wiley & Sons, reproduced with permission.]

The shoreface can be subdivided into an upper and lower section. The **upper shoreface**, referred to as the littoral zone by Hallermeier (1981), is defined as the region in which erosion and accretion result in significant (*i.e.*, measurable) changes in bed elevation during a typical year. The **lower shoreface**, referred to as the shoal zone by Hallermeier (1981), experiences sediment transport processes under typical wave conditions, but significant morphological changes only occur during extreme storm events. The upper and lower shoreface therefore represent distinctly different morphodynamic regions.

## 8.2.1 Depth of closure

The boundary between the upper and lower shoreface is referred to as the **depth of closure** (or **closure depth**) and can be used to infer a seaward limit to significant cross-shore sediment transport (Nicholls *et al.*, 1998). The concept is useful for applications such as estimating coastal budgets, numerical models of coastal change, beach nourishment design and the disposal of dredged material.

The depth of closure can be derived with relative ease if high-quality, repetitive morphological surveys of the shoreface are available. Plotting all cross-shore profiles from one location at different times together creates a profile bundle (Figure 8.2). The **sweep zone** can then be defined as the area between the minimum and maximum heights of the profile bundle, and the thickness of the sweep zone is an indication of the morphological variability over the studied time span. The morphological variability is greatest over the surf zone region and decreases progressively in the seaward direction. The profile bundles converge to a depth at which morphological change is insignificant (*i.e.*, within the measurement error) and this is the closure depth.

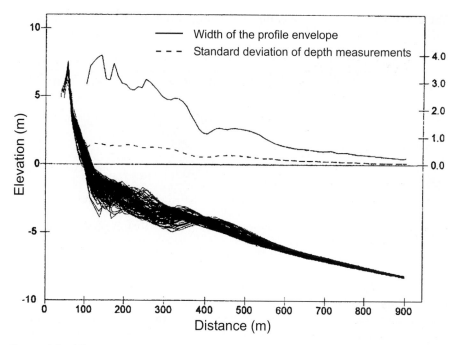

**Figure 8.2** – The envelope of a large number of upper shoreface profiles surveyed at the Field Research Facility, Duck, North Carolina. The upper part of the graph shows the maximum vertical change in the depth and the standard deviation of the depth variations. [From Komar, 1998.]

More often than not, high-quality survey data such as shown in Figure 8.2 are not available and the depth of closure has to be determined indirectly. Hallermeier (1981) used morphological data from a large number of shorefaces and related the observed closure depths to the incident wave conditions. He proposed that the depth of closure $h_c$ can be predicted using the annual wave climate according to

$$h_c = 2.28 H_{sx} - 68.5 \frac{H_{sx}^2}{gT_{sx}^2} \tag{8.1}$$

where $H_{sx}$ is the nearshore storm wave height that is exceeded only 12 hours each year and $T_{sx}$ is the associated wave period. The wave parameters $H_{sx}$ and $T_{sx}$ may not always be available and for practical purposes Hallermeier (1981) provided the simpler approximation

$$h_c = 2\bar{H}_s + 11\sigma \tag{8.2}$$

where $\bar{H}_s$ is the mean annual significant wave height (defined as the annual mean height of the highest one-third of waves measured each day) and $\sigma$ is the standard deviation of $\bar{H}_s$. Typical values for $h_c$ are listed in Table 8.1.

The transition from upper to lower shoreface is generally not

**Table 8.1** Depth of closure computed using equation (8.1) for a number of sites; see text for explanation

| | $\bar{H}_s$, m | $\sigma$, m | $\bar{T}_s$, s | $D_{50}$, mm | $h_c$, m | $h_i$, m | $L_o/4$, m |
|---|---|---|---|---|---|---|---|
| US Gulf of Mexico: Naples, FL | 0.3 | 0.2 | 4.6 | 0.12 | 2.8 | 4.4 | 8.3 |
| US Atlantic: Nags Head, NC | 1.0 | 0.5 | 8.8 | 0.11 | 7.5 | 31.0 | 30.2 |
| US Pacific, La Jolla, CA | 1.2 | 0.5 | 12.0 | 0.11 | 7.9 | 52.4 | 56.1 |
| Netherlands | 1.2 | 0.8 | 5.0 | 0.19 | 11.2 | 13.4 | 9.7 |
| Southeast Australia | 1.5 | 1.2 | 9.5 | 0.16 | 16.2 | 36.7 | 35.2 |

(data derived from Hallermeier (1981) and Cowell et al. (1999))

characterized by a change in morphology. However, the theoretical closure depth may correspond to a distinct break in sediment characteristics (Niedoroda et al., 1985). Upper shoreface sands are usually well-sorted and similar to beach sediments, and tend to display a seaward-fining trend. In many places, the seaward fining of sediments occurs down to an abrupt transition at a depth similar to $h_c$, seaward of which coarser, poorly sorted sand may be found. On prograding, deltaic shorefaces, however, the coarse sand facies is usually not found. Here, the sandy sediments become finer from the upper to the lower shoreface, grading to clays on the continental shelf.

### 8.2.2 Wave base

The seaward limit of the lower shoreface is less straightforward to define than the depth of closure because it depends on the time scale of interest (Figure 8.1). The longer the time scale under consideration, the greater the potential for extreme wave events to occur, and therefore the further the lower shoreface extends seaward. In the absence of other clear-cut definitions, it is useful to turn to Hallermeier (1981) once more, who estimated that the limiting depth $h_i$ for significant cross-shore sediment transport of sand by waves throughout a typical year is given by

$$h_i = (\bar{H}_s - 0.3\sigma)\bar{T}_s \left(g/5000D_{50}\right)^{0.5} \tag{8.3}$$

where $\bar{T}$ is the mean annual significant wave period (defined as the annual mean period of the highest one-third of waves measured each day) and $D_{50}$ is the median sediment size (in m) determined from a sand sample taken in a water depth of $h = 1.5h_c$. Typical values for $h_i$ are listed in Table 8.1.

One of the difficulties in applying Equation 8.3 is the availability of reliable sediment data. Even if a sediment sample is available, it remains to be established how representative the sample is considering the variable nature of the sediment distribution on the lower shoreface. An alternative approach to defining the seaward limit of the lower shoreface is through the application of the wave base concept, where **wave base** is defined as the water depth beyond which wave action ceases to stir bed sediment. The wave base can be taken as the water depth at which ocean waves start interacting with the sea bed. This corresponds to the transition from deep-water to intermediate-water waves (Chapter 4). Conventionally this limit occurs where the water depth is equal to half the deep-water wave length ($h = L_o/2$). However, according to Komar (1976) this limit is too stringent for most practical cases and he suggests a limit of $h = L_o/4$. Redefining the wave base in this way provides water depths for the seaward limit of sediment disturbance by waves that are similar to $h_i$ computed using Equation 8.3 (Table 8.1). It should be noted, however, that the wave base approach does not consider the height of the wave, nor the sediment characteristics.

## 8.2.3 Equilibrium profile

Assuming that the shoreface is in equilibrium with the hydrodynamic conditions and that sufficient sediment is available to accomplish profile equilibrium, the concave-upward shape of the upper shoreface profile can be described by a simple function

$$h = Ax^m \tag{8.4}$$

where $h$ is the still-water depth at a horizontal distance $x$ from the shoreline, $A$ is a dimensional parameter and $m$ is a dimensionless exponent. The parameter $A$ is a scaling coefficient that controls the overall steepness of the profile, while the exponent $m$ determines the profile shape. Convex, straight and concave profiles have $m$-values of larger than 1, equal to 1 and smaller than 1, respectively. The profile described by Equation 8.4 is used extensively to describe the cross-shore profile of the upper shoreface.

Elaborate fitting of Equation 8.4 to natural upper shoreface profiles by Bruun (1954, 1988) and Dean (1977, 1991) has suggested an average value of $m = 0.67$. However, the value for $m$ is subject to quite a large variability. For example, values for individual profiles within Dean's (1977) data base of 502 US east coast and Gulf coast beach profiles ranged from about 0 to 1.4. According to Cowell et al. (1999), the value for $m$ depends on the beach type and is relatively low ($m \approx 0.4$) for reflective beaches and relatively high ($m \approx 0.8$) for dissipative beaches. The steepness parameter $A$ of the resulting profile has been empirically related by Dean (1987) to the sediment fall velocity $w_s$

$$A = 0.067w_s^{0.44} \tag{8.5}$$

where $w_s$ is in cm s$^{-1}$. As expected, the larger the sediment size, the greater the value of $A$ and the steeper the beach profile. For sandy sediments, $A$ = 0.05–0.25.

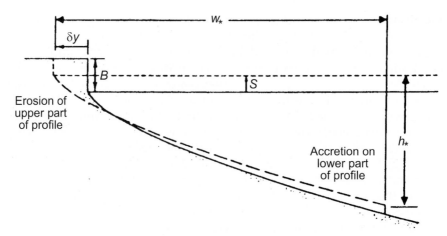

**Figure 8.3** – Response of shoreface profile to rising sea level according to the Bruun rule. Erosion of the subaerial beach is balanced by deposition on the shoreface. $\delta y$ = shoreline retreat; $S$ = rise in sea level; $w_*$ = width of the shoreface; $h_*$ = depth of the shoreface; and $B$ = height of the subaerial beach. [Modified from Dean, 1991.]

The attractiveness of the equilibrium shoreface profile as a concept is that it can be used to predict shoreline change due to a rise in sea level (Figure 8.3). The hypothesis is that when sea level rises, the shoreface profile adjusts by rising and moving landward, whilst keeping its equilibrium shape. Bruun (1962) proposed that the retreat of the shoreline $\delta y$ can be derived by assuming that the amount of erosion occurring on the upper part of the profile is equal to the amount of deposition on the lower part

$$\delta y = - S \frac{w_*}{h_* + B} \tag{8.6}$$

where $S$ is the rise in sea level, $w_*$ is the width of the shoreface, $h_*$ is the depth of the shoreface (taken as the depth of closure $h_c$ or the wave base $h_i$) and $B$ is the height of the subaerial beach. Equation 8.6 is known as the **Bruun rule** and is the most widely used approach in predicting the effects of sea-level rise on sandy shorelines.

It should be mentioned that the existence of an equilibrium shoreface profile and the validity of the Bruun rule is strongly questioned (*e.g.*, Pilkey *et al.*, 1993). The three main objections are to the assumptions that: (1) the underlying geology does not play a role in determining the shoreface shape; (2) shoreface sediment is moved only by waves; and (3) there is no significant movement of sediment beyond the depth of closure $h_c$. Pilkey *et al.* (1993) acknowledge that there may be regionally consistent shoreface slopes that are in equilibrium with a number of oceanographic and geologic factors. They believe, however, that such an equilibrium shape is controlled by factors more numerous than those described by Equation 8.4 and conclude that a fundamental re-examination of the methods of determining shoreface evolution is needed.

## 8.2.4 Waves and currents

The seaward limit of the shoreface is formed by the wave base ($L_o/2$ or $L_o/4$) and this places the entire shoreface in the shallow- and intermediate wave region (Chapter 4). Therefore, wave motion extends to the sea bed and has the potential for entraining and transporting sediments. In addition to wave-orbital flows, a large variety of currents operate in this region. Surf zone currents operate on the upper part of the shoreface and include long-shore wave-driven currents, offshore-directed rip currents and bed return flow. These surf zone currents will be discussed later in this chapter. The

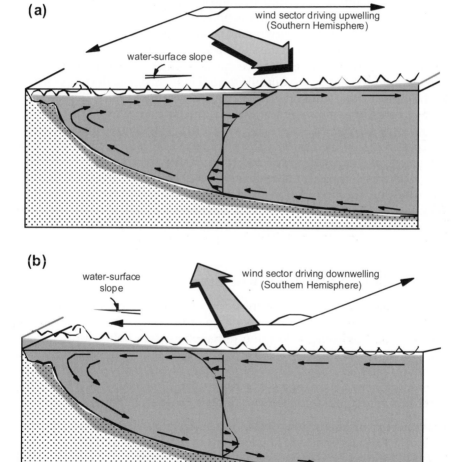

**Figure 8.4** – Wind-driven currents on the shoreface in the Southern Hemisphere: (a) upwelling with onshore near-bottom flow and (b) downwelling with offshore flow near the bed. [From Cowell et al., 1999.] [Copyright © 1999 John Wiley & Sons, reproduced with permission.]

two main types of currents that operate on the shoreface seaward of the surf zone are tidal currents and wind-driven currents.

**Tidal currents** generally flow in an alongshore direction. Only near estuaries and tidal inlets do they have an important cross-shore component. Similar to incident wave motion, tidal currents are also characterized by flow reversal, albeit on a longer time scale. In microtidal environments, tide-driven currents on the shoreface are generally weak, but on macrotidal shorefaces they may exceed $0.25$ m s$^{-1}$. Usually, flood current velocities are not balanced by those of the ebb and this creates residual tidal currents. Although these currents are generally weak ($< 0.25$ m s$^{-1}$), they may contribute significantly to net sediment transport on the shoreface.

Water responds rapidly to wind action and there is a strong tendency in shallow water for surface currents to be aligned with the direction of the wind. However, over most of the shoreface the situation becomes complicated because the wind-driven surface current deviates from the wind direction by up to 45° due to the Coriolis force (Figure 8.4). The deviation will be to the right in the Northern Hemisphere and to the left in the Southern Hemisphere. If the movement of water is directed away from the coastline, the water level near the coast will be lowered. This causes **upwelling** near the coast whereby the near-bed flow is directed onshore. Conversely, if the movement of the water is directed towards the coastline, the water level near the coast is raised. The elevated water level near the coast induces **downwelling** currents that result in offshore flow near the sea bed. Flow velocities associated with up- and downwelling are usually less than $0.25$ m s$^{-1}$ (Niedoroda and Swift, 1991). The upwelling and downwelling are driven by the pressure gradient force, which is described in Box 3.1.

## 8.2.5 Sediment transport processes

At first glance, it appears that there is only limited potential for the net transport of sediments on the shoreface. Tide- and wind-driven current velocities are generally below the entrainment threshold and hence too weak to move sediment. Wave-orbital velocities are usually strong enough to entrain sediment, but the reversing nature of wave-driven flows reduces their effectiveness in causing net transport. However, there are three factors that contribute significantly to net sediment transport on the shoreface:

- **Wave-current interaction** – Most net sediment transport on the shoreface is accomplished by waves and currents working in concert (*e.g.*, Bagnold, 1963). Stresses exerted by the wave motion support and suspend sediments above the bed, but may not cause a net transport due to the reversing behaviour of the motion. However, if a unidirectional current is superimposed on this to-and-fro motion, a net transport of sediment will result in the direction of the current. Since the waves have already supplied the power to put the sand into motion, the unidirectional current can cause a net transport no matter how weak the flow.

- **Wave asymmetry** – During wave shoaling, waves become increasingly non-linear and asymmetric as they enter shallow water. This change in wave shape is reflected in the wave orbital velocities such that the onshore stroke of the wave is stronger, but of shorter duration than the offshore stroke (onshore asymmetry). Sediment transport is related to the flow velocity raised to a power of at least 3. Therefore, the onshore transport under the wave crest is likely to exceed the offshore transport under the wave trough (Figure 8.5). The potential for onshore sediment transport under asymmetric waves is particularly large for coarse sediment particles. These particles tend to be 'nudged' landward during the onshore stroke of the wave, while remaining at rest during the offshore stroke.
- **Bed morphology** – Sediment transport under waves is generally considered to be in phase with the wave motion. In other words, maximum onshore (offshore) sediment transport rates coincide with maximum onshore (offshore) flow velocities. In reality, the response of the sediment to the flow is not instantaneous since there are time lags between the flow field and the sediment transport pattern. In other words, maximum flow velocities may not coincide with maximum sediment transport rates. These time lags are particularly significant when suspended sediment transport occurs over a rippled sea bed and may induce net onshore/offshore sediment transport, despite the presence of symmetrical wave motion (Figures 5.20 and 5.21).

## 8.3 BEACHES

A **beach** can be defined as a wave-lain deposit of sand or gravel found along marine, lacustrine and estuarine shorelines. Generally speaking, the beach relates to that part of the shore profile that extends from the spring low tide level to some physiographic change such as a cliff or dune field, or to the point where permanent vegetation is established. From a morphological point of view, the part of the nearshore profile that is regularly affected by

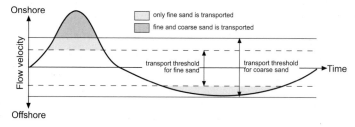

**Figure 8.5** – The effect of onshore wave asymmetry on sediment transport. Onshore sediment transport, especially of coarse sediment, is promoted because the onshore wave orbital velocity is stronger than the offshore velocity. [Modified from Cowell *et al.*, 1999.]

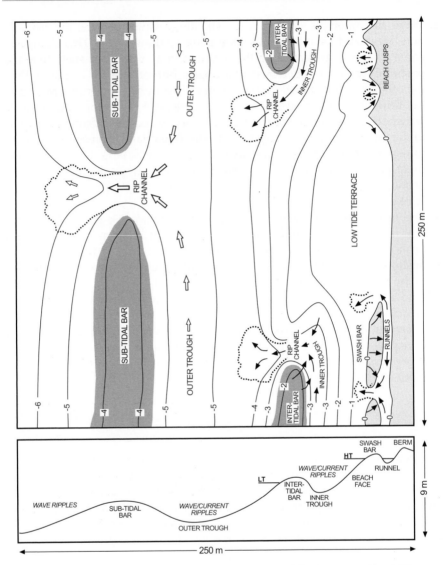

**Figure 8.6** – Schematic showing dominant morphological features on a beach. The top panel represents a contour plot of the beach morphology, whereas the bottom panel shows a typical cross-shore beach profile. It is unlikely that all the morphological features shown in the diagram are present at the same time on the same beach, but they may be present at different times on a beach. High tide level on this beach is at an elevation of 0 m and the tidal range is 2 m.

surf zone processes should also be considered part of the beach. The beach therefore constitutes the upper part of the shoreface. As such, it complies with the concave-upward shape that characterizes the shoreface profile.

On most beaches, perturbations to the concave-upward profile may occur in the form of smaller-scale, secondary morphological features such as beach cusps, berms and nearshore bars (Figure 8.6). It is the presence of these secondary morphological features that gives beaches their distinctive morphology and, in combination with differing morphodynamic process regimes, allows for the classification of distinct beach types.

Before discussing the different types of secondary morphology, their dynamics and beach classification, we will first introduce the characteristics of wave-driven, nearshore currents. These currents are of paramount importance because they give direction to the movement of beach sediments and as a result act to shape the beach morphology.

## 8.3.1 Nearshore currents

In the surf zone, incident waves progressively dissipate their energy due to wave breaking. A significant proportion of this energy is used for

**Figure 8.7** – Quasi-steady currents in the surf zone: (a) shore-parallel longshore currents due to obliquely-incident waves, (b) vertically-segregated bed return flow or undertow and (c) horizontally-segregated rip currents as part of the nearshore cell circulation system.

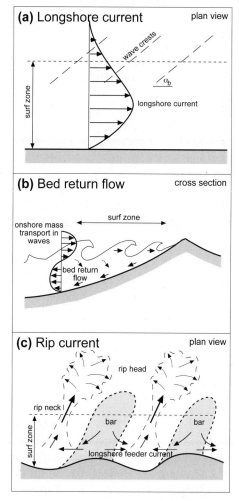

generating nearshore currents and sediment transport, ultimately resulting in the formation of distinct morphology. **Nearshore currents** are due to cross-shore and/or longshore gradients in the mean water surface that arise due to variations in the wave breaker height. This implies that when waves are not breaking, these currents will not be generated. In addition, the intensity of the currents increases with increasing incident wave energy level and hence the strongest currents are encountered during storms. Nearshore currents are capable of transporting large quantities of sediment. This is partly due to their often significant flow velocities, but also because sediment entrainment is considerably enhanced by the stirring motion of the breaking waves.

There are three types of wave-induced currents and these systems dominate net water movement in the nearshore (Figure 8.7): longshore currents, bed return flow and rip currents. All three currents are quasi-steady currents, meaning that they flow with a relatively constant velocity for a given set of wave/tide conditions. Longshore currents flow predominantly parallel to the shore, while bed return flow and rip currents are principally directed perpendicular to the beach.

**Longshore currents** are shore-parallel flows within the surf zone (Figure 8.7a). They are primarily driven by waves entering the surf zone with their crests aligned at oblique angles to the shoreline. Longshore currents may reach velocities in excess of 1 m s$^{-1}$. On planar beaches, maximum longshore current velocities occur around the mid-surf zone position where most of the waves are breaking. On barred beaches, the longshore current is mainly confined to the trough between the bar and the shoreline, or in troughs between multiple bars. The strength of the longshore current increases with the incident wave energy level and also with the angle of wave approach. According to Komar and Inman (1970), the longshore current velocity at the mid-surf zone position $\bar{v}$ can be given by

$$\bar{v}_l = 1.17 \sqrt{gH_b} \sin\alpha_b \cos\alpha_b \qquad (8.7)$$

where $H_b$ is the root mean square breaker height and $\alpha_b$ is the wave angle at the break point. For example, a breaker height and breaker angle of 1 m and 10°, respectively, will produce a longshore current with a velocity of 0.63 m s$^{-1}$ at the mid-surf zone position. Longshore currents can be modulated by tidal changes in the nearshore water level (Thornton and Kim, 1993) and may also be affected by alongshore winds (Whitford and Thornton, 1993). For strong winds blowing in the direction of the longshore current, measured current velocities may be significantly higher than that predicted using Equation 8.7.

The **bed return flow**, or **undertow**, is an offshore-directed mean flow near the bed. The current is part of a vertically-segregated circulation of water characterized by onshore flow in the upper part of the water column and seaward flow near the bottom (Figure 8.7b). Measured bed return velocities are typically 0.1–0.3 m s$^{-1}$, but under extreme wave conditions may reach values of up to 0.5 m s⁻¹. On planar beaches, maximum bed return velocities occur around the mid-surf zone position, but on barred beaches the strongest flows are encountered at, or slightly landward of, the bar crest. The bed return flow is clearly fed by the water carried to the shore by breakers and bores, but a more precise explanation for the occurrence of bed return flow involves the set-up gradient across the surf zone produced by wave energy dissipation (Svendsen, 1984). The set-up produces a seaward-directed pressure gradient of water. This pressure gradient is, on average, balanced by the momentum of the waves directed to the shore (Box 4.4). However, close to the bed, the offshore-directed pressure gradient is greater than the onshore-directed force due to the wave momentum. This results in a net offshore-directed force on the near-bed water particles, which drives the bed return current.

**Rip currents** are strong, narrow currents that flow seaward through the surf zone in channels, and present a significant hazard to swimmers

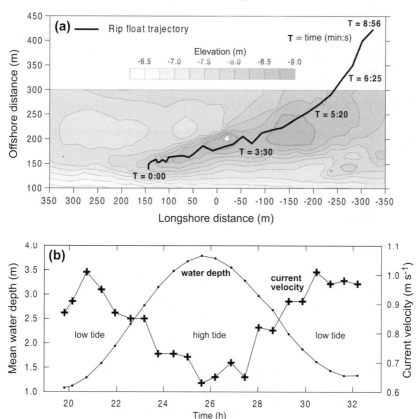

Figure 8.8 – Flow velocities in a rip channel on Muriwai Beach, New Zealand: (a) trajectory of a rip floater showing current velocities in excess of 1 m s⁻¹ and (b) temporal variation in water depth and current velocity measured in the rip feeder indicating tidal modulation of the rip system. [Modified from Brander and Short, 2000.]

(Figure 8.7c). They are an integral part of the nearshore cell circulation system (Shepard and Inman, 1950) and consist of: (1) onshore transport of water between rip currents; (2) longshore **feeder currents** that are fully contained within the surf zone and carry water into the rip; (3) a fast-flowing **rip neck** that extends from the confluence of two opposing feeder currents and transports water seaward through the surf zone; and (4) the **rip head**, which is a region of decreasing velocity and flow expansion seaward of the surf zone. Maximum current velocities associated with cell circulation are encountered in the rip neck and may reach up to 2 m s$^{-1}$ under extreme storm conditions when so-called 'mega-rips' form (Short, 1985). Typical rip current velocities are in the order of 0.5–1 m s$^{-1}$. Rips 'work' best at low tide and current velocities in the rip during low tide are generally stronger than during high tide (Figure 8.8; Brander, 1999). Cell circulation is usually associated with and controlled by nearshore bar/rip morphology (Sonu, 1972). Wave energy dissipation and wave set-up is larger across the barred areas than in the rip channels. This results in a longshore pressure gradient from bar to rip that drives feeder currents and therefore the cell circulation.

The three wave-induced current systems discussed above do not occur separately. When waves break at small angles to the shoreline, longshore currents as well as a cell circulation system with rip currents may be present. In addition, the bed return flow is always present under breaking waves although its role is expected to be limited in the case of pronounced cell circulation.

## 8.3.2 Swash morphology

The **swash zone** is the upper part of the beach that is alternately wet and dry (Figure 8.9). Sediment transport in the swash zone is performed by wave uprush (onshore flow) and backwash (offshore flow). Field measurements demonstrate that swash motion is asymmetric – the backwash is not simply the reverse of the uprush (Hughes *et al.*, 1997). Generally, onshore flow velocities during the uprush are larger, but of shorter duration than the offshore velocities during the backwash. Maximum onshore velocities occur at the start of the uprush and then decrease, whereas offshore velocities increase to a maximum at the end of the backwash. Swash hydrodynamics are fundamentally different from surf zone hydrodynamics. First, because the swash zone is alternately wet and dry, quasi-steady currents are absent. Second, interactions take place between the swash flow and the beach groundwater table in the form of infiltration into the beach and exfiltration out of the beach. Typical morphological features associated with swash motion include the beachface, berm, beach step and beach cusps.

The **beachface** is the planar, relatively steep upper part of the beach profile that is subject to swash processes. Large amounts of sediment are transported up and down the beachface by the uprush and backwash, respectively. When the net transport averaged over several swash cycles is zero, the beachface can be defined as being in equilibrium. In general terms,

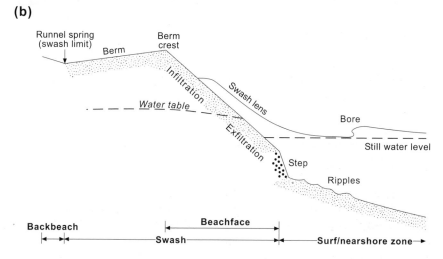

**Figure 8.9** – (a) Freshly deposited berm on Warriewood Beach, New South Wales. [Photo A.D. Short.] (b) Schematic diagram of swash zone and beachface morphology showing terminology and principal processes. [From Hughes and Turner, 1999.] [Copyright © 1999 John Wiley & Sons, reproduced with permission.]

the equilibrium slope of the beachface reflects the balance between the onshore asymmetry of the near-bed velocities in the swash zone tending to move sediment landward (uprush is stronger than backwash), and the opposing force of gravity acting to move sediment seaward. Onshore swash asymmetry arises mainly from energy losses due to bed friction and infiltration of water into the beach during the uprush, and is promoted by coarse and permeable sediments. The equilibrium beachface gradient is therefore positively correlated with the beachface sediment size (Figure 8.10). The steepest beachfaces occur on gravel beaches, where gradients in excess of 1/5 are not uncommon.

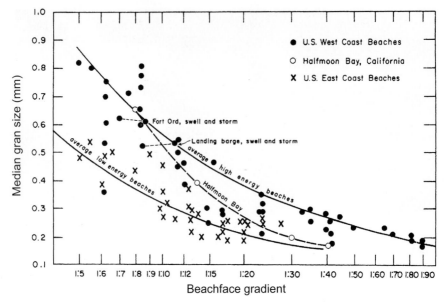

**Figure 8.10** – Relation between median sediment size and beachface gradient and angle. [From Komar, 1998; modified from Wiegel, 1964; based on data from Bascom, 1951.] [Copyright © 1964 Wiegel. Reprinted by permission of Pearson Education, Inc., Upper Saddle River, NJ.]

The **berm** is the nearly planar section landward of the beachface and is an accretionary feature that results from the accumulation of sediment at the landward extreme of wave influence (Figure 8.9). It forms the first line of defence of the beach and protects the backshore and coastal dunes from erosion under mild wave conditions and during the early phase of a storm. The berm is separated from the beachface by the berm crest. Some beaches, in particular gravel beaches, can have multiple berms at different elevations. The berm crest tends to be most distinct on beaches composed of coarser sediment, because on these beaches the beachface is typically much steeper than the near-horizontal backbeach region. On fine-grained beaches, the berm may be indistinct due to similar backbeach and beachface gradients. The height of the berm is controlled by the maximum elevation to which sediment is transported during wave uprush. According to Takeda and Sunamura (1982), the berm height $Z_{berm}$ can be predicted using

$$Z_{berm} = 0.125 H_b^{5/8} (gT^2)^{3/8}$$
(8.8)

The larger the wave height and/or wave period, the larger the vertical wave runup and hence the higher the berm.

Berms and beachfaces are highly dynamic and respond rapidly to changing wave conditions (Figure 8.11). Berm/beachface accretion and the formation of a convex beach profile occur under low-energy swell conditions. Storm conditions, on the other hand, result in berm/beachface erosion and the development of a concave and sometime scarped beach profile. Vertical

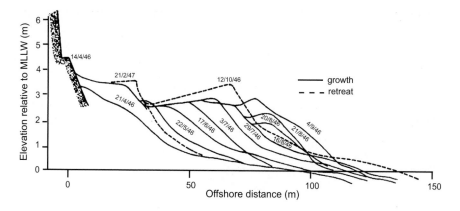

**Figure 8.11** – The progressive growth and retreat of the berm at Carmel, California, from 1946–1947. MLLW refers to mean lowest low water. [From Komar, 1998; modified from Wiegel, 1964; based on data from Bascom, 1953.] [Copyright © 1964 Wiegel. Reprinted by permission of Pearson Education, Inc., Upper Saddle River, NJ.]

accretion of the berm requires overtopping of the berm crest, followed by deposition of sediment from the ponded water behind the berm crest. However, erosion of the berm also requires overtopping of the berm crest. There thus seems to be a threshold separating berm erosion and accretion. This threshold will be discussed in Section 8.3.5.

The **beach step** is a small, submerged scarp located at the base of the beachface and can range in height from several centimetres to over a metre (Figure 8.9). Beach steps generally comprise the coarsest material found on a beach and their morphology is therefore most pronounced on steep beaches consisting of coarse sand and gravel material (Bauer and Allen, 1995). They are fairly insignificant features, but their formation and maintenance highlights the importance of positive feedback in coastal morphodynamics. Beach step formation is causally linked to the vortex that develops when the backwash interacts with the oncoming incident wave (Figure 8.12). This **backwash vortex** forms immediately seaward of the still water shoreline during the backwash phase of the swash cycle. The rotation of the vortex is such that the flow is directed landward near the bed. The backwash vortex may erode the bed resulting in the formation of the step and also provides the hydrodynamic regime necessary to sort the sediment and concentrate the coarsest material in the step. The size of the step is related to the size (radius) and the angular velocity of the backwash vortex, which in turn depends on the wave height at the base of the beachface and the wave period. By combining field and laboratory data, Hughes and Cowell (1987) proposed the following predictive equation for the step height $Z_{step}$

$$Z_{step} = 0.55 \sqrt{H_b T w_s} \tag{8.9}$$

where $w_s$ is the sediment fall velocity. Thus, step height increases with increasing wave height, period and sediment size.

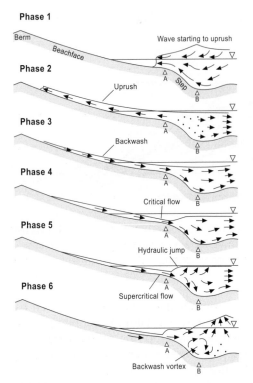

Figure 8.12 – Characteristic swash flow patterns over the beachface and step during uprush (phases 1 and 2) and backwash (phases 3 to 6). The length of the arrows shows the relative flow velocity. [From Hughes and Turner, 1999; modified from Larson and Sunamura, 1993.] [Copyright © 1999 John Wiley & Sons, reproduced with permission.]

**Beach cusps** are rhythmic shoreline features formed by swash action and may develop on sand or gravel beaches. They are typically (quasi-) regularly spaced and exhibit a crescentic planform (Figure 8.13). Beach cusp morphology is characterized by gentle-gradient, seaward-facing cusp embayments separated by steep-gradient, seaward-pointing cusp horns. The spacing of the cusps (distance between consecutive cusp horns) is related to the horizontal extent of the swash motion and may range from *c*. 10 cm on lake shores to 50 m on exposed ocean beaches. On some beaches, different sets of cusps may be present across the beach profile with different

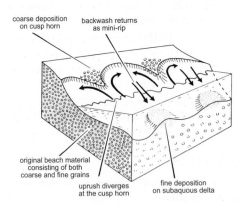

Figure 8.13 – Schematic showing beach cusp morphology. Divergence of the uprush occurs at the cusp horns and the backwash converges in the cusp embayments. [From Pethick, 1984.] [Copyright © 1984 Edward Arnold, reproduced with permission.]

spacings. Cusp horns generally contain coarser, more permeable, sediments than the embayments.

Swash motion on a beachface with cusp morphology is distinctly three-dimensional. Most commonly, the wave uprush is deflected from the horns into the embayments where the water flows offshore in concentrated back-wash streams or mini-rips. Equilibrium conditions on the beachface are to a large extent determined by the relative strengths of the wave uprush and backwash (Masselink and Pattiaratchi, 1998). On cusp horns, the wave uprush is considerably stronger than the backwash because a significant volume of water is diverted to the two adjacent embayments during uprush. As a result, relatively steep gradients can be achieved due to the preferential onshore sediment transport across the horns. In the embay-ments, the backwash volume is significantly larger than the uprush volume due to the contributions of the two adjacent cusp horns. In addition, the wave uprush in the embayment is impeded by the strong backwash at that location. Consequently, offshore transport dominates, resulting in gentle beachface gradients.

Beach cusps are in themselves not very important features, however, their formation has intrigued coastal researchers for over a century. Although the presence of three-dimensional swash flow circulation may help to explain the development of subtle cusps into pronounced cusps, it cannot account for the formation of cusps on a planar beach. At present, only two theories appear to provide an adequate explanation for the formation of regularly-spaced beach cusps on natural beaches: standing edge waves and self-organization. Field verification of the two theories so far seems to favour the self-organization hypothesis (Masselink, 1999).

The **standing edge wave model**, put forward by Guza and Inman (1975), suggests that swash from incident waves is superimposed upon the motion of standing edge waves to produce a systematic longshore variation in swash height which results in a regular erosional perturbation (edge waves are a special class of waves that travel along the beach, rather than towards the beach). The theory indicates that cusp embayments are scoured out at the locations of edge wave antinodes, whereas cusp horns occur at edge wave nodes. The distance between cusp horns is equal to half the wave-length of the prevailing edge wave. Although several different types of edge waves may occur on beaches, the sub-harmonic edge wave, which has a period twice that of the incident waves, is the most easily excited and is generally implicated in the formation of beach cusps. The sub-harmonic edge wave model predicts a beach cusp spacing $\lambda$ of

$$\lambda = \frac{g}{\pi} T^2 \tan \beta \qquad (8.10)$$

where $T$ is incident wave period and $\tan \beta$ is beach gradient. Beach cusp for-mation by standing edge waves is a self-limiting process because the forma-tion of cusps results in the reduction of the edge wave amplitude. Hence, it appears that although edge waves may initiate a longshore perturbation in the foreshore topography, an additional process, such as positive feedback between incident waves and the perturbed morphology, needs to be invoked to explain the further evolution of the cusps.

The **self-organization model**, proposed by Werner and Fink (1993), suggests that beach cusps develop through a combination of positive feedback between beach morphology and swash flow that can operate to enhance existing topographic irregularities, and negative feedback that inhibits accretion or erosion on well-developed cusps. Morphological regularity arises from the internal dynamics of the system, and hence the term 'self-organization' or 'self-emergence' is employed (Box 1.2). Positive and negative feedback processes between foreshore morphology and swash hydrodynamics have been observed by several previous researchers. However, only recently have the computational resources and sediment transport formulations become available to demonstrate that such feedback processes may produce stable and rhythmic morphological features. According to the self-organization theory of beach cusp formation, cusp spacing $\lambda$ is proportional to the horizontal extent of the swash motion $S$ according to

$$\lambda = fS \qquad (8.11)$$

where the constant of proportionality $f$ is $c.$ 1.5.

### 8.3.3 Surf zone morphology

Water motion in the surf zone is dominated by breaking incident waves, standing infragravity wave motion and quasi-steady currents. These water

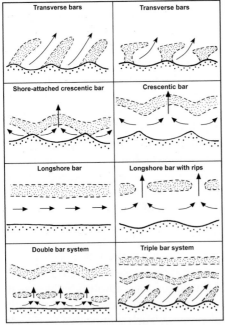

Figure 8.14 – Examples of different types of nearshore bar morphology.

motions induce sediment transport at a range of frequencies and in different directions. The morphology of the surf zone can be planar, but usually bars and troughs are present. **Nearshore bars** are the main expression of hydro-dynamic and sediment transport gradients, with bars forming as a result of sediment convergence and troughs forming in areas of sediment diver-gence. Nearshore bar morphology can assume a large variety of configura-tions (Figure 8.14), including transverse bars, crescentic bars and longshore bars. Rip channels often dissect the bars. In addition, multiple bars may be present on a beach. When the bar morphology is characterized by a domi-nant alongshore wave length (*i.e.*, spacing of rips), this is referred to as **rhythmic morphology**.

The mechanism of **bar formation** remains unresolved despite consider-able research and modelling efforts. The most likely hypothesis is that nearshore bars form near the breakpoint as a result of onshore sediment transport outside the surf zone due to wave asymmetry and offshore trans-port in the surf zone due to bed return flow (Figure 8.15a; Roelvink and Stive, 1989). The bar location thus represents a point of sediment conver-gence. However, bar formation, particularly for rhythmic and multi-bar morphology, has also been attributed to net sediment transport patterns associated with standing infragravity waves (Figure 8.15b; Holman and Bowen, 1982). According to the latter hypothesis, bars form at the node positions if sediment transport is predominantly by bedload and at the antinodal positions if suspension is the main mode of transport. Regardless, the locations of the bars represent points of sediment convergence.

Nearshore bars are highly dynamic and tend to move in response to changing wave conditions. Following formation, nearshore bars can migrate both in an onshore and offshore direction. Offshore **bar migration** is gener-ally observed during high wave events (storms) when sediment transport due to off-shore directed bed return flow is more important than that driven by the onshore asymmetry of the waves (Figure 8.16a). Onshore bar

Figure 8.15 – Two models of bar forma-tion: (a) breakpoint mechanism and (b) standing wave model. [Modified from Komar, 1998.]

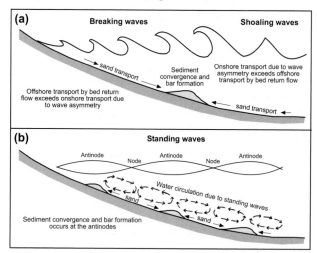

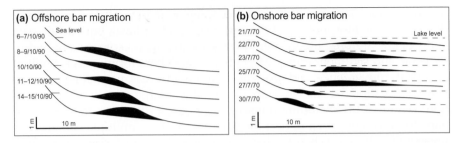

Figure 8.16 – (a) Offshore bar migration over a 10-day period at Duck, North Carolina. [Modified from Thornton *et al.*, 1996.] (b) Onshore bar migration sequence on a Lake Michigan beach over a 10-day period. [Modified from Davis *et al.*, 1972.]

migration is generally observed during low- to moderate-energy wave conditions when the shoreward-directed sediment flux due to wave asymmetry dominates over the seaward-directed sediment flux by the bed return flow (Figure 8.16b). At some stage the nearshore bars may merge with the beachface, resulting in a welded bar. Bar migration rates vary widely and are affected by a range of factors, including wave energy level, tidal range, bar size and water depth over the bar (Sunamura and Takeda, 1984; Orme, 1985). Typical rates of bar migration are 1–10 m day$^{-1}$, but under extreme conditions may reach up to 30 m day$^{-1}$. Offshore bar migration rates are generally faster than onshore rates on the same beach. Over a time span of years, the nearshore bar morphology may show a cyclic behaviour with a generation phase just below the low-tide water line, a net seaward migration phase through the nearshore to finally a destruction phase of the feature at the seaward end of the nearshore (Box 8.1).

## Box 8.1 – Bar Dynamics on the Dutch Coast

The coast of the Netherlands experiences a mean significant wave height of 1.2 m and a mean period of 5 s. The mean spring tidal range is 1.6 m. The beach morphology is characterized by a double- or triple-bar system (Short, 1992). Long-term monitoring of the nearshore morphology along the Dutch coast has demonstrated that these bars exhibit long-term cyclic behaviour (Figure 8.17). Over periods of many years, all bars migrate in a seaward direction. As the outer bar migrates in progressively deeper water, it decays while at the same time a new bar is generated near the shoreline.

The offshore movement of the bars does not represent net offshore sediment transport. Rather, the sediment is redistributed across the nearshore zone, because the sediment comprising the decaying outer bar is transported back onshore. The cyclicity in the long-term bar behaviour is most likely explained by morphologic feedback in the breaker bar system, rather than by cyclicity in the wave conditions, with the outer bar playing a central role. As long as the outer bar remains well developed, the inner bar(s) will only move to and fro within a limited

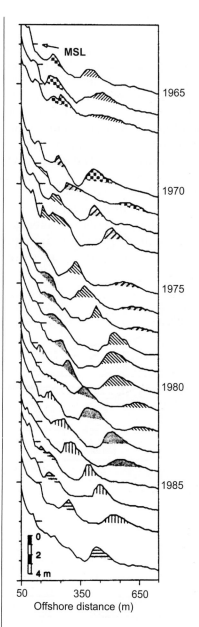

**Figure 8.17** – Offshore migration of the bar system on the Dutch coast. Bars are shaded to emphasise the offshore progression cycle. [Modified from Wijnberg, 1996.]

cross-shore range. Once the outer bar disappears, however, the inner bar(s) start migrating in the offshore direction. The behaviour of the outer bar appears to be controlled by the local wave conditions – breaking waves maintain the outer bar, while very asymmetric, shoaling waves promote the decay of the outer bar (Wijnberg, 1996). The offshore migration of the nearshore bars along the Dutch coast is not unique, but also occurs along many other multi-barred coastlines (Shand et al., 1999).

## 8.3.4 Intertidal morphology

So far we have only considered beach morphological features formed by either swash or surf zone processes and have ignored the effect of tides on beach morphology. Tide-induced changes in nearshore water levels cause the shoreline to shift across the intertidal beach profile with the rate of shifting being dependent on the tidal range and the beach gradient. Two important consequences are the migration of different hydrodynamic zones (swash zone, surf zone and shoaling wave zone) and the occurrence of beach watertable fluctuations (Masselink and Turner, 1999). The morphological effects of these tide-related processes are now discussed.

On tidal beaches, swash, surf zone and shoaling wave processes shift both vertically and horizontally with the rise and fall of the tide. If the tide range is sufficiently large, as on macrotidal beaches, parts of the intertidal beach profile may be subjected to the entire suite of processes during a single tidal cycle. Masselink and Short (1993) used a numerical model to

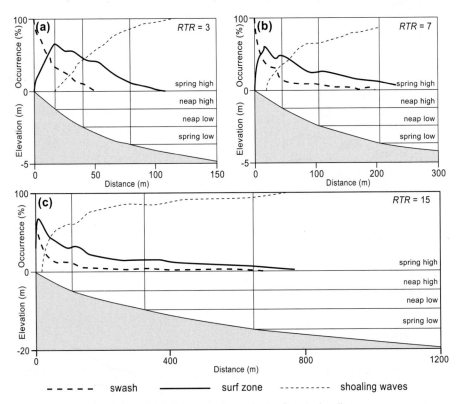

Figure 8.18 – The relative occurrence of swash, surf and shoaling wave processes across the beach profile calculated over a lunar tidal cycle for three different values of the relative tide range *RTR*: (a) *RTR* = 3, (b) *RTR* = 7 and (c) *RTR* = 15. [From Masselink and Turner, 1999; modified from Masselink and Short, 1993.] [Copyright © 1999 John Wiley & Sons, reproduced with permission.]

determine the amount of time, expressed as a percentage of the total submergence time, that different parts of the intertidal profile are subjected to swash, surf or shoaling wave processes for different wave-tide conditions over a lunar tidal cycle (Figure 8.18). It was found that the relative occurrence of the different hydrodynamic processes over the intertidal profile is primarily dependent on the ratio between the mean spring tide range MSR and the breaker wave height $H_b$

$$RTR = \frac{MSR}{H_b} \tag{8.12}$$

where $RTR$ is referred to as the **relative tide range**. Swash and surf zone processes prevail around the high tide level, but their relative occurrence decreases with increasing $RTR$. For $RTR \geq 15$, shoaling wave processes dominate over practically the entire intertidal zone.

The model results shown in Figure 8.18 indicate that on beaches experiencing large tidal ranges, swash and surf zone processes are relatively unimportant. In addition, due to the rapid **tidal translation** rates, stationary water level conditions are only encountered around the high and low tide levels. As a result, there is generally insufficient time for the development of secondary morphological features, such as berms, beach cusps, beach steps, bars and rip currents. Therefore, the intertidal zone of such beaches is often flat and featureless.

The water table within beaches oscillates in response to the ocean tide. A major feature of tide-induced **beach water-table** fluctuations is their asymmetry, characterized by the water table rising faster than it falls. During the flooding tide, the rising water table will keep up with the rising ocean tide level. However, during the ebbing tide, the water table may decouple from the ocean because it cannot fall as fast as the ocean tide level. This results in the formation of a **seepage face** where groundwater outcrops onto the beachface (Figure 8.19). The intersection of the beach groundwater table and the beach is referred to as the **exit point**.

The moment of decoupling and the characteristics of the seepage face are largely dependent on the drainage capabilities of the beach, which are a

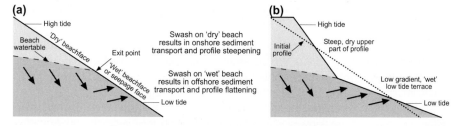

**Figure 8.19** – (a) Definition sketch of the exit point and the seepage face. [Modified from Turner, 1993]. (b) Development of a steep upper beach with low-gradient low tide terrace as a result of interactions between swash hydrodynamics and the beach groundwater table.

# BOX 8.2 – MACROTIDAL RIDGE AND RUNNEL BEACHES

In macrotidal coastal environments, some beaches are characterized by the presence of a number of intertidal bars and troughs referred to as 'ridge and runnel' morphology. Conditions conducive to the formation of ridge and runnel morphology are a macrotidal tide range, a gentle beach gradient, a short fetch and a surplus of sediment (King, 1972). The north Lincolnshire coast in east England satisfies all these criteria and is characterized by well developed ridge and runnels (Figure 8.20).

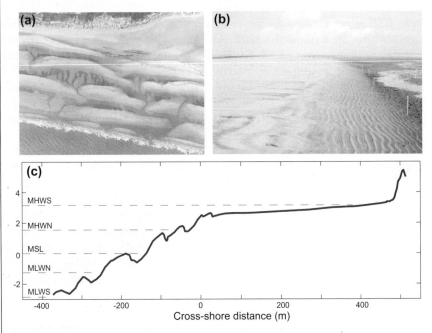

**Figure 8.20** – Ridge and runnel morphology on the north Lincolnshire coast, England: (a) aerial photograph [Aerial photo supplied by Environment Agency, Peterborough, UK.]; (b) individual ridge and runnel [photo G. Masselink]; and (c) beach profile. ODN refers to Ordnance datum Newlyn which is approximately mean sea level in Britain, MHWS and MLWS refer to mean high and low water spring, respectively.

According to King (1972), the ridges are swash bars that develop in an attempt by the waves to form an equilibrium swash zone gradient on a beach that is naturally much flatter. King (1972) further notes that the ridge locations roughly correspond to the positions on the intertidal profile where the water level is stationary for the longest time. However, a recent investigation by Masselink and Anthony (2001) demonstrated that the ridge locations are unrelated to positions of relatively stationary water-level conditions, which may suggest that the ridges are breaker bars, rather than swash bars.

function of beach gradient and sediment properties (Turner, 1993). Low-gradient, fine-grained macrotidal beaches ($D < 0.3$ mm) do not drain very well and decoupling occurs shortly following high tide. This results in the formation of a wide seepage face, occupying most of the intertidal zone. Steep-gradient, coarse-grained macrotidal beaches ($D > 0.3$ mm) drain efficiently and decoupling occurs closer to low tide, resulting in a narrow seepage face.

The occurrence of a seepage face defines two distinct morphodynamic domains: (1) an upper intertidal region that alternates between saturated and unsaturated over the tidal cycle; and (2) a lower intertidal region that remains in a permanently saturated state. Turner (1995) postulated that the intersection of these contrasting intertidal zones marks a point of divergent sediment transport with onshore transport occurring over the dry, upper intertidal region and offshore transport taking place over the wet, lower intertidal region. Such sediment divergence may, in turn, provide an explanation for the presence of a distinct break in slope often observed on large tide range beaches and the development of a low tide terrace (Figure 8.19). On the upper profile, steepening is enhanced by swash infiltration on the rising tide, which in turn promotes profile drainage through subsequent tidal cycles. In contrast, offshore transport is promoted across the low tide terrace, maintaining this region in a low-gradient form.

Despite the fact that tide-induced water level fluctuations inhibit bar formation, **intertidal bars** may occur on tidal beaches. Such bars are thought to be primarily controlled by swash processes and are referred to as **swash bars**. However, it is likely that they are equally, if not more so, affected by breaking wave processes. The intertidal bars generally migrate onshore under low-energy conditions and prolonged low-energy wave conditions may result in the merging of the bars to the supra-tidal beach. High-energy wave conditions interrupt the onshore bar migration and may flatten the bar morphology within a period of several hours. On macrotidal beaches, numerous intertidal bars may be present on the intertidal profile and such morphology is referred to as **ridge and runnel morphology** (Box 8.2). The ridges are rather immobile features, but flattening of the ridges occurs during storm periods, whereas ridge building occurs under calm weather conditions (King, 1972).

## 8.3.5 Seasonal and cyclic beach change

Beach morphology clearly responds to changing wave conditions and of greatest significance is the exchange of sediment between the berm and bar (Figure 8.21). Long-term monitoring of beaches on the west coast of the United States has resulted in the identification of a 'winter' and a 'summer' profile, and a seasonal cycle of beach morphology (Hayes and Boothroyd, 1969). Wave conditions during winter storms remove sand from the berm and result in the formation of a narrow beach with nearshore bar morphology. Calm conditions during the summer induce landward migration of the

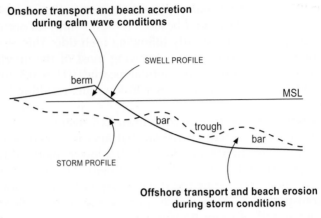

**Figure 8.21** – Idealized barred and non-barred beach profiles. Storm conditions induce offshore transport, beach erosion and the formation of a nearshore bar. Calm wave conditions result in onshore sediment transport, beach accretion and the formation of a berm. [From Aagaard and Masselink, 1999; modified from Komar, 1998.] [Copyright © 1999 John Wiley & Sons, reproduced with permission.]

bar, followed by welding of the bar to the beachface resulting in the formation of a wide, non-barred beach profile with a berm. Subsequent research indicates that the terms 'winter' and 'summer' are somewhat misleading in that beach response is cyclic, as opposed to seasonal (Short, 1979; Nordstrom, 1980). Several sequences of barred/non-barred profiles may occur within a given year depending on the storminess of the wave climate.

Early laboratory investigations correlated the cross-shore sediment transport direction, and hence the occurrence of bar morphology, with the deep water wave steepness ($H_o/L_o$). Steep waves (large wave height and/or small wave period; $H_o/L_o > 0.02$), such as occur during storms, were found to promote offshore transport and the formation of a bar-type profile. Waves with a low wave steepness (small wave height and/or large wave period; $H_o/L_o < 0.02$), which are typical of calm wave conditions, produced onshore transport and the formation of a berm-type profile. Subsequent research has demonstrated that the critical wave steepness is not constant, but increases with the size of the beach sediment. On a coarse-grained beach, a larger wave steepness is required to produce a barred profile than on a fine-grained beach.

Accounting for the effects of both wave steepness and sediment size, Gourlay (1968) and Dean (1973) presented heuristic models for the occurrence of onshore sediment transport and berm-type profiles, and offshore transport and bar-type profiles. Both workers proposed the following parameter

$$\Omega = \frac{H_o}{w_s T} \qquad (8.13)$$

where $H_o$ is the deep water wave height, $T$ is the wave period, $w_s$ is the sediment fall velocity and $\Omega$ is referred to as the **dimensionless fall velocity**. Small-scale laboratory experiments indicated that offshore sediment transport occurs and bar-type profiles form when $\Omega > 1$, whereas onshore transport takes place and berm-type profiles develop when $\Omega < 1$. Subsequent large-scale experiments and field verification showed that, due to scale effects, the threshold value of 1 was too small and suggested a more appropriate $\Omega$-value of 2–2.5 separating bar-type and berm-type beaches.

The value of Equation 8.13 is based on its consideration of the trajectory of a suspended sand particle during its fall to the sea bed whilst under the influence of wave-driven horizontal flow. The numerator $H_o$ can be considered as a measure of how high the sediment particle will be lifted from the bed due to wave action – the larger the wave, the higher the sediment particle will be lifted from the bed and the longer it will remain in suspension. The denominator $w_s T$ reflects the vertical distance over which a sediment particle will fall over one wave period assuming it is entrained at the start of the wave cycle. If the fall of the grain to the bed requires a short time relative to the wave period, either due to small waves and/or a large sediment fall velocity, then the sediment particle will be acted upon primarily by the onshore wave velocities. In this case, $\Omega$ will have a relatively small value. On the other hand, if the fall of the grain to the bed requires a long time relative to the wave period, either due to large waves and/or a small sediment fall velocity, then the sediment particle will remain in suspension a long time and is likely to move offshore. In this case, $\Omega$ will have a relatively large value.

## 8.3.6 Beach classification

Classification of beaches into distinct groups or types can provide a useful framework within which beach morphodynamics and morphological change can be studied. At the most elementary level, sandy beaches can be classified on the basis of their overall slope into steep and gentle-gradient beaches. The simplicity of this classification, however, is deceiving, because these two types of beaches represent two fundamentally different morphodynamic process regimes. On steep beaches, a surf zone is generally absent and incident waves break directly by surging or plunging on the beachface. A significant part of the incoming wave energy is reflected back from the shoreline and hence these beaches are referred to as **reflective beaches**. On gentle-gradient beaches, the surf zone is wide with multiple lines of spilling breakers. The majority of the incoming wave energy is dissipated during the wave breaking process and these beaches are known as **dissipative beaches**. A useful parameter to determine the relative importance of reflection and dissipation is the **surf scaling parameter** $\varepsilon$

$$\varepsilon = \frac{4\pi^2 H_b}{gT^2 \tan^2 \beta} \tag{8.14}$$

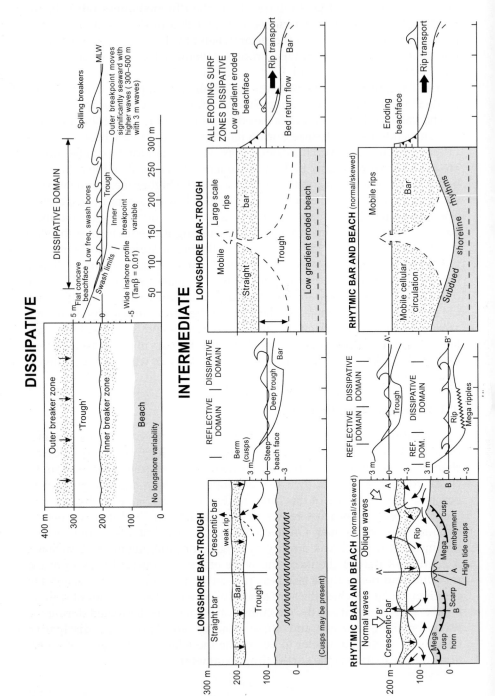

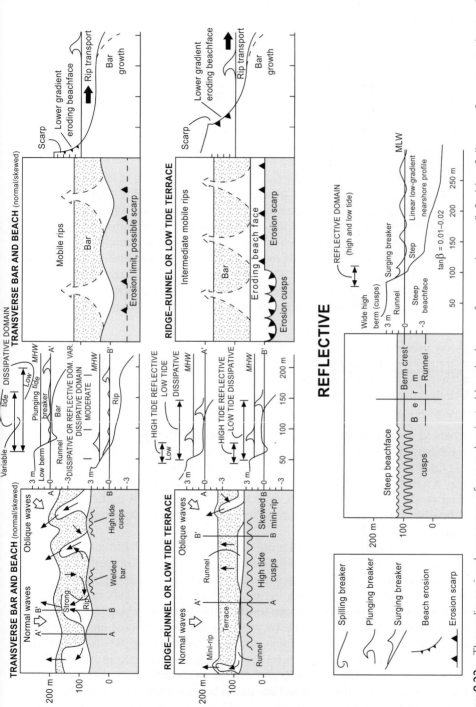

**Figure 8.22** – Three-dimensional sequence of wave-dominated beach changes for accretionary (left side of diagram, from top to bottom) and erosional (right side of diagram, from bottom to top) wave conditions. [From Short, 1999.] [Copyright © 1999 John Wiley & Sons, reproduced with permission.]

where $H_b$ is breaker height, $g$ is gravity, $T$ is wave period and $\tan\beta$ is beach gradient (determined across the swash/surf zone). Reflective conditions prevail when $\varepsilon < 2.5$ and dissipative conditions occur if $\varepsilon > 20$ (Guza and Inman, 1975). When $\varepsilon = 2.5{-}20$ both reflection and dissipation occur and these conditions are referred to as intermediate.

The secondary morphological features superimposed on the main beach profile are strongly related to the morphodynamic regime. A steep beach-face with beach cusp and/or beach step morphology is characteristic of reflective beaches. Nearshore bar morphology dissected by rip channels is typical of intermediate beaches. Multi-barred morphology is generally associated with dissipative beaches.

**Beach type** can be predicted to some degree on the basis of environmental parameters and a widely used predictive parameter for this purpose is the dimensionless fall velocity (Equation 8.13). This parameter, which was originally developed as a predictor for on/offshore sediment transport and bar occurrence has been implemented in the '**Australian beach model**' proposed by Wright and Short (1984). This model, of which the most elaborate version is shown in Figure 8.22, specifically relates to medium to high wave energy, micro-tidal sandy coastlines and identifies six main beach types whose occurrence can be predicted on the basis of the dimensionless fall velocity:

- reflective beach ($\Omega < 1.5$)
- low tide terrace beach ($\Omega \approx 2$)
- transverse bar and rip beach ($\Omega \approx 3$)
- rhythmic bar and beach ($\Omega \approx 4$)
- longshore bar-trough beach ($\Omega \approx 5$)
- dissipative beach ($\Omega > 5.5$)

where $\Omega$ is calculated using the breaker height, rather than the deep water wave height. It is noted that the model shown in Figure 8.22 does not apply to gravel beaches. Although gravel beaches are reflective beaches, they display a number of distinctive morphodynamic attributes that sets them apart from sandy reflective beaches (Carter and Orford, 1993). In particular, gravel beaches never develop bar morphology, not even under extreme storm conditions.

The beach model shown in Figure 8.22 is useful in explaining spatial differences between sandy beach environments and can also be used to predict how beach morphology changes under the influence of rising and falling wave conditions. An increase in wave height causes offshore sediment transport and the beach will move towards the dissipative beach state (right side of Figure 8.22). A decrease in wave height induces onshore sediment transport and the beach will move towards the reflective beach state (left side of Figure 8.22). The end members of the beach type continuum shown in Figure 8.22 (*i.e.*, the reflective and dissipative beaches) are relatively stable over time, while the intermediate beach types are the most dynamic.

The effect of tides is not considered in Figure 8.22, but they can be significant, particularly in macrotidal environments. With increasing tidal range,

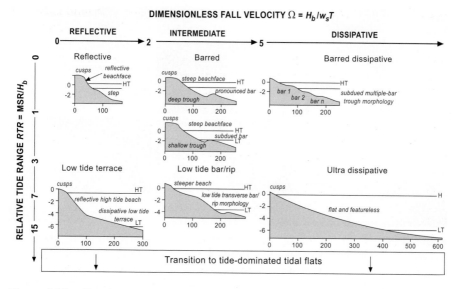

**Figure 8.23** – Beach classification model for the prediction of beach morphology on the basis of the dimensionless fall velocity ($\Omega = H_b/w_sT$, where $H_b$ is breaker height, $w_s$ is sediment fall velocity of mid-beachface sediment and $T$ is wave period) and the relative tide range ($RTR = MSR/H_b$, where MSR is the mean spring tide range). [From Masselink and Short, 1993.] [Copyright © 1999 John Wiley & Sons, reproduced with permission.]

the role of tides in shaping sandy beaches increases relative to wave effects. However, the importance of tides also increases as wave importance decreases due to a reduction in incident wave energy level. Quantification of wave versus tide effects may be achieved using the relative tide range parameter $RTR$ discussed in Section 8.3.4. Tidal effects are relatively insignificant when $RTR < 3$, but dominate sediment transport processes and morphology when $RTR > 15$.

Masselink and Short (1993) formulated a beach classification model that takes into account the three most important environmental constraints (waves, tides and sediment) by combining the relative tide range parameter $RTR$ and the dimensionless fall velocity $\Omega$ (Figures 8.23 and 8.24). Reflective, intermediate and dissipative beach types occur when tidal effects are insignificant, as reflected by the low value of the relative tide range. Tidal effects become important when $RTR > 3$ and three types of mixed wave-tide beaches can be identified:

- low tide terrace beach ($\Omega < 2$ and $3 < RTR < 15$)
- low tide bar/rip beach ($2 < \Omega < 5$ and $3 < RTR < 7$)
- ultra-dissipative beach ($\Omega > 5$ and $3 < RTR < 15$)

When the relative tide range increases even more ($RTR > 15$), beaches grade into tidal flats.

**Figure 8.24** – Examples of: (a) reflective beach, Pearl Beach, Australia [photo P.J. Cowell]; (b) intermediate beach, Palm Beach, Australia [photo G. Masselink]; (c) dissipative beach, Goolwa, Australia [photo G. Masselink]; (d) low tide terrace beach, Westward Ho!, England [photo G. Masselink]; (e) low tide bar-rip beach, Nine Mile Beach, Australia [photo A.D. Short]; and (f) ultra-dissipative beach, Farnborough Beach, Australia [photo A.D. Short].

The conceptual models illustrated in Figures 8.22 and 8.23 can be used to classify sandy beaches in a wide range of coastal environments. However, application of the models to sandy beaches subject to low mean wave energy levels ($H < 0.25$ m) is considered inappropriate. In these environments morphological response times are long and therefore the morphology may be dominated by features inherited from high-energy events and may not reflect modal wave conditions. In addition, for low wave energy levels, minor absolute changes in the wave height induce large changes in the values of $\Omega$ and $RTR$, and consequently the predicted beach types.

# 8.4 COASTAL DUNES

**Coastal dunes** are ubiquitous features in wave-dominated coastal environments. Their formation requires an ample supply of sand and an energetic wind climate. The prevailing wind direction does not necessarily have to be onshore, but onshore winds capable of inducing sediment transport must occur for a significant amount of time. Coastal dunes are most commonly found in temperate climates where they can reach heights of up to 100 m. They are generally less developed at high and low latitudes (Davies, 1980). The lack of extensive coastal dune development at high latitudes is due to the lack of sandy sediments and the action of frost and sea ice. Their relatively modest development at low latitudes has been ascribed to a variety of factors, including the effect of dense backbeach vegetation, unsuitable sediments (too coarse or too fine), relatively low wave energy levels and reflective beaches, continuous dampness, the presence of salt crusts and the lack of suitable sand-trapping vegetation.

Dunes play a crucial role in protecting the coast from erosion by providing a buffer to extreme waves and winds. Extreme storm activity inevitably results in elevated water levels and beach erosion, and may lead to coastal flooding. However, if a well-developed dune system is present behind the beach, the storm waves will dissipate their energy through dune erosion, preventing flooding from occurring. The sand eroded from the dune system will be transported offshore, but will eventually return to the beach under fair weather conditions. As the sediment is returned to the beach, aeolian processes may result in renewed dune development. Maintenance of coastal dune systems is an important component of coastal protection and management.

## 8.4.1 Aeolian sand transport

Although air is almost a thousand times less dense than water, many of the hydrodynamic principles discussed in Chapter 5 also apply to wind. Sediment transport by currents and wind arises from the shear stress imposed on the bed by the flow. The wind shear stress $\tau_0$ is related to the shear velocity $u_*$ according to

$$\tau_0 = \rho_a u_*^2 \qquad (8.15)$$

where $\rho_a$ is the air density. The shear velocity $u_*$ cannot be measured directly, but can be derived from the vertical velocity profile using the Law of the Wall (Chapter 5).

If a sandy surface is exposed to an increasing wind field, there will be a point when aeolian sediment transport commences. The velocity at which this occurs is the threshold velocity and according to Bagnold (1941) it can be given by

$$u_{*t} = A \left[ gD \left( \frac{\rho_s - \rho_a}{\rho_a} \right) \right]^{1/2} \qquad (8.16)$$

where $u_{*t}$ is the threshold shear velocity, $A$ is an empirical coefficient usually taken as 0.1, $D$ is sediment size and $\rho_s$ is the sediment density. The threshold wind velocity at height $z$ above the surface (rather than the threshold shear velocity) can be derived by inserting the threshold shear velocity $u_{*t}$ into Equation 5.15 (i.e., the Law of the Wall) and assuming a value for the roughness length $z_o$. Figure 8.25a shows the threshold wind velocity at 2 m above the sand surface as a function of sediment size derived using Equations 5.15 and 8.16 and assuming $z_o = D/30$. Coastal dunes are commonly made up of fine to medium sand and such sediments are characterized by a threshold wind velocity of 4–8 m s$^{-1}$. It is clear that for wind speeds less than 4 m s$^{-1}$ no sediment is expected to be transported.

Once the threshold of movement is exceeded and sand grains are entrained, they travel in three distinct ways. In increasing order of velocity

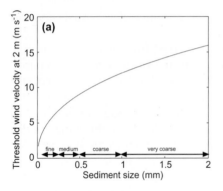
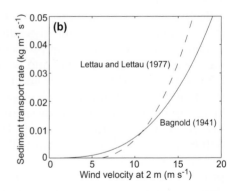

**Figure 8.25** – (a) Relation between threshold wind velocity (measured 2 m above sand level) and sediment size predicted following Bagnold (1941; Equation 8.17). (b) Predicted aeolian sediment transport as a function of wind speed (measured 2 m above sand) for a sediment size of 0.2 mm according to Bagnold (1941; Equation 8.18) and Lettau and Lettau (1977; Equation 8.19).

these are: (1) **creep** (rolling/sliding grains that remain in contact with the bed); (2) **reptation** (grains making small 'hops'); and (3) **saltation** (grains making large 'leaps'). Although it is difficult to allocate the relative contributions of the different sediment transport modes to the total sediment transport, saltation is generally considered the most important process. Once sediment transport has started, the process has the tendency to drive itself due to the ballistic impact of sand grains returning to the bed. There is an interesting implication of this self-forcing nature of aeolian sediment transport. If sediment transport occurs and the wind velocity is reduced to below the initial threshold wind velocity, aeolian sediment transport does not cease. In fact, the velocity required to maintain sediment movement through collision is about 80% of that required to initiate sediment transport (Livingstone and Warren, 1996).

Numerous empirical and semi-empirical equations have been used to predict **aeolian transport** rates (refer to Sherman *et al.*, (1998) for a recent evaluation of aeolian transport equations). The model of Bagnold (1941) is probably the most widely used and represents one of the simplest approaches

$$q = C \left(\frac{D}{D_r}\right)^{1/2} \frac{\rho_a}{g} u_*^3 \qquad (8.17)$$

where $q$ is the sediment transport rate (in kg m$^{-1}$ s$^{-1}$), $C$ is a constant (1.8 for naturally graded sand), $D$ is the mean grain diameter and $D_r$ is a reference grain diameter of 0.25 mm. A major weakness of Equation 8.17 is that it does not include a threshold velocity. Therefore, it predicts sediment transport even for very low, below threshold wind velocities. The model of Lettau and Lettau (1977) does take into account the threshold velocity and is also widely implemented

$$q = C \left(\frac{D}{D_r}\right)^{1/2} \frac{\rho_a}{g} (u_* - u_{*t})u_*^2 \qquad (8.1$$

where $C$ has a value of 4.2 and $u_{*t}$ is given by Equation 8.16. Figure 8.25b compares aeolian sediment transport rates according to Equations 8.17 and 8.18.

Practically all aeolian sand transport equations consider the sediment transport rate proportional to the shear velocity cubed ($q \propto u^3$). This implies that, once sediment is moving, a small increase in the wind velocity causes a large increase in the sediment transport rate. For example, only a 25% increase in the wind speed will induce a doubling of the sediment transport rate.

Aeolian sand transport equations are generally derived and validated under 'ideal' conditions, and therefore provide an estimate of the maximum transport rate that may occur. In other words, they predict the potential sediment transport rate. There are many factors that inhibit sediment transport under natural conditions, including sand moisture content, crusts, vegetation and slope effects. Of particular relevance to coastal environments is the fact that beaches are often too narrow for winds to become fully saturated with sand (Nordstrom and Jackson, 1993). The sediment transport can then be said to be fetch-limited. In addition, the equations which use time-averaged wind velocity measures do not account for wind unsteadiness (turbulence and gustiness), which may be a fundamental shortcoming (Bauer et al., 1998).

Wind flow is greatly affected by topography over all spatial scales, ranging from small obstacles, such as clumps of vegetation, driftwood and bottles, to individual dunes. Whatever the scale of the topographic feature, the flow of air around them alters the surface shear velocity and therefore the sediment transport rates and patterns. Flow acceleration generally occurs on the up-wind side and along the flanks of the obstacle. The increased shear stress at these locations may result in scour. Flow deceleration takes place in the lee of the obstacles and may induce sediment deposition. If a dune ridge is discontinuous and characterized by gaps, the wind flow will tend to concentrate on these gaps, causing increased flow velocities and shear stresses at these locations. The ensuing scour may widen and deepen the gap, exemplifying a positive feedback process. With a continuous dune ridge and an onshore wind, the air has to be 'lifted' over the dune (Figure 8.26). The lines of equal wind velocity (isovels) tend to be compressed near the top of the

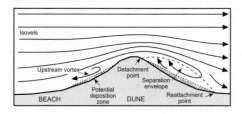

**Figure 8.26** – Movement of air over a dune as indicated by isovels (lines of equal wind speed). [Modified from Carter, 1988.]

erosion at this location. A region of flow separation and
develop in the lee of the dune and here sediment may
pending on the upwind topography, a small region of flow
ay also be found at the front of the dune, leading to sediment

## 8.4.  ...stal dune morphology

Coastal dunes can be broadly classified into two types:

- **Primary dunes** are found closest to the shoreline and are significantly affected by wave processes (*e.g.*, overwashing, storm erosion) and have a dynamic interrelationship with the beach. They include the ephemeral dunes present on the backshore and foredunes.
- **Secondary dunes** are located further inland and have been dissociated from wave processes through coastal progradation. They include mature dune ridges, blowouts, parabolic dunes and sand sheets.

The morphodynamics of secondary dunes are very similar to those of terrestrial dunes (*e.g.*, desert dunes), but primary dunes are fundamentally different. In fact, Bauer and Sherman (1999) contend that the only distinctive coastal dune is the primary dune because its form is integrally coupled to the nearshore processes fronting the beach–dune system. In addition, the primary dune is geomorphologically conditioned by the germination, colonization and succession of vegetation assemblages characteristic of coastal environments.

Coastal dune formation can be initiated in various ways, but dunes generally begin to develop around the drift line above the spring high tide line. Here, tidal litter (seaweed, driftwood) represents an obstacle to the wind, promoting the formation of **shadow dunes** with tails stretching-out downwind (Figure 8.27a). Shadow dunes cannot reach elevations higher than that of the obstacle, but ongoing accretion can occur following the establishment of **pioneer plant** species. Examples of such plants are Marram Grass, Leyme Grass, Saltwort, Sea Couch Grass, Sea Rocket and Spinifex. Pioneer plants are all characterized by a high tolerance to salt, elaborate root systems that

**Figure 8.27** – (a) Shadow dunes [photo J.E. Bullard]. (b) Well-developed foredune fronted by a wide intertidal zone [photo G. Masselink].

can reach down to the freshwater table and rhizomes that grow parallel to the upper dune surface. In addition, pioneer plants can withstand, or may actually thrive on being buried by sand, as long as vertical accretion rates are not excessive (less than *c.* 0.5 m yr$^{-1}$).

The sand trapping ability of the pioneer plants enables the shadow dune to grow upward and outward into **incipient foredunes** (also termed **embryo dunes**), which are low (1–2 m high) vegetated mounds of sand. Given suitable conditions (onshore winds and adequate sand supply) and sufficient time, the incipient foredunes will coalesce, forming a **foredune** or **foredune ridge** (Figure 8.27b). Exactly how and why this happens is not known, but it appears that vegetation again plays a crucial role. The growth of plants into unoccupied sand will induce sand deposition at these locations such that the dunes tend to expand and advance in the direction of vegetation growth. Foredunes may grow at quite noticeable rates and can reach a height of several metres over a period of 5–10 years (Figure 8.28). Foredune height ranges from 1 m to several tens of metres and depends on a range of factors, of which sediment supply to the dune system and shoreline progradation are the most important.

The secondary dune system is found behind the primary dune ridge. On some coasts there are multiple dune ridges that have formed as successive foredunes behind a prograding beach. The age of the dune ridges increases in the landward direction and this is reflected by a concomitant increase in vegetation maturity and soil development. Sediment supply to the secondary dunes is limited due to increased distance to the source area (the beach) and the presence of the primary dune providing an effective barrier to landward aeolian transport. This does not imply, however, that the morphology

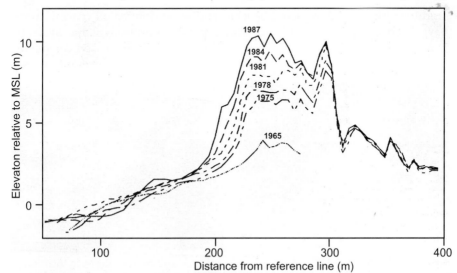

**Figure 8.28** – Example of foredune growth in the Netherlands. [Modified from Arens, 1994.]

of the generally well-vegetated secondary dunes is arrested. Unstable, at least partly unvegetated dune forms are often present within the secondary dunes and there are two main types of coastal, unvegetated dune forms (Hesp, 1999):

- **Blowouts** – These are saucer-, cup- or trough-shaped depressions that can develop as a result of disturbance of the vegetation by natural (*e.g.*, storm erosion, weather cycles) or unnatural causes (footpath erosion, grazing) (Figure 8.29a). The airflow and sediment transport pattern associated with blowouts is such that a small blowout can be enhanced through positive feedback. Specifically, onshore winds are funnelled into the blowout, resulting in flow acceleration and scouring of the base and walls of the depression. The eroded sediment is deposited in rims or parabolic-shaped lobes along their leeward ends and the blowout can grow and migrate landward.
- **Transgressive dune fields** – These consist of small to extensive dune fields that migrate landward or alongshore (Figure 8.29b). Transgressive dune fields occur on many of the world's coasts, but are best developed in high wind- and wave-energy environments with an ample sediment supply. Some may consist of broad, flat-to-undulating sand sheets, whereas others contain pronounced dune ridges and **parabolic dunes**. On the landward side of active transgressive dune fields there is often a pronounced slip face, referred to as a **precipitation ridge**. Transgressive dune fields are very similar to terrestrial dunes in arid regions.

### 8.4.3 Beach/dune interactions

Traditionally, wave-dominated beaches and wind-dominated coastal dunes have been examined as distinct and separate morphodynamic systems. During the last two decades, however, there has been increased consideration of the presence of **beach–dune interaction** with the recognition that

**Figure 8.29** – (a) Multiple blowouts of varying sizes and morphologies in Warnboro Sound, Australia. (b) Small developing transgressive dune field, south of Perth, Australia. [From Hesp, 1999] [Copyright © 1999 John Wiley & Sons, reproduced with permission.]

these two environments are strongly coupled and mutually adjusted (Sherman and Bauer, 1993). This notion has led to the formulation of a number of conceptual models, two of which will be discussed below.

Short and Hesp (1982) were among the first to explore the interaction between beaches and dunes and proposed that the morphodynamics of coastal dunes are strongly related to the morphodynamic state of the nearshore (Table 8.2). Their model considers that landward aeolian sand transport is greater for dissipative nearshore systems than for reflective nearshore systems. This is partly attributed to the greater width (fetch) of dissipative beaches, but also due to their finer and more abundant sediments. In addition, dune erosion by waves is thought to occur less frequently on dissipative beaches than on reflective beaches. Intermediate nearshore systems are characterized by an accordingly intermediate sediment supply, but dune erosion tends to be focused opposite the rip currents that characterize these beaches. Such discontinuous erosion promotes the formation of blowouts and parabolic dunes on intermediate beaches. Although the model proposed by Short and Hesp (1982) makes intuitive sense, it is important to point out that it was derived on the basis of observations on the high-wave energy, microtidal coast of south-eastern Australia, and was not intended to be universally applicable. In addition, the model is designed for stable beach systems and is not really readily applicable to prograding or eroding beaches.

Psuty (1992) proposed a model of beach-dune interaction that considers the longer-term sediment budget of the beach. His model comprises a series of combinations of sediment budgets of the beach and dune portrayed in a simple matrix to show the morphological outcome of each of the combinations (Figure 8.30). Maximum foredune development occurs when the beach

**Table 8.2** Summary of the Short and Hesp (1982) model relating beach morphodynamic state to dune system dynamics

| Morphodynamic beach state | Frequency: type of dune scarping | Potential aeolian transport: foredune size | Probability of foredune destruction, per 100 years | Nature of dominant dunes |
|---|---|---|---|---|
| Dissipative | Low: continuous scarp | High: large | Moderate | Large-scale transgressive dune sheets |
| Intermediate | Moderate: scarps in rip embayments | Moderate: moderate | Low to high | Parabolic dunes and blowouts |
| Reflective | High: continuous | Low: small | Low | Foredune scarping and small blowouts |

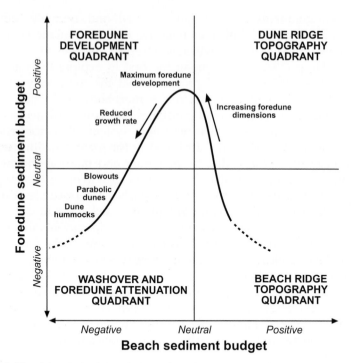

**Figure 8.30** – Morphologic evolution of the foredune as a function of beach and fore-dune sediment budget. [Modified from Psuty, 1992.]

sediment budget is neutral and the dune sediment budget is positive. Under such conditions, sand can accumulate for long periods of time at the same location, resulting in the formation of a well-developed, single foredune ridge. A slightly negative beach budget results in modest scarping of the dunes and landward migration. An increasingly negative beach budget inhibits the formation of foredunes, and promotes the development of blowouts, parabolic dunes and dune hummocks. For a positive beach sediment budget, the continuous progradation of the shoreline provides a constraint to the amount of sand that can accumulate on a foredune before it is left 'stranded' too far landward of the shoreline. The morphological end product of such rapid progradation is beach ridge topography with low foredune ridges.

## 8.5 BARRIERS AND RELATED ACCUMULATION FEATURES

In the three previous sections we have discussed the morphodynamics of shorefaces, beaches and coastal dunes. Collectively, these three sub-environments make up large-scale coastal accumulation features known as barriers. This section will discuss how barriers can be described and classified by their planform shape and cross-sectional morphology/stratigraphy. We will first, however, briefly discuss the role of longshore sediment transport processes in affecting the morphology of wave-dominated coasts.

## 8.5.1 Longshore sediment transport processes

The transport of sediment along coasts is accomplished by longshore currents and is very important from a sediment budget point of view. The net removal of material from a stretch of coast by longshore currents may result in coastal erosion, while the net influx of sediment by longshore currents may cause coastal accretion. Knowledge of longshore transport rates is therefore very useful for understanding coastal change.

The rate of **longshore sediment transport** or **littoral drift** can be predicted using the so-called CERC equation (CERC, 1984)

$$Q_l = KP_l = K(ECn)_b \sin \alpha_b \cos \alpha_b \qquad (8.19)$$

where $Q_l$ is the volumetric transport rate in $m^3 \, day^{-1}$, $K$ is a constant, $P_l$ is the longshore component of the wave energy flux, $(ECn)_b$ is the wave energy flux evaluated at the breaker point and $\alpha_b$ is the wave breaker angle (refer to Chapter 4 for more information on the different wave parameters). Equation 8.19 can be simplified considerably by inserting the appropriate equations and constants

$$Q_l = 89429 \, H_b^{5/2} \sin \alpha_b \cos \alpha_b \qquad (8.20)$$

which considers the littoral drift rate is a sole function of breaker height $H_b$ and angle $\alpha_b$. As an example, the littoral drift for a day during which the breaker height is 2 m and the angle of wave incidence is 12° amounts to just over 15,000 $m^3$ of sediment. Along ocean shores, net annual littoral drift rates are generally of the order of 100,000 $m^3 \, yr^{-1}$. Due to the often reversing nature of longshore currents, the total littoral drift rate along ocean shores is likely to exceed the net rate many times.

The relation between coastal morphology and longshore transport can be illustrated by the predictive model of May and Tanner (1973). This model was designed to predict and explain the development of an **equilibrium bay** (Figure 8.31), but the underlying principles have a much wider applicability. Wave refraction around the headland causes wave convergence on the headland and wave divergence in the adjacent bay. This results in a longshore variation in the wave energy level, with wave energy decreasing progressively from headland to bay. The angle of wave approach also exhibits a longshore variation. The largest wave angles are found at the transition from headland to bay, while smaller angles of wave approach occur at the headland and in the bay. From the longshore variation in wave energy (or wave height) and wave angle, the change in the longshore wave energy flux $P_l$ can be computed. The largest value for $P_l$ occurs somewhere between the headland and where the wave angle is maximum. Subsequently, the longshore variation in the littoral drift rate $Q_l$, which is proportional to $P_l$, can be determined.

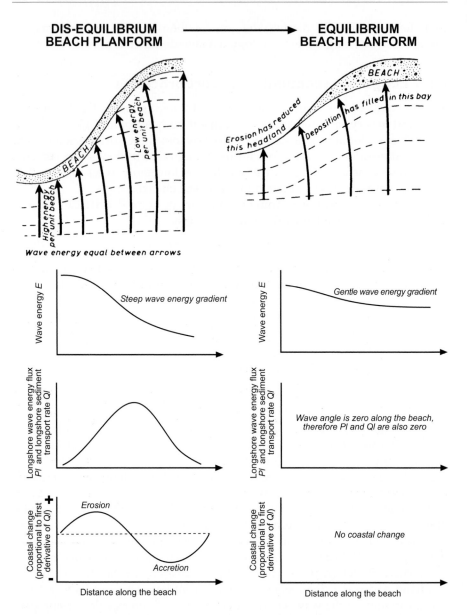

**Figure 8.31** – Relation between wave refraction, littoral drift and coastal change in a headland-bay system. [Modified from Pethick, 1984; modified from May and Tanner, 1973.]

From the longshore variation in $Q_l$ we can now predict changes in the coastline. If the littoral drift rate increases along a stretch of coast, progressively more sediment will be entrained. This sediment will have to be provided by erosional processes, so that the coastline is expected to retreat. Similarly, if the littoral drift rate decreases, sediment will progressively be deposited, causing coastal progradation. Therefore, it is the rate of change in the littoral drift rates, which is proportional to the first derivative of the longshore gradient of $Q_l$ (*i.e.*, $\delta Q_l/\delta x$), that controls coastline changes. In our example of the headland/bay system, the coastline from the headland to where $P_l$ is maximum exhibits a progressive increase in the littoral drift rate and hence will experience erosion. The coastline from where $P_l$ is maximum to the bay is characterized by a progressive decrease in the littoral drift rate and will therefore undergo deposition. The result of this analysis therefore shows that deposition occurs in the low-energy zones of the bay and erosion at the headlands. These morphological changes will continue until the morphology is such that everywhere along the coast the wave angles are zero and littoral drift ceases.

## 8.5.2 Planform shape of barriers

We will see later in this section that a large variety of barrier planforms exists. However, it is useful at this stage to distinguish between two fundamentally different types of coastline:

- **Swash-aligned coasts** are oriented parallel to the crests of the prevailing incident waves (Figure 8.32a). They are closed systems in terms of longshore sediment transport and net littoral drift rates are zero.
- **Drift-aligned coasts** are oriented obliquely to the crest of the prevailing waves (Figure 8.32b). The shoreline of drift-aligned coasts is primarily controlled by longshore sediment transport processes. Drift-aligned coasts are open systems in terms of longshore sediment transport and are sensitive to changes in littoral drift rates.

It is important to appreciate that the planform shape of many coasts is a combination of drift- and swash alignment. In addition, along a large stretch of coast there may be alternate swash and drift-aligned sections depending on coastal geology and local sediment supplies.

**Embayed beaches** are generally swash-aligned because there is little or

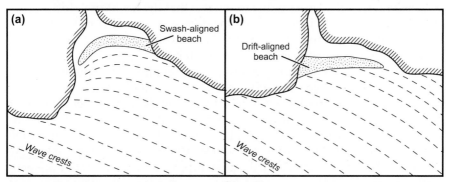

**Figure 8.32** – Two main types of coastal alignment: (a) swash-alignment and (b) drift-alignment. [Modified from Davies, 1980.]

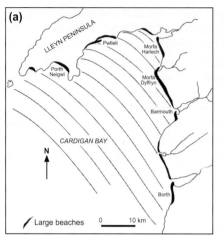
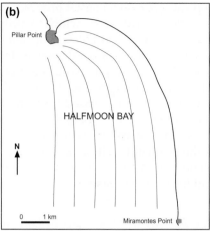

**Figure 8.33** – Examples of swash-aligned coastlines. (a) Wave refraction pattern for southwesterly waves and orientation of large beaches within Cardigan Bay, Wales. All large beaches are oriented parallel to the refracted wave crests. Note that the spits at the end of the beaches at Pwllheli and Morfa Dyffryn are not swash-aligned. [From Hansom, 1988.] [Copyright © 1988 Cambridge University Press, reproduced with permission.] (b) Wave refraction pattern and shoreline curvature of Halfmoon Bay, California. [Modified from Yasso, 1965.]

no additional sediment supplied to the beach and essentially no losses. Therefore, the equilibrium planform shape of the beach depends solely on the wave conditions, in particular the pattern of refraction and diffraction. The resulting planform shape is curved with the shoreline aligned to the crests of the prevailing waves (Figure 8.33a). Beaches found down-drift from large headlands capable of blocking the littoral drift are also usually swash-aligned (Figure 8.33b). Their shape can be described by a logarithmic spiral or modifications thereof, and beaches conforming to such a curved shape have been termed **log-spiral bays**, **crenulate bays** and **zeta bays** (Hsu *et al.*, 1987). The long-term average shoreline configuration of swash-aligned beaches is relatively constant due to the fact that the net littoral drift is zero. However, short-term changes in the incident wave conditions, for example seasonal or inter-annual changes in the prevailing wave direction, will induce minor adjustments in the planform shape.

Tombolos, salients and (cuspate) forelands (Figure 8.34) represent a hierarchy of sedimentary coastal landforms that have build out into the prevailing wave direction (Zenkovich, 1967). They may develop under a range of conditions and settings, but their formation is usually associated with offshore geological features (*e.g.*, islands, reefs) that provide shelter from the waves and hence form loci for deposition. **Tombolos** comprise a relatively thin strip of beach linking an offshore island to the mainland. They are generally swash-aligned features and have resulted from the convergence of longshore currents and sediment transport in the shadow zone behind the island due to wave refraction and diffraction. Where islands extend for some distance along the coast, tombolos may form at each end of the shadow zone, forming double tombolos. The development of a tombolo is strongly dependent on the alongshore length of the island $I$ and its distance to the mainland $J$. Sunamura and Mizuno (1987) found that tombolos develop when $J/I < 1.5$, whereas the island does not exert a significant influence on the coast if $J/I > 3.5$. For $J/I = 1.5–3.5$, a salient forms in the lee of the island. The terms 'salient' and (cuspate) 'foreland' have been used rather inconsistently in the coastal literature. **Salients** are relatively inconspicuous protuberances behind submerged reefs or offshore islands, whereas **forelands** extend a considerable distance out from coasts where two dominant swells are in opposition. Examples of well-known forelands are Dungeness

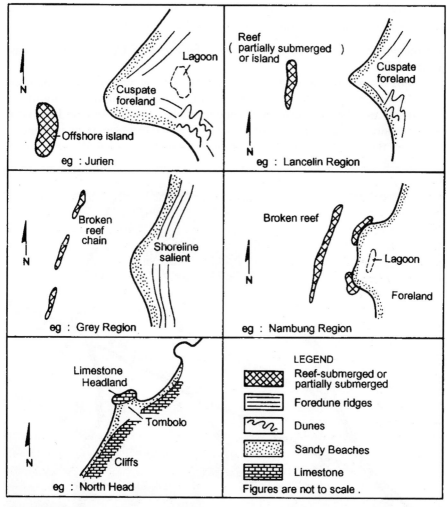

**Figure 8.34** – Types of tombolos, salients and forelands along the central west coast of Western Australia. [From Sanderson and Eliot, 1996.] [Copyright © 1996 Coastal Education & Research Foundation, reproduced with permission.]

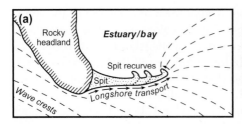

**Figure 8.35** – (a) Formation of a (recurved) spit. The longshore current transports sediment along the coast. When the current enters deeper water near the tip of the spit, it spreads out and loses its sediment transporting capacity, resulting in the deposition of sediment and the growth of the spit. Curvature of the end of the spit may occur due to refraction of the waves around the tip of the spit. (b) Oblique aerial photo of spit and tidal inlet on the Oregon coast, USA. [Photo A. Kroon.]

on the southeast coast of England and the capes of Carolina on the east coast of the US.

A **spit** is a narrow accumulation of sand or gravel, with one end attached to the mainland and the other projecting into the sea or across the mouth of an estuary or a bay (Figure 8.35). They may also develop where there is an abrupt change in the direction of the coast. Coastal dunes may be present on spits, but they tend to be low. The simplest type of spit is a linear spit, which

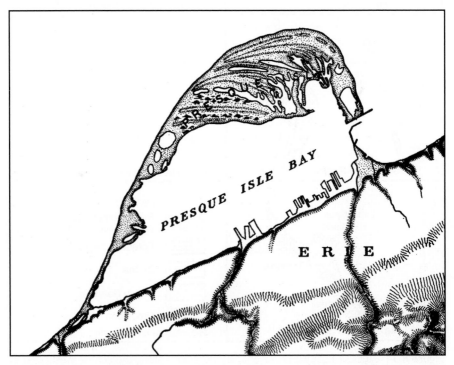

**Figure 8.36** – Compound recurved spit with lagoon, Presque Isle, Lake Erie. [From Johnson, 1919.]

is predominantly a straight feature, but they often have a curved body of sand at their distal end pointing in the landward direction. This curved distal end may be the result of wave refraction processes or the interplay of waves having different angles of approach. If the curved distal end is very pronounced, the spit is called a **recurved spit**. Many spits have developed over extended periods of time (hundreds of years) and are characterized by complex morphologies reflecting different stages in their evolution. One example of such complex spit systems is the **compound recurved spit**, which is characterized by a narrow proximal end and a wide, recurved distal end that often encloses a lagoon-type environment (Figure 8.36). The wide distal end usually consists of several dune/beach ridges associated with older shorelines demonstrating seaward migration of the shoreline. Examples of well-known spits are Spurn Head, Blakeney Point and Hurst Castle in England and Sandy Hook in New Jersey.

Spits grow in the direction of the predominant littoral drift and are a classic example of a drift-aligned feature that can only exist through the continued longshore supply of sediment (Orford *et al.*, 1991). If the longshore sediment supply ceases, the spit will eventually subsume itself and disappear. Erosion will commence at the proximal end of the spit to supply ongoing spit growth at the distal end. However, thinning of the proximal end of the spit will increase the chances of overwashing and breaching, followed by disintegration of the spit complex. Spits are highly dynamic morphological features and rapidly respond to changes, positive or negative, in the longshore sediment supply.

## 8.5.3 Barrier morphology and stratigraphy

**Barriers** are the basic depositional element for wave-dominated coasts and can be defined as elongated, shore-parallel sand bodies that extend above sea level (Roy *et al.*, 1994). Generally a back-barrier environment (estuary or lagoon) is situated between the barrier and the mainland. Barriers consist of a number of facies, including beach, dune, shoreface, tidal delta, inlet and washovers (Figure 8.37). Together they form a continuum of barrier morphologies, ranging from **barrier islands** with wide shallow lagoons on low-gradient coasts to **mainland beaches** with negligible back-barrier morphologies on steep coasts. Between these end members lie **bay barriers**, which are located within embayments, confined by rocky headlands.

Most contemporary barriers were established sometime during the final stages of the post-glacial marine transgression at the start of the Holocene and migrated landward over the remainder of the Holocene during conditions of decelerating sea-level rise. The evidence for an 'offshore' origin for barriers has been provided by sediment cores taken seaward and landward of current barrier systems (Hoyt, 1967; Sanders and Kumar, 1975). Cores that encountered relict estuarine sediments in present-day shelf environments demonstrate the existence of barrier lagoons when sea level was lower (Figure 8.38a). In addition, the absence of open marine sediments on

**(a)**

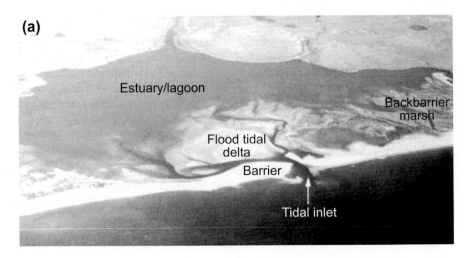

**(b)**

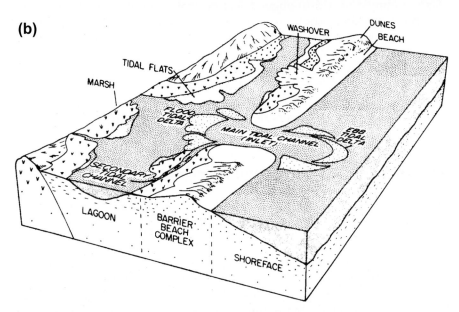

**Figure 8.37** – (a) Oblique aerial view of a small barrier island system on the northeast coast of New Brunswick, US, showing a linear barrier-beach, the tidal inlet through the barrier, the lagoon behind the barrier, the flood tidal delta and the back-barrier marsh developed on abandoned flood delta deposits. (b) Block diagram illustrating the various sub-environments in a barrier island system. [From Reinson, 1984.] [Copyright © 1984 Geological Association of Canada, reproduced with permission.]

the mainland shore behind barriers indicates that, rather than emerging in place as offshore bars, most barriers migrated landward during the post-glacial transgression from an initial position on the outer shelf. Finally, many present-day barrier systems are underlain by estuarine or lagoonal deposits, demonstrating that the barrier at some stage during its

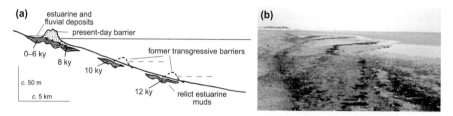

**Figure 8.38** – (a) Coring investigations on the continental shelf of southeastern Australia encounter subsurface relict estuarine sediments with ages indicating formation as the shelf was inundated by the postglacial marine transgression. Estuarine and lagoonal environments only exist in the protection of barriers, which means that these latter features must also have existed at times of lower sea levels. [From Roy et al., 1994.] [Copyright © 1994 Cambridge University Press, reproduced with permission.] (b) Exposed salt marsh deposits on the intertidal beach at Gibraltar Point, east coast of England. [Photo G. Masselink.]

development has migrated into its own back-barrier environment. One can sometimes find these estuarine deposits exposed on the beach after major storms (Figure 8.38b).

As sea level rises, a barrier may undergo one of three responses (Figure 8.39):

- **Barrier erosion** – The cross-sectional geometry of the shoreface is maintained, but the entire profile moves upward by the same amount as the rise in sea level. Material eroded from the shoreface is deposited below the wave base in the nearshore zone, hence this model involves offshore sediment transport. This model is also known as the Bruun rule (Section 8.2.3).
- **Barrier translation** – The entire barrier migrates across the substrate gradient without loss of material. This is accomplished through erosion of the shoreface and deposition of this sediment behind the barrier in the form of **washovers**. Barrier translation represents onshore sediment transport and is also referred to as the **roll-over model**.
- **Barrier overstepping** – Under some conditions sea level may rise too fast for the barrier morphology to respond. In this case overstepping of the barrier may occur, leaving it drowned on the sea bed as a relict feature.

Evidence presented in the literature seems to suggest that the barrier translation is the most common response of barriers to rising sea level (Davis, 1994). The barrier translation response appears to be particularly effective in gravel barriers (Orford et al., 1995), strongly wave-dominated barriers (Cooper, 1994) and on relatively gentle substrate slopes (< 0.8°; Roy et al., 1994).

Rates of shoreline recession due to **barrier retreat** depend on a number of factors with the rate of sea-level rise being the most important (Roy et al., 1994). Barrier retreat is also affected by the gradient of the underlying substrate and rates of retreat increase with decreasing substrate slope. Longshore sediment transport processes can also be important and barrier

Figure 8.39 – Models of barrier response to rising sea level. [Modified from Carter, 1988.]

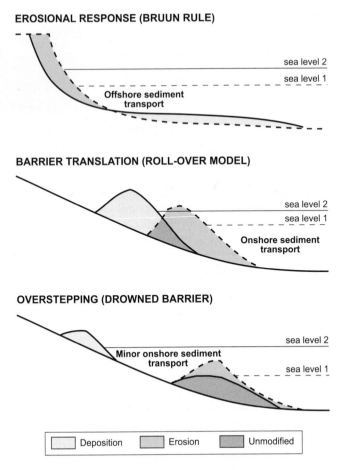

**EROSIONAL RESPONSE (BRUUN RULE)**

sea level 2

sea level 1

Offshore sediment transport

**BARRIER TRANSLATION (ROLL-OVER MODEL)**

sea level 2

sea level 1

Onshore sediment transport

**OVERSTEPPING (DROWNED BARRIER)**

sea level 2

Minor onshore sediment transport

sea level 1

☐ Deposition   ▨ Erosion   ▨ Unmodified

retreat is promoted (inhibited) by a negative (positive) longshore sediment budget. Depositional processes in the back-barrier lagoon or estuary also need to be considered. If the back-barrier sedimentation rates are high (*i.e.*, similar to or higher than the rate of sea-level rise), then the barrier will transgress over an increasingly thick back-barrier deposit that will significantly slow down barrier retreat. Finally, foredune development may affect the rate of barrier retreat under rising sea level. While sea level is rising rapidly, aeolian processes only have time to build low, discontinuous foredunes on the barrier surface that provide little impediment to storm washovers. As the rate of sea-level rise slows, however, larger dunes build above storm surge levels and the frequency of storm washovers is reduced thus retarding barrier translation. It is likely that landward barrier translation recommences when the sea rises to a level that effectively narrows the barrier and once again allows storm washovers (Leatherman, 1983). Thus it follows that barrier translation is fairly continuous while sea level is rising rapidly, but may become rather intermittent when it slows down.

Barriers that migrate landward under the influence of rising sea level are

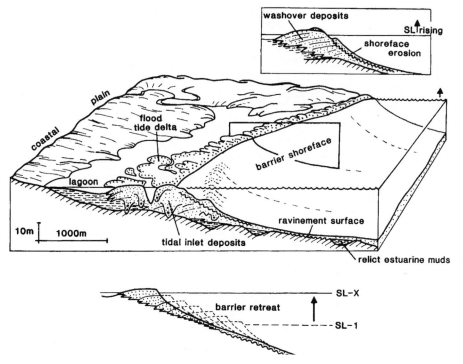

Figure 8.40 – Model of transgressive barrier during sea-level rise. Transgressive barriers are almost entirely composed of tidal delta and washover deposits. The barrier migrates into estuarine/lagoonal environments as sea level rises. Landward migration is through a process of erosional shoreface retreat as the barrier adjusts to changing sea level. [From Roy *et al.*, 1994.] [Copyright © 1994 Cambridge University Press, reproduced with permission.]

termed **transgressive barriers** (Figure 8.40). Their stratigraphy is such that sediments deposited in seaward environments stratigraphically end up on top of sediment that originated in more landward environments (transgressive sequence). Transgressive barriers consist mainly of tidal delta and/or washover deposits, and are underlain by back-barrier estuarine or lagoonal deposits. **Regressive barriers** are those that develop under the influence of a falling sea level (Figure 8.41). The stratigraphy of a regressive barrier is such that the landward sediments are deposited on top of more seaward ones (regressive sequence). The barrier is capped by aeolian sand, below which is beach sand, while at still greater depths within the barrier is sand that accumulated in progressively deeper water, finally giving way to silt and clay that had been deposited on what was formerly the continental shelf.

Many barriers experienced a transgressive phase during the first half of the Holocene when sea level was rising rapidly. The end of the Holocene transgression occurred around 5,000–6,000 years BP, after which sea level has remained more or less stable. Under the influence of a relatively stable

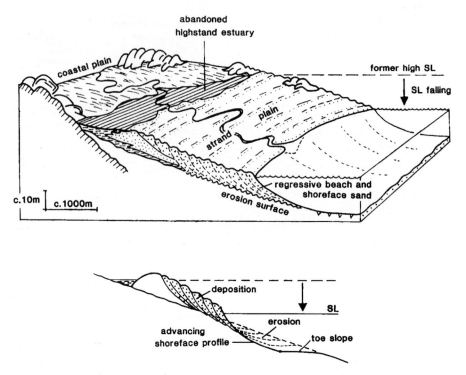

**Figure 8.41** – Model of regressive barrier during sea-level fall. Regressive barriers consist of a tabular, gently seaward-inclined sand deposit 10–20 m thick with an erosional base. As sea level falls, the inner shelf surface erodes and sand moves onto the shoreface. The surface of the regressive barrier forms a wide strandplain, generally without estuaries. [From Roy *et al.*, 1994.] [Copyright © 1994 Cambridge University Press, reproduced with permission.]

sea level and a large sediment supply from the shelf, the coast started accreting through **barrier progradation**. This second phase of Holocene barrier development is often referred to as regressive, but strictly speaking the term should only be used for barriers that have developed under the influence of a falling sea level. Coastal progradation during the second half of the Holocene resulted in the development of extensive coastal plains known as **strand plains** of which there exist four main types (Figure 8.42):

- **Prograded barrier** – The general term 'prograded barrier' is used to indicate a barrier system that has prograded seaward under the influence of a positive sediment budget, and is not a foredune ridge, beach ridge or a chenier plain. Prograded barriers are characterized by extensive dune development and may attain elevations of tens of metres above mean sea level. A clear example of a prograded barrier is the coast of the Netherlands.
- **Beach ridge plain** – Beach ridges can be defined as wave-deposited, linear alongshore, triangular to convex ridges, formed of sand, gravel or shelly

sediments (Hesp, 1999). They form by fairweather swash processes during beach progradation, by storm wave/swash processes or a combination of both (Taylor and Stone, 1996). Beach ridges generally prograde when an abundance of sediment exists and the offshore gradient is low, resulting in the development of a beach ridge plain (Box 8.3). An aeolian capping may be present on beach ridges, but the role of aeolian processes is secondary to that of wave and swash processes. Well-developed beach ridge plains occur, for example, along the northern shore of Lake Michigan.

- **Foredune ridge plain** – Foredunes are shore-parallel, convex dune ridges formed on the top of the backshore by aeolian deposition within vegetation (Hesp, 1999). Systematic beach progradation over time frames of tens to thousand of years may lead to the development of foredune plains. The faster the rate of coastal progradation, the lower and more numerous the foredune ridges are. A clear example of an extensive foredune ridge plain is the shore of Warnboro Sound in Western Australia.

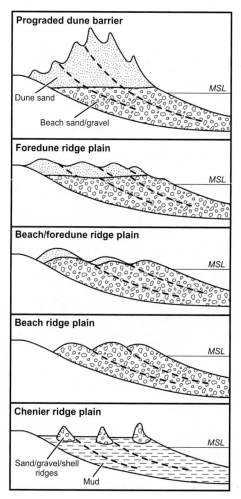

**Figure 8.42** – Five main types of wave-dominated, coastal plains: (a) prograded dune barrier, (b) foredune ridge plain, (c) beach/foredune ridge plain, (d) beach ridge plain and (e) chenier plain. The different types of coastal plain are similar with respect to their planform and morphology, but can be distinguished on the basis of their stratigraphy.

## BOX 8.3 – BEACH RIDGES ALONG THE WEST COAST OF AFRICA

Anthony (1995) investigated beach-ridge development along the west coast of Africa. Here, beach ridges form the dominant mode of Holocene progradation. The West African coast is exposed throughout the year to constant, low to moderate energy ($H_s$ = 0.5–1.5 m), long-period ($T_s$ = 10–16 s), southwesterly swells from the Atlantic. Shoreface gradients vary from very low (1:1200) at low wave-energy portions of the coast of Sierra Leone, to moderately steep (up to 1:120) off the exposed coasts of the Bight of Benin and southern Sierra Leone. Mesotidal tide ranges prevail from the central coast of Sierra Leone northwards, while the rest of the west coast of Africa experiences microtidal ranges.

There are four areas along the west coast of Africa that show massive beach-ridge development and progradation (Figure 8.43). These are southern Sierra Leone, the Ivory coast, the Bight of Benin and the Niger Delta. The beach ridge

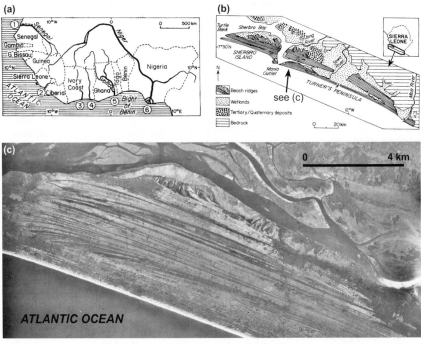

Figure 8.43 – (a) Map showing the west coast of Africa. Major beach ridge plains have developed in southern Sierra Leone, the Ivory coast, the Bight of Benin and the Niger delta. Numbers refer to the following river deltas: 1 = Senegal; 2 = Moa; 3 = Comoé; 4 = Bandama; 5 = Volta; and 6 = Niger. (b) Map showing beach ridge plains in southern Sierra Leone. (c) Aerial photograph showing beach ridge plain on Sherbro Island. Old recurved ridges fed by longshore drift from the southwest are succeeded seaward by linear parallel to sub-parallel ridges constructed with sands from the nearshore zone. [From Anthony, 1995.] [Copyright © 1995 Elsevier Science, reproduced with permission.]

plains range in width from a few hundred metres to over 10 km, with lengths exceeding 100 km. The ridges only contain minimal amounts of aeolian sands, thus are almost wholly wave-formed. Beach-ridge development probably started following the eustatic highstand at around 5500 BP.

Anthony (1995) found that the most important factor controlling the spatial and temporal variation in beach-ridge development was sediment supply (sea level was relatively stationary). The simplest ridges were found to be related to just one or two sources of sand supply from a nearby source (cliff erosion or nearshore zone). The major beach ridge plains in southern Sierra Leone and in the Bight of Benin exhibit complex patterns of development related to large sediment supplies from the nearshore zone and rivers. Here, early phases of progradation were generally driven by sediment supply from the nearshore, while later phases appeared to have been largely dominated by supplies from aggraded river mouths and associated deltaic sediments. In many areas, beach-ridge formation has ceased because the sediment supply has dwindled.

■ **Chenier plain** – A chenier is a coastal ridge composed of sand and shell overlying muddy or marsh deposits, surrounded by mudflats or marshes. Chenier plains consist of at least two cheniers separated by a progradational muddy unit. There has been considerable debate over the conditions responsible for the episodic deposition of fine- and coarse-grained sediments on chenier plains (Augustinus, 1989). Mud flat deposition occurs during periods of abundant sediment supply, whereas cheniers form during periods of diminished supply, as a result of waves winnowing and concentrating the sand fraction from the eroding mud, or from sand carried by longshore transport from nearby sediment sources. Development of a chenier plain then requires alternation of these two conditions (Otvos and Price, 1979), for example due to switching of delta lobes or variations in storminess. Chenier plains occur mainly in tropical and subtropical regions and are well known along the coasts of Louisiana, Guyana, Suriname and north Australia.

Beach ridge, foredune ridge and chenier plains are rather similar in planform and morphology, and it is clearly important to establish the genesis of the landform by stratigraphic and process studies prior to ascribing an origin or term to the feature.

Barrier morphology is affected by a large number of factors including sea level, sediment supply, substrate gradient, geological inheritance, wave energy level, tides and wind (Hesp and Short, 1999). Aside from eustatic sea level, these factors are all local to regional in extent and explain to a large degree the great variety in barrier types that occurs in nature. In terms of barrier evolution and stratigraphy, the two most important factors are sea level and sediment supply. Depending on the combination of these two factors, barriers may be prograding, stationary or retrograding (Table 8.3). Thus, **prograding barriers** generally occur with a stationary to falling sea

**Table 8.3** Primary dependence of barrier type on the balance between sea-level change and the sediment budget

| | Falling sea level, regressive | Stationary sea level | Rising sea level, transgressive |
|---|---|---|---|
| Positive sediment budget | Prograding barrier | Prograding barrier | Indeterminate |
| Neutral sediment budget | Prograding barrier | Stationary barrier | Retrograding barrier |
| Negative sediment budget | Indeterminate | Retrograding barrier | Retrograding barrier |

level and a neutral to positive sediment budget. However, a barrier may still be prograding under rising sea-level conditions when there is a large positive sediment budget. Similarly, **retrograding barriers** occur when sea level is stationary to rising with a neutral to negative sediment budget. However, a barrier may also be retrograding under falling sea-level conditions when there is a large negative sediment budget. **Stationary barriers** occur under conditions of stationary sea level and a neutral sediment budget, or where sea level rise (fall) is exactly balanced by a positive (negative) sediment budget.

## 8.6 SUMMARY

- The three main wave-dominated sub-environments are the shoreface, beach and coastal dunes. Collectively, they make up coastal barriers and these landforms can be considered the basic depositional elements of wave-dominated coasts.
- The shoreface is the upper part of the continental shelf that is affected by contemporary wave processes. It has a concave-upward shape that can be described by a simple function – the equilibrium shoreface profile. A variety of currents are active on the shoreface, including wave-, tide- and wind-driven currents. Most net sediment transport on the shoreface is accomplished by waves and currents working in concert. Important factors that affect shoreface sediment transport are wave asymmetry and the presence of bedforms.
- Beaches are defined as wave-lain deposits of sand- or gravel-sized material found along marine, lacustrine and estuarine shorelines. The three main factors that control beach morphology are waves, tides and sediments. Beach morphology is highly variable (spatially and temporally) and a variety of secondary morphological features can be found on beaches (*e.g.*, beach cusps, nearshore bars, rip channels). It is the presence of these secondary morphological features that gives a beach its distinctive morphology. Furthermore, it is the assemblages of secondary

morphological features on a particular beach that, in combination with the morphodynamic process regime (reflective versus dissipative), allow beaches to be classified into distinct beach types.

■ Coastal dunes can be broadly classified into two types. Primary dunes are found closest to the shoreline and are significantly affected by wave processes (*e.g.*, overwashing, storm erosion). Their dynamics are directly linked to that of the beach. Secondary dunes are located further inland and have disassociated themselves from wave processes through coastal progradation.

■ The planform shape of wave-dominated coastlines is strongly controlled by wave refraction processes. Two fundamentally different types of coastline can be distinguished. Swash-aligned coasts are oriented parallel to the crests of the prevailing incident waves (*e.g.*, equilibrium bays), while drift-aligned coasts are oriented obliquely to the crest of the prevailing waves (*e.g.*, spits).

■ Barriers are defined as elongated, shore-parallel sand bodies that extend above sea level. Most contemporary barriers were established sometime during the final stages of the post-glacial marine transgression at the start of the Holocene and migrated landward during the remainder of the Holocene under conditions of decreasing rates of sea-level rise. Barrier morphology is affected by a large number of factors, including sea level, sediment supply, substrate gradient, geological inheritance, wave energy level, tides and wind. In terms of barrier evolution and stratigraphy, the two most important factors are sea level and sediment supply. Depending on the combination of these two factors barriers may be prograding, stationary or retrograding.

## 8.7 FURTHER READING

Komar, P.D., 1998. *Beach Processes and Sedimentation* (2nd Edition). Prentice Hall, New Jersey. [Chapters 7–11 of this text provide an advanced and up-to-date review of beach processes and morphology.]

Short, A.D. (editor), 1999. *Handbook of Beach and Shoreface Morphodynamics*, Wiley, Chichester. [Very recent book with various up-to-date reviews on shoreface, beach, dune and barrier morphodynamics.]

Roy, P.S., Cowell, P.J., Ferland, M.A. and Thom, B.G., 1994. Wave-dominated coasts. In: R.W.G. Carter and C.D. Woodroffe (editors), *Coastal Evolution*, Cambridge University Press, Cambridge, 87–120. [Comprehensive book chapter on the morphodynamics of wave-dominated coasts. The emphasis is on barrier morphology and processes.]

Trenhaile, A.S., 1997. *Coastal Dynamics and Landforms*. Oxford University Press, Oxford. [Chapters 4, 5 and 6 of this text give brief, but comprehensive overviews of beaches, barriers and coastal dunes.]

# 9

# ROCKY COASTS

## 9.1 INTRODUCTION

The wave-, tide- and river-dominated coastal environments discussed in the previous chapters are depositional environments that have developed over thousands of years through accretionary processes. Rocky coasts, on the other hand, are continually being cut back by the sea and are characterized by erosional features. The erosive nature of rocky coasts makes it very difficult to deduce their evolutionary history because the different evolutionary stages are not preserved in the stratigraphy. Another factor that confounds our understanding of rocky coast development is the very slow rate of morphological change. Along clastic depositional coasts we can usually observe morphological changes using observations, measurements, maps and aerial photographs. Rocky coast morphology, on the other hand, is characterized by extremely long relaxation times and is greatly affected by inheritance. As a result of these confounding factors, we can only speculate on the mode and rate of development of rocky coasts (Trenhaile, 1987; Sunamura, 1992).

Sunamura (1992) categorizes rocky coast morphology into three types (Figure 9.1): sloping shore platform, sub-horizontal shore platform and plunging cliff. Naturally, as with all classifications, such a general categorization of rocky coasts is an oversimplification. However, it does provide us with a useful framework for discussing rocky coast morphodynamics and will form the basis of this chapter.

## 9.2 PROCESSES

Rocky coasts are affected by a range of processes, including mechanical wave erosion, physical and chemical weathering, bio-erosion and mass movements (for a comprehensive review of these processes, refer to Trenhaile, 1987). The different processes often occur concurrently and one of the main problems associated with understanding rocky-coast

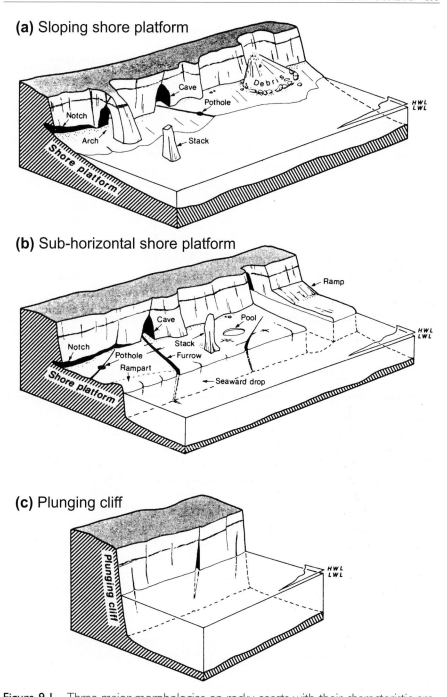

**Figure 9.1** – Three major morphologies on rocky coasts with their characteristic erosional features: (a) sloping shore platform, (b) sub-horizontal shore platform and (c) plunging cliff. [From Sunamura, 1992.] [Copyright © 1992 John Wiley & Sons, reproduced with permission.]

**Table 9.1** – Summary of the main erosional processes on rocky coasts

| Process | Description | Conditions conducive to the process |
|---|---|---|
| **Mechanical wave erosion:** | | |
| Erosion | Removal of loose material by waves | Energetic wave conditions and microtidal tide range |
| Abrasion | Scouring of rock surfaces by wave-induced flow with mixture of water and sediment | 'Soft' rocks, energetic wave conditions, a thin layer of sediment and microtidal tide range |
| Hydraulic action | Wave-induced pressure variations within the rock causes and widens rock capillaries and cracks | 'Weak' rocks, energetic wave conditions and microtidal tide range |
| **Weathering:** | | |
| Physical weathering | Frost action and cycles of wetting/ drying causes and widens rock capillaries and cracks | Sedimentary rocks in cool regions |
| Salt weathering | Volumetric growth of salt crystals in rock capillaries and cracks widens these capillaries and cracks | Sedimentary rocks in hot and dry regions |
| Chemical weathering | A number of chemical processes remove rock material. These processes include hydrolysis, oxidation, hydration and solution | Sedimentary rocks in hot and wet regions |
| Water-layer levelling | Physical, salt and chemical weathering working together along the edges of rock pools | Sedimentary rocks in areas with high evaporation |
| **Bio-erosion:** | | |
| Biochemical | Chemical weathering by products of metabolism | Limestone in tropical regions |
| Biophysical | Physical removal of rock by grazing and boring organisms | Limestone in tropical regions |
| **Mass movements:** | | |
| Rock falls and toppling | Rocks falling straight down the face of the cliff | Well jointed rocks, undercutting of cliff by waves |
| Slides | Deep-seated failures | Deeply weathered rock, undercutting of cliff by waves |
| Flows | Flowing of loose material down a slope | Unconsolidated material, undercutting of cliff by waves |

morphodynamics is to determine the relative importance of these processes. Table 9.1 presents a summary of the major erosional processes on rocky shores. These processes will be discussed in more detail below.

## 9.2.1 Mechanical wave erosion

Mechanical wave action is the main erosional agent in most swell and storm-wave environments. Under less energetic conditions, waves have limited erosive powers but may still play an important role in removing the products of weathering. In addition to the removal of loose material by waves, the two principal wave effects are abrasion of rock surfaces due to wave-induced currents and wave-generated pressure variations in the rock.

**Abrasion** is the scouring action of wave-induced currents and includes the sweeping, rolling, or dragging of rocks and sand across gently-sloping rock surfaces, and the throwing of coarse material against steep surfaces. The scouring effect of water is greatly enhanced if sand and/or gravel are in transport. For example, Robinson (1977) found that erosion rates at the base of the cliff were 15–20 times higher in places where there was a beach at the foot of the cliff, compared to where there was no beach. Abrasion is most efficient if only a thin layer of sediment is present (< 0.1 m) because a thick layer protects the underlying rock surface from abrasion.

The impact of waves on rocks induces pressure variations that weaken the rock by causing and widening capillaries and cracks. These pressure variations consist of a dynamic (impact of the moving water) and a hydro-static component (due to the weight of the water column). Breaking waves exert the greatest pressures (particularly plunging breakers), followed by broken waves and reflected waves (Sunamura, 1992). Pieces of rock that have been dislodged due to **hydraulic action** can subsequently be removed from the matrix in a process known as **quarrying**. The presence of large, angular debris and fresh rock scars along many rocky shores has convinced many researchers that wave quarrying is the dominant erosive mechanism along rocky coasts.

Most studies argue that **mechanical wave erosion** operates most effectively at, or slightly above, the still water level. Whilst the waves, and particularly breaking or broken storm waves, perform the erosive work, tides control the water level and therefore also the level where mechanical wave erosion occurs. This 'level of greatest wear' must therefore be associated with the elevation most frequently occupied by the water surface (Carr and Graff, 1982). By computing the frequency distribution of the water level across the intertidal zone (Figure 9.2), the level of greatest wear can be determined. From such distributions we can deduce that the elevations most prone to mechanical erosion are at and between the mean tide levels, because it is at these levels that the water level occurs most frequently. The period that the tide will occupy certain levels decreases with increasing tide range. Therefore, the potential for mechanical wave erosion is greatest in microtidal environments. Usually, the vertical distribution of wave energy

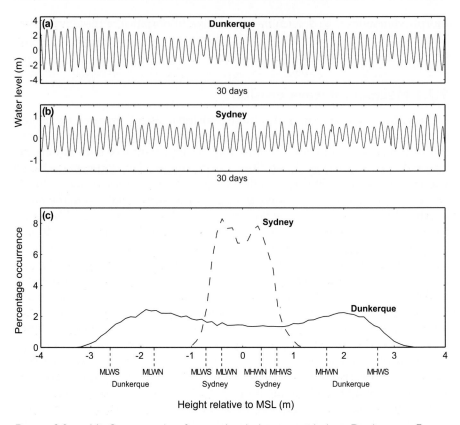

**Figure 9.2** – (a) One month of water-level data recorded at Dunkerque, France (macrotidal range). (b) One month of water-level data recorded at Sydney, Australia (microtidal range). (c) Relative frequency distribution of the water level for Dunkerque and Sydney. The frequency distribution was computed from one year of data using a bin width of 0.1 m. MLWS = mean low water spring, MLWN = mean low water neap, MSL = mean sea level, MHWN = mean high water neap, MHWS = mean high water spring.

and mechanical wave erosion is somewhat skewed towards the upper portions of the tidal range (Trenhaile, 1987). This is because waves are generally more energetic at high tide than low tide. In addition, the most erosive waves occur during storms when nearshore water levels are raised due to storm surge and wave set-up.

## 9.2.2 Weathering

Cliffs and intertidal shore platforms are subjected to alternate wetting and drying by salt spray, wave swash, tides and rain. They therefore represent suitable environments for many physical and chemical weathering processes. In sheltered areas and on particularly susceptible rocks, **weathering** is probably the major erosive mechanism. Here, wave action may be merely responsible for the removal of the weathered material. In storm and high-energy swell wave environments, weathering can also be significant some distance above the high tide level out of reach of the waves. Like mechanical wave action, weathering works most effectively in, and slightly above, the intertidal zone. Weathering processes are strongly controlled by climatic factors (temperature and rainfall) and rock type (Table 9.1).

**Physical weathering** can result from frost action and alternating cycles of wetting/drying. The breakdown of the rock is accomplished through the formation and subsequent widening of capillaries and cracks in the rock. If a frozen rock surface is submerged by the rising tide, the ice within the rock will thaw. Therefore, intertidal rocks may experience two freeze/thaw cycles during a single day when the tide is semi-diurnal and sub-zero temperatures occur throughout the day. The mid-intertidal zone of any rocky surface subjected to semi-diurnal tides experiences almost 700 wetting/drying cycles per year. Frost action and wetting/drying cycles can only cause significant rock weathering if the rock attains high levels of saturation (Trenhaile and Mercan, 1984). Therefore these processes are particularly important in material that can take up and hold large quantities of water, such as fine-grained sedimentary rocks. The mechanical breakdown of rocks is most important along rocky coasts in cold coastal regions because here the rocks are able to attain high levels of saturation and also experience a large number of freeze-thaw cycles per year.

Considerable quantities of salt are deposited in areas near the coast. The presence of salts in the capillaries of coastal rocks facilitates a physical weathering process known as **salt weathering**. The salt inside the capillaries undergoes volumetric changes due to adsorption of water to salt crystals, temperature-induced expansion of salt crystals and crystal growth from solution (Trenhaile, 1987). These volumetric changes result in a widening of the capillaries and cracks in the rock, and a subsequent weakening of the rock. Salt weathering therefore operates in a similar manner to frost action, and the efficiency of salt weathering also increases with rock permeability and porosity. Sandstone is particularly vulnerable to salt weathering. Climatic conditions conducive to salt weathering are high temperatures and low rainfall. Salt weathering is therefore most significant in (semi-) arid climates and Mediterranean regions.

**Chemical weathering** of rocks is usually the result of a number of chemical reactions working together, including hydrolysis, oxidation, hydration and solution. The efficiency of these weathering processes is mainly determined by the amount of water available for the chemical reactions and for the removal of the soluble products. If the weathered product remains in the system, chemical equilibrium may be attained, and this will inhibit further weathering. It is very difficult to rank the susceptibilities of rocks to weathering, but igneous rocks are generally considered to be less susceptible to chemical weathering than sedimentary rocks. Chemical weathering is most significant in hot, wet climates. Cold climates are characterized by relatively slow chemical weathering rates, although the lack of liquid water may be more important in explaining this than the low temperatures (Trenhaile, 1987).

Physical, salt and chemical weathering often work in concert and operate most effectively when and where the rocky substrate experiences frequent wetting and drying. Erosion due to weathering is particularly active around the margins of rock pools in the supra- or intertidal zone, where alternate wetting and drying occurs as the water is replenished by spray or tides. This type of weathering process, which is referred to as **water-layer levelling** (Matsukura and Matsuoka, 1991), cuts into the divides between the rock pools, enlarging them and eventually causing them to merge. Water-layer levelling may play a significant role in the lowering and modification of shore platforms in areas with high evaporation levels, but is less likely to be important in cool, wet climates.

### 9.2.3 Bio-erosion

**Bio-erosion** is the removal of rock by organisms. This process is probably most important in tropical regions due to the enormously varied marine biota and the abundance of calcareous substrates, which are very susceptible to biochemical and biophysical processes (Spencer, 1988). The crucial factor regarding the efficiency of bio-erosion is the spatial distribution of marine organisms across the rock surface, which is largely controlled by the availability of moisture and thus depends on tidal characteristics and wave energy level.

Marine organisms remove rock in various ways. Algae form dense mats on the rock surface and their products of metabolism cause the chemical erosion of the rock surface underneath the algal mats. As a result, the algal colonies can be etched into the rock surface. Grazers effectively abrade rock surfaces as they feed on micro-flora (algae, lichen and fungi) adhered to the rock surface. Chemical or mechanical borers directly remove rock material, but also weaken the surrounding rock making it more susceptible to wave erosion and weathering. In addition, borers enhance a rock environment for algal colonization and increase the area of rock surface exposed to other physical and chemical processes. Marine flora and fauna may also assume a protective role on rock coasts. Some organisms may contribute to the formation of organic crusts that protect the underlying rock from wave erosion and physical/chemical weathering. Dense organic algal mats can also prevent the drying out of a rock surface, thereby protecting a surface that would otherwise be susceptible to water-layer levelling.

### 9.2.4 Mass movements

The steep slopes of rocky coasts indicate that they are unstable features prone to mass movement. A spectrum of **mass movements** can occur on rocky coasts (Figure 9.3), depending primarily on the properties of the rock (lithology and structure).

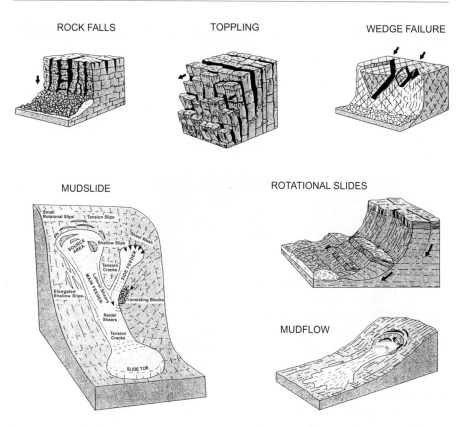

**Figure 9.3** – Types of mass movement affecting cliffs. [From Brunsden and Goudie, 1997; modified from Allison, 1990.] [Copyright © 1997 Geographical Association, reproduced with permission.]

**Rock falls** and **toppling** are characteristic of hard rocks, especially where the rocks are well-jointed and cliffs have been undercut by waves at their base. They are also common on consolidated clay coasts (*e.g.*, glacial tills), although strictly speaking the material is not 'rocky'. Fresh rock faces and the presence of talus at the base of cliffs attest to the importance of rock falls on many coasts. Rock falls and toppling occur most frequently in the winter months and this can be ascribed to frost action, rainfall and increased basal undercutting due to storm waves.

**Landslides** are deep-seated failures that occur when the compressive strength of a rock is exceeded by the load on it. Failure begins when the stress at some point within the rock mass exceeds the strength of the rock, so that the cohesion at that point becomes zero. A distinction is usually made between translational slides, rotational slides and mudslides. Translational slides (or slips) involve movement along a straight plane and are often structurally controlled. Rotational slides fail along a concave-upward surface and typically occur in thick, fairly homogenous deposits of

# Box 9.1 – Black Ven: A Coastal Mudslide System

The Black Ven mudslide complex is situated between Lyme Regis and Charmouth on the Dorset coast, UK, and is one of the largest of such systems in Europe (Brunsden and Goudie, 1997). The uppermost cliffs collapse in single and multiple rotational landslides to form a thick debris slope. When this debris material is saturated due to groundwater seepage, it moves across the upper terrace and falls 20 m onto the next terrace. The muddy material is then funnelled into mudslide

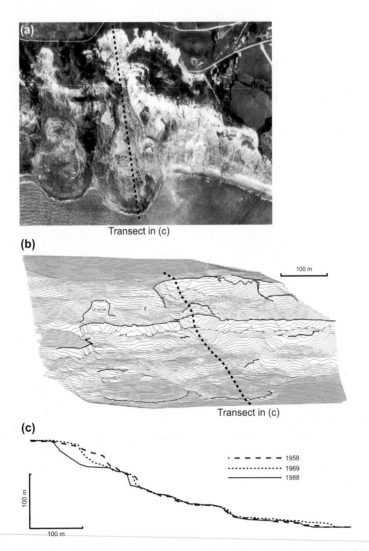

Figure **9.4** – The Black Ven mud slide complex: (a) aerial photograph, (b) digital elevation model (DEM) of 1958 and (c) cliff profiles derived from DEMs of 1958, 1969 and 1988. [Modified from Brunsden and Goudie, 1997.]

tracks. When very active, intermixing streams of mud pour continuously over the benches to merge into broad lobes, which push steadily into the sea. When activity ceases or is slowed down, the toes are then eroded back by the sea to leave behind arcuate areas of boulders on the foreshore. Important feedback processes occur between mass movement and nearshore processes (Figure 9.4a). The presence of vast amounts of sediment at the base of the cliff inhibits further cliff failure, whereas when beach sediment becomes depleted in front of a vulnerable cliff, this leads to an acceleration of cliff erosion.

Black Ven is very active and since the 1920s there has been intermittent failure almost every year. Very pronounced activity occurred in the winter of 1957/1958, when the mudslides extended nearly 100 m across the beach (Figure 9.4a). Cliff retreat of 5–30 m yr$^{-1}$ is characteristic of the active periods, while between major events the coastal erosion of the toe is 15–40 m yr$^{-1}$. Chandler and Brunsden (1995) applied a technique known as archival analytical photogrammetry to produce digital elevation models (DEMs) from historical photographs (Figure 9.4b, c). From these DEMs they determined that between 1958 and 1988 over 200,000 m$^3$ of sediment was transported from the cliff to the sea over the staircase of terraces.

Systems such as Black Ven are gigantic conveyor belts transporting sediments from the land into the nearshore zone and can be of fundamental importance with respect to the littoral drift budget (Bray, 1997).

clay, shale or marl. They also occur in deeply weathered granite and are therefore common in the humid tropics. The slumping of a rock mass removes the lateral support of the material behind, inducing further slides inland, giving rise to multiple rotational land slide systems (Box 9.1). Mudslides occur when fine-grained sediment slides relatively slowly on a shear surface producing a lobate form. The occurrence of all types of landslides is promoted by three main factors: (1) addition of material, possibly from an up-slope mass movement causing an increase in the load; (2) steepening of the slope angle due to basal undercutting by marine processes; and (3) reduction in the compressive strength of the rock due to an increase in moisture content and/or weathering. Landslides are often triggered by a change in moisture conditions and/or increased wave action. They are therefore more common in winter than in summer.

Mass movements that involve movement of material with a high liquid content are referred to as **flows**. They may be divided according to the sediment size into debris flows (coarse) and mudflows (fine). Flows only occur under exceptional rainfall and groundwater conditions.

## 9.3 CLIFFS

Coastal **cliffs** can be defined as 'steep slopes that border ocean coasts' and occur along *c.* 80% of the world's coastline (Emery and Kuhn, 1982). A bewildering variety of cliff profiles are found in nature (Figure 9.5) and this

**Figure 9.5** – Examples of rocky coastlines: (a) stacks at the Twelve Apostles, Victoria, Australia [photo G. Masselink]; (b) rugged coastline of Hartland Quay, Devon, UK [photo G. Masselink]; (c) slope-over-wall cliffs at Baggy Point, Exmoor, UK [photo P. Keene]; and (d) cliff fronted by shore platform at Boulby, Yorkshire, UK [photo G. Masselink].

reflects the large number of factors involved in the development of coastal cliffs. These factors include marine processes, subaerial processes (weathering, mass movements, run-off), rock type (lithology and structure) and sea-level history.

## 9.3.1 Cliff morphology

The importance of the relative roles of marine or subaerial processes in determining **cliff morphology** can be illustrated in a simple morphodynamic model (Figure 9.6). Subaerial processes result in the movement of sediments down the cliff slope to the sea. If the ability of marine processes to remove the debris exceeds the supply of material by mass-wasting processes, sediment will not accumulate at the base of the cliff. In this case, the angle of the cliff profile depends mainly on the structure and lithology of the rock. Vertical cliffs form in massive uniform rocks, whereas if the rock strata dip seaward, the resultant slope will tend to follow this dip. In any case, the predominance of debris removal over supply will maintain a bare rock face at a constant angle. If the supply of debris exceeds the capacity of removal at the base of the cliff, the material accumulates into a talus slope. The resulting angle of the cliff profile is the angle of repose of the debris. In between these two extremes lies an infinite range of slope forms, each of

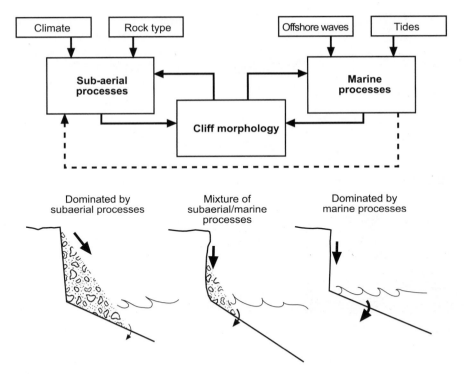

**Figure 9.6** – Generalized representation of the coastal cliff morphodynamic system. Typical cliff morphologies can be identified depending on the relative roles of marine and subaerial processes.

which depends on the relative rates of sediment supply and removal at the shoreline (Pethick, 1984). The relative roles of marine and subaerial processes may vary over time and hence cliff morphology may change accordingly. For example, following a period of intense down-slope sediment transport, a large amount of talus may be present at the base of the cliff that may require a significant amount of time before it is removed by marine processes.

The roles of marine and subaerial processes in determining cliff morphology are exemplified by Hutchinson's (1973) process/form classification of the London Clay cliffs of southeastern England. This classification is generally considered representative of slopes in stiff fissured clay:

- Type 1 occurs where the rate of marine erosion is balanced by the rate of sediment supply to the toe of the slope by shallow mud sliding. The slope undergoes parallel retreat and waves only remove the loose material collected at the base of the slope. Removal of the mudslide debris promotes further sliding and a dynamic equilibrium condition may be attained.
- Type 2 occurs where the erosion by waves exceeds the down-slope transport of weathered material. Waves remove all the material delivered by mudslides and additional undercutting of *in situ* clay results in a

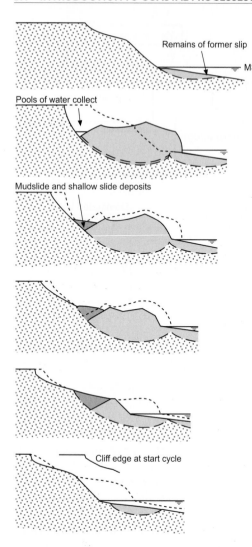

Remains of former slip

MSL

Pools of water collect

Mudslide and shallow slide deposits

Cliff edge at start cycle

**Figure 9.7** – Stages in the cyclic behaviour of London Clay cliffs experiencing strong wave erosion. [Modified from Hutchinson, 1973.]

steepening of the profile. Eventually, this steepening will cause a deep-seated failure, typically a rotational slide. Wave processes subsequently remove the slump debris and basal erosion occurs until another failure occurs. This type of cliff therefore exhibits cyclical behaviour (Figure 9.7).

■ Type 3 coastal cliffs develop where the sea has abandoned the cliff or where coastal defences have been constructed at the base of the cliff. After an initial period of rotational land sliding, the slope angle will be reduced by hill wash and soil creep until it is at the ultimate angle of stability against land sliding (c. 8° in London Clay).

An important aspect of the cliff morphodynamic system is the feedback between cliff erosion processes and the presence/absence of material at the base of the cliff (Figure 9.6). This is elegantly demonstrated by laboratory

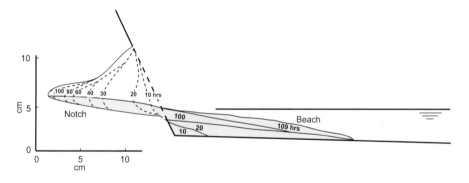

**Figure 9.8** – Erosion by waves of a cliff composed of cement plus sand in a laboratory experiment. The eroded notch extends obliquely landward, while the released sand forms a protective beach at the base of the cliff. [Modified from Sunamura, 1976.]

experiments conducted by Sunamura (1976) who constructed artificial cliffs of loosely cemented sand and subjected these to wave action (Figure 9.8). Initially there was no loose material present at the base of the cliff and erosion of the cliff was relatively modest due to the limited abrasive powers of the water. Cliff erosion made available sediments that accumulated at the base of the cliff in the form of a beach. This initially increased the erosion rate because a mixture of sand and water has a greater abrasive capacity than water alone (positive feedback). However, at a later stage the presence of the beach acted as a buffer, causing the waves to break on the beach rather than against the cliff (negative feedback). Sunamura's (1976) experiments highlight the importance of mainland beaches fronting sea cliffs in inhibiting cliff erosion.

So far we have not considered the role of the geology in determining cliff morphology. Emery and Kuhn (1982) proposed a classification of marine cliffs, based on the relative importance of marine and subaerial erosion, and the effects of variations in rock hardness on the efficiency of these erosive agencies (Figure 9.9). This classification shows that in homogeneous rocks, steep cliffs tend to form where marine processes are important, whereas convex profiles develop if subaerial processes are most important. Where both subaerial and marine processes are important, the cliff will consist of a concave lower part and a convex upper part. The classification also demonstrates that strata differing greatly in their resistance to erosion impose marked variations on these basic forms. The role of the dip of the bedding and joint planes in the rock is also important in determining cliff morphology. For example, there is the tendency for cliffs to be very steep in rocks either horizontally or vertically bedded, whereas slopes are more moderate in rocks dipping landward or seaward.

It must be evident by now that the relative roles of debris-producing and debris-removing processes are of fundamental importance in determining cliff morphology. These relative roles roughly follow a global distribution through the control of climate on wave conditions and weathering.

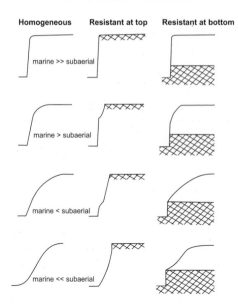

**Figure 9.9** – Generalized coastal cliff profiles according to variations in rock resistance and in the relative efficiency of marine and subaerial processes. It is assumed that the cliffs are cut into plateaux and are in steady state equilibrium. The more resistant rock strata are shaded. [Modified from Emery and Kuhn, 1982.]

Therefore, cliff form can be tentatively related to latitude (Pethick, 1984). In the tropics, the low wave energy levels and high chemical weathering rates generally produce low-gradient coastal cliffs. Coastal cliffs in high latitudes are also characterized by relatively gentle gradients because in these regions periglacial processes produce large amounts of cliff-base material and the wave climate is not very energetic. It is consequently in temperate regions that steep coastal cliffs are most common. Here, rapid debris removal promoted by high wave energy prevents talus formation while active cliff erosion causes steepening of the slope by undercutting. It should be emphasized that the latitudinal distribution of cliff form is a generalization. Variations in local factors, such as rock structure, lithology and wave exposure, can produce greater differences in cliff morphology within the same region than between different regions.

### 9.3.2 The role of inheritance

Besides the balance between debris-supplying and debris-removing processes, most cliffs are also controlled by their evolutionary history. They are therefore greatly affected by inheritance. In fact, it is sometimes difficult to decide whether a given cliff is contemporary (*i.e.*, the result of present-day processes), fossil (*i.e.*, the result of past processes) or a mixture of the two. Many coastal cliffs are so-called **composite cliffs** consisting of more than one slope element (Figure 9.5c). These include multi-storeyed cliffs, with two or more steep faces separated by more gentle slopes, and cliffs with a convex or straight seaward-facing slope above a near-vertical, wave-cut rock face (this latter cliff type has been variously termed 'slope over wall', 'bevelled' or 'hog's back' cliffs). To understand the formation of such

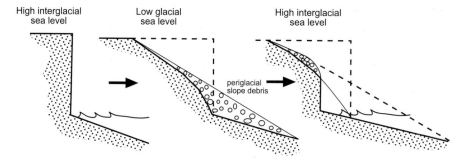

**Figure 9.10** – Development of a bevelled cliff commonly found in regions that have experienced periglacial conditions during the last glacial. The first stage represents a sea cliff during the last interglacial with sea level close to present. During the subsequent glacial, the abandoned cliff is subjected to periglacial conditions and becomes mantled by weathered material. Marine erosion of the base of the weathered slope occurs during present sea level conditions, leading to a bevelled cliff. [Modified from Trenhaile, 1987.]

composite cliff profiles we must consider climatic and sea-level change during the Quaternary.

The formation of **bevelled cliffs** is easiest to explain and requires three stages (Figure 9.10).

- A vertical cliff is formed due to marine processes during the last interglacial period when sea level was high and comparable to the present-day level.
- During the subsequent glacial period, sea level falls and the cliff is abandoned. Periglacial processes progressively degrade the abandoned cliff and mantle it with debris.
- When the sea level rises again during the following interglacial, renewed wave attack removes the debris and steepens the base of the cliff, leaving the regolith-covered remnant above.

Explanations are less straightforward for multi-storeyed cliffs requiring a number of sea level fluctuations for their formation (Griggs and Trenhaile, 1994).

A composite cliff does not necessarily imply a polycyclic origin. It can also be the one-cycle product of differential erosion brought about by the juxtaposition of lithologies with different resistance to erosion. For example, a coastal cliff formed by a resistant rock overlain by a less resistant rock can have a form that is similar to that of a bevelled cliff (Figure 9.9).

### 9.3.3 Cliff erosion rates

Rocky coasts are eroding coasts and it is of importance to identify the factors that affect the rates of recession. The main factor that controls **cliff erosion** is the hardness of rock, in other words its resistance to subaerial and marine erosion. A rock's resistance to erosion is, to a large degree, determined by its lithology and cliff erosion rates can be loosely correlated to lithology as follows (Sunamura, 1992):

- $< 0.001$ m yr$^{-1}$ for granitic rocks
- $0.001–0.01$ m yr$^{-1}$ for limestone
- $0.01–0.1$ m yr$^{-1}$ for flysch and shale
- $0.1–1$ m yr$^{-1}$ for chalk and Tertiary sedimentary rocks
- $1–10$ m yr$^{-1}$ for Quaternary deposits
- $> 10$ m yr$^{-1}$ for volcanic ejecta.

It is telling to translate some of these recession rates to predictions of cliff retreat since the sea level has reached its present position (*c.* 6,000 years ago). For example, granitic rocks can be expected to have retreated since this time by about 6 m, while glacial till cliffs may have been cut back by 6–60 km! The number for the glacial till cliffs may appear quite unbelievable. However, the cliffs along the Holderness coast in eastern England have retreated by almost 3 km since Roman times.

Other factors are important as well. For example, areas of structural weakness due to intense faulting and fracturing may provide focal points of accelerated erosion. Exposure to wave action is also important, not only because waves actively erode the cliff, but also because they may remove material at the base of the cliff. The presence of a beach in front of a coastal cliff generally reduces the rates of cliff erosion by protecting the base of the cliff. Additionally, the height of the cliff is often considered important, with low cliffs eroding faster than high cliffs, because less material needs to be removed to accomplish cliff recession. Human activity may also significantly affect cliff erosion rates. In particular, coastal engineering structures may lead to the disruption of the littoral drift and the removal of a protective beach at the base of the cliff resulting in increased erosion. In addition, the building of structures, such as residences, on cliff tops will increase the bearing weight, making the cliffs more prone to failure.

Cliff evolution and recession are episodic processes with most morphological change and erosion occurring during events with large rainfall and waves. Interpretation of short-term cliff erosion rates should therefore be considered with the utmost care. Cliff erosion rates are often site-specific and difficult to predict other than from past records of sufficient length.

## 9.4 SHORE PLATFORMS

**Shore platforms** are horizontal or gently sloping rock surfaces found in the intertidal zone. In the past, they were referred to as **wave-cut platforms**, but this term should no longer be used because it assumes that shore platforms are the result of wave action, which is not always true. Shore platforms are clearly erosional features that develop when erosion of a rocky coast and the subsequent removal of the debris by waves and currents leaves behind an erosional surface – the shore platform. Shore platform formation is therefore intrinsically linked with cliff erosion, although the processes acting to lower the shore platform are not necessarily the same as those causing recession of the cliff. The junction between the shore platform and the cliff is usually close to the high tide level and sometimes a high tide beach is present at this location. Shore platforms also form along cohesive coasts where they front clay cliffs. These shore platforms are generally covered by a thin layer of sand/gravel material, but become exposed when severe storms remove this veneer.

## 9.4.1 Shore platform morphology

Two major morphologies of shore platforms have been identified (Figure 9.1):

- **Sloping platforms** (1°–5°) extend from the cliff-platform junction to below low tide level, without any major break of slope or abrupt seaward terminus (Figure 9.11a).
- **Sub-horizontal platforms** can be supra-, inter- or subtidal, and generally terminate in a low tide cliff or ramp (Figure 9.11b).

According to Trenhaile (1987), sloping platforms are most common in macrotidal environments and sub-horizontal platforms in meso- and microtidal regions. However, Sunamura (1992) believes that the development of the different types of platforms is more related to the balance between wave action and rock resistance. He found that sloping platforms tend to form in relatively weak rocks and under relatively energetic conditions, while sub-horizontal platforms develop in more resistant rocks and less energetic conditions. Sunamura (1992) further demonstrated that shore platforms are not present when the rocks are too resistant and/or the waves are too weak to cause platform erosion. Therefore, shore platforms are notably absent along coasts consisting of very resistant igneous rocks (*e.g.*, granite), even in vigorous environments, unless the rock has been intensely weathered.

Figure 9.11 – Examples of shore platforms: (a) sloping platform in a macrotidal environment (Bude, Cornwall, UK) and (b) sub-horizontal platform in a microtidal environment (Wollongong, New South Wales, Australia). [Photos G. Masselink.]

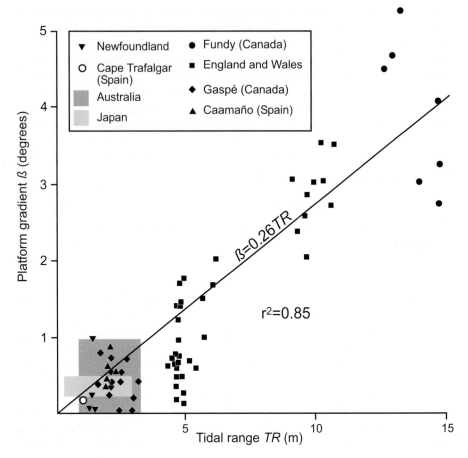

**Figure 9.12** – Relationship between shore platform gradient β and tidal range TR. Each data point represents the regional mean of a large number of platform profiles. The data from Japan and Australia were not included in the regression equation. [From Trenhaile, 1999.] [Copyright © 1999 Coastal Education & Research Foundation, reproduced with permission.]

Trenhaile (1999) conducted a comprehensive investigation into the factors that control shore platform morphology by combining data from south Wales (UK), Gaspé (Canada) and southern Japan. He found a strong, positive relationship between platform gradient and tidal range (Figure 9.12). This relationship may to some extent explain the morphological difference between the seaward sloping platforms along the macrotidal UK coastline and the predominantly sub-horizontal platforms found along the microtidal coasts of Australasia. Trenhaile (1999) further found that the width of the platform was largely independent of tidal range, but increases with wave intensity and decreases with rock hardness. This can be explained by the increased cliff erosion rates with increasing wave intensity and decreasing rock strength.

It is important to emphasize that the strength of the rock is not merely a matter of the lithology, but often more related to the strike and dip of rock strata, joint density and degree of weathering (Trenhaile, 1987). Rock strength is therefore very difficult to determine and quantify. Another major obstacle in relating platform morphology to environmental and geological conditions is the complexity of the relationships and the difficulty of isolating the individual effects. For example, the morphology of shore platforms (*i.e.*, gradient, width and elevation) associated with headlands generally differs from that found within embayments. Headlands experience higher wave energy levels, but also consist of the strongest rock, making it virtually impossible to ascribe the difference in platform morphology to either wave action or rock hardness.

## 9.4.2 Formation of shore platforms: The 'wave erosion *versus* weathering' debate

The formation of shore platforms has been the subject of a more than a century-old scientific debate. This ongoing debate is mainly centred on the question of whether the platforms are the result of wave erosion or weathering. Dana (1849) proposed that platforms are cut by waves at the level of maximum wear, in cliffs that are weathered down to the point at which the rocks are saturated with seawater. In contrast, Bartrum (1916) considered that the platforms he referred to as 'Old Hat platforms', are not the result of wave attack concentrated upon a definite zone of weathered rock. He agreed with Dana that coastal cliffs are weathered down to the level at which they become permanently saturated with seawater, but believed that the platforms develop at this saturation level, rather than at the level of maximum wave erosion. Bartrum proposed that the platforms form about 0.5 m below high tide level in sheltered areas where wave action is weak. Wave action only removes the weathered material, but is by itself not considered capable of actively cutting into the rock. Most of the subsequent theories of platform formation are intermediate between these two early mechanisms attributing varying importance to the roles of waves and weathering.

The weathering *versus* wave action debate is ongoing but the argument seems somewhat academic. Shore platforms occur in a variety of localities and different processes may operate depending on where they occur. Clearly, platforms found in sheltered environments are unlikely to have been cut by waves. Similarly, on platforms subjected to energetic wave conditions it is probable that wave action plays the dominant role, although the weakening of rock by weathering may facilitate the erosive action of the waves. In addition, the relative roles of weathering and wave action are expected to vary across the platform profile and also with time. On most platforms, both wave action and weathering require consideration. The role of weathering, including frost action, wetting and drying, salt weathering and chemical weathering, is primarily to weaken the rock. The role of the

## Box 9.2 – Shore Platforms on Kaikoura Peninsula, New Zealand

Well-developed shore platforms are present on Kaikoura Peninsula, South Island, New Zealand (Figure 9.13). These platforms are exposed to an energetic wave climate, where long periods of relatively calm seas (< 0.5 m) are interrupted by high-energy storms (> 1.5 m). The tides in the region are predominantly microtidal. Stephenson and Kirk (1996) have monitored the platforms on Kaikoura Peninsula over a 20-year period. They used for this purpose the micro-erosion meter (MEM), a device designed to give precise measurements of erosion rates on rock surfaces. Measurement of the erosion rates using the MEM at a large number of sites has revealed an average rate of platform lowering of 1.4 mm yr$^{-1}$.

In a number of studies the relative importance of waves versus subaerial weathering on platform development on Kaikoura Peninsula has been addressed (Stephenson and Kirk, 1998, 2000a, b). Interestingly, it was found that despite high offshore wave energy levels, particularly during storms, the amount of wave energy delivered to the platforms is low. This was due to the offshore topography, most of the offshore waves break before arriving on the platform surface. It was concluded

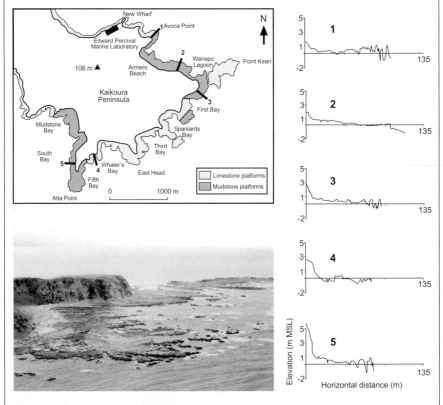

**Figure 9.13** – Shore platform morphology on Kaikoura Peninsula, South Island, New Zealand. [Modified from Stephenson and Kirk, 2000b; photo W. Stephenson.]

that wave forces are not capable of causing erosion on the shore platforms. Strong indications were found that subaerial weathering plays the dominant role. Evidence of weathering came from a number of distinctive surface morphologies and rock strength tests. Furthermore, it was found that platform erosion rates were strongly correlated to the number of wetting-drying cycles. It was therefore concluded that the development of shore platforms on Kaikoura Peninsula results mainly from subaerial weathering caused by repeated wetting and drying.

waves is to remove weathered material and actively erode the platform surface through inducing pressure variations and abrasion. To assess the dominant platform-forming processes for a specific shore platform requires an integrated research approach, such as demonstrated in Box 9.2. An alternative approach to determine the relative roles of waves and weathering is computer modelling (Trenhaile, 2000, 2001).

There are now many field data on rates of platform lowering and most rates average between 0.1 and 2 mm yr$^{-1}$ (Trenhaile, 1987). Not surprisingly, it appears that there is a general tendency for the rate of lowering to increase as the hardness of the rock diminishes (Sunamura, 1992). Rates of platform lowering are also linked to cliff recession rates and field studies suggest that the rate of vertical lowering of the cliff base (top of the shore platform) is 2–5% of the horizontal cliff recession (Goudie, 1995).

## 9.4.3 The role of inheritance

It has been proposed that wave action is presently modifying ancient erosional surfaces inherited from periods when sea level was similar to today's. This appears to be the case in many igneous and metamorphic rocks where present rates of platform lowering (and cliff erosion) seem far too low to account for the formation of shore platforms since the sea reached its present level (c. 6,000 years ago). In addition, these platforms front high cliffs with composite profiles suggesting that they are also ancient features that are slowly being modified by contemporary marine and subaerial processes. The role of inheritance appears less important for shore platforms cut in less resistant rocks where contemporary rates of platform lowering range from 0.1 to 2 mm yr$^{-1}$ (Trenhaile, 1987). These rates are not necessarily representative of the processes and erosion rates at an earlier period when the platforms were narrower and steeper, and when wave quarrying may have been a more significant process than today. Nevertheless, contemporary rates of cliff recession and platform lowering in the weaker sedimentary rocks are generally within the range of values necessary to explain the formation of shore platforms over the last 6,000 years. Platforms cut in rocks of intermediate strength may reflect a mixture of inherited and contemporary features.

## 9.5 OTHER GEOMORPHOLOGICAL FEATURES ON ROCKY COASTS

In addition to cliffs and shore platforms, a number of other features can be found along rocky coasts. Some of these are typical of limestone environments and are discussed first.

### 9.5.1 Limestone coasts

In some parts of the world, specifically the mid-latitudes of the Northern Hemisphere, processes and landforms along **limestone coasts** are similar to those of other rock types. However, limestone coasts found at higher latitudes experience a rather unique suite of physical/chemical/biological processes, resulting in the formation of characteristic limestone coastal geomorphology. These characteristic processes and landforms are brought about by the generally low energy wave conditions, higher temperatures and an enormously varied marine biota. Limestone features are typically corrosion features, resulting from the removal of limestone by physical weathering, solution and bio-erosion.

Guilcher (1953) proposed that the coastal morphology in limestone regions can be classified on the basis of temperature and tidal regime, and has described four main types of coastal limestone landforms (Figure 9.14). Each of these types is characterized by a distinct zonation of a number of typical limestone features.

- **Lapies** (or marine karren) are small ridges that coexist with pool structures and have developed as a result of the solution of rock by running or standing water. Lapies are found in the intertidal zone at mid-latitudes but extend into the spray and splash zone at higher latitudes.
- **Vasques** are wide (up to several 10's of cm), shallow pools with flat bottoms, which form a network of a terrace-like series of steps. The pools are separated from each other by sinuous, narrow lobed ridges, between 1 and 20 cm in height and running continuously for tens of metres. They develop in the intertidal zone, especially in inter-tropical and Mediterranean climatic regions.
- **Corniches** are organic protrusions, generally formed of calcareous algae that grow out from steep rock surfaces in the intertidal zone. They form a narrow pavement (0.5–2 m wide) or sidewalk-like path at the foot of sea cliffs.
- **Trottoirs** are rock ledges caused by erosional processes and coated with a veneer of organic material.
- **Notches** are often found at the base of a cliff around high tide level (Figure 2.2). They may develop along all rocky coasts, but are ubiquitous on limestone coasts. Depending on the environment in which they occur, the erosional process that results in their formation may be wave erosion, weathering and/or bio-erosion.

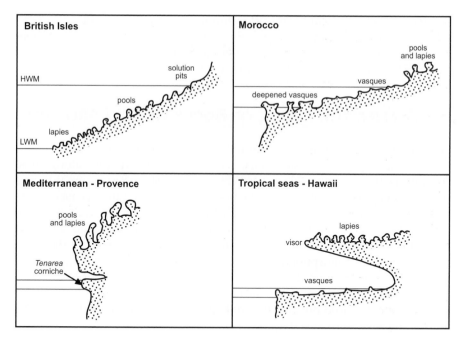

**Figure 9.14** – Four main types of limestone coast distinguished on the basis of temperature and tidal regime. [Modified from Guilcher, 1953.]

## 9.5.2 Coastal embayments, headlands, stacks and arches

The ultimate shape and character of a rocky coast is a reflection of the efficacy of erosional processes integrated over long time periods (thousands to millions of years). Recession rates depend on the balance between the force of the eroding processes (wave action and weathering) and the resistance of the rock to these erosive forces. This balance is expected to vary significantly along coasts mainly through longshore variations in rock resistance. As a result, recession rates vary along the coast resulting in an irregular coastline.

Differences in lithology are often reflected by the occurrence of wide coastal embayments and prominent headlands. Rocks that are resistant to wave attack, such as most igneous rocks, commonly form headlands. Embayments are generally cut in less resistant material such as shale and sandstone. Swash-aligned beaches are often present in the embayments (Chapter 8). Embayments are, however, not solely the result of wave erosional processes. Usually, bays form where rivers and streams enter the sea and hence occupy drowned river valleys. Embayment size, shape and configuration are therefore largely inherited properties (Bishop and Cowell, 1997).

Mechanical wave erosion is particularly sensitive to variations in rock structure. Weaknesses in the rock (joints, faults, fracturing) are focal points for wave erosion, resulting in small bays and headlands. On an even smaller scale, gorges, stacks (free-standing pinnacles of rock; Figure 9.5a), arches (bridge of rock) and caves are often found in close association with each other on coasts which have well-defined planes of weakness. For such features to form, the rock must be strong enough to stand as high, near-vertical slopes. They are therefore less likely to form in very weak rock types.

## 9.6 A Tentative Model of Rocky Coast Evolution

Sunamura (1992) proposed a comprehensive model for rocky coast evolution during a prolonged stationary sea level (as experienced along most coastlines over the past 6,000 years). The model is shown in Figure 9.15 and predicts rocky coast evolution on the basis of the type of initial coast and the relative strength of the rock. Although the model has some rather simple assumptions (rocks are made of insoluble uniform rocks and no sediment accumulation takes place in the nearshore zone), it may provide the rudimentary modes of morphological evolution occurring on most rocky coasts in the world. In addition, the model represents the first attempt to quantify the processes involved in rocky coast evolution.

The model is based on five different types of initial profiles (**I** to **V**) thus accounting for inheritance (Figure 9.15):

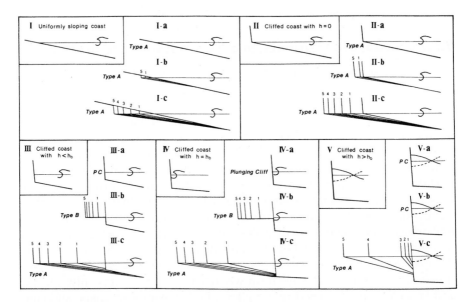

Figure 9.15 – Model for rocky coast evolution beginning with five types of initial landform (**I** through **V**) with different degrees of rock hardness (**a, b** and **c**). 'Type A' refers to sloping shore platforms and 'Type B' indicates sub-horizontal platforms. The water depth at the base of the cliff is indicated by $h$ and the breaker depth is represented by $h_b$. [From Sunamura, 1992.] [Copyright © 1992 John Wiley & Sons, reproduced with permission.]

- **Coast I** – A uniformly sloping coast with a low gradient. Incident waves break some distance offshore and the coastline is only affected by broken waves.
- **Coast II** – A coast with a cliff and a platform. The water depth at the base of the cliff is zero ($h = 0$). Incident waves break offshore on the subtidal platform and this coast is also only affected by broken waves.
- **Coast III** – A coast with a cliff and a platform. The water depth at the base of the cliff is between 0 and the depth of wave breaking ($h < h_b$). Wave breaking occurs offshore of the cliff face, but broken waves with more energy than compared to the former two coastal types hit the cliff.
- **Coast IV** – A coast with a cliff and a platform. The water depth at the base of the cliff is equal to the breaker depth ($h = h_b$). Incident waves break directly against the cliff face.
- **Coast V** – A coast with a plunging cliff. The water depth at the base of the cliff is greater than the breaker depth ($h > h_b$). Incident waves do not break, but are reflected from the cliff face, resulting in standing waves.

The five different types of coasts thus represent different wave conditions at the base of the cliff.

The relative resistance of the rock to wave erosion is categorized on a qualitative basis into three classes:

- **a** – very strong and highly resistant to weathering
- **b** – moderately strong and slightly susceptible to weathering
- **c** – very weak and vulnerable to weathering

The fact that these classes are qualitative, rather than quantitative makes the model difficult to apply. However, Sunamura (1992) does make an attempt at quantifying the different strength classes:

$$\text{for } \mathbf{a:} \quad \frac{S_c}{\rho g H_l} \geq 590$$

$$\text{for } \mathbf{b:} \quad \frac{S_c}{\rho g H_l} = 77 - 590 \tag{9.1}$$

$$\text{for } \mathbf{c:} \quad \frac{S_c}{\rho g H_l} \leq 77$$

where $S_c$ is the compressive strength of the rock (a measure of rock strength), $\rho$ is density of seawater, $g$ is gravity and $H_l$ is the height of the largest waves occurring in the area under consideration. Equation 9.1 is very tentative, but represents the first attempt at quantifying the force balance between wave erosion and rock strength.

The morphological evolution of rocky coasts can now be traced depending on the initial type of coast (**I** to **V**) and the relative resistance of the rock to wave erosion (**a** to **c**). From the model it can be seen that, given sufficient time, sloping shore platforms (Type A) develop on coasts **I** and **II**, regardless of the strength of the rock, and on coasts **III** and **IV** if the rock is very weak. Sub-horizontal shore platforms (Type B), on the other hand, only seem to develop in moderately strong rock on coasts **III** and **IV**. Plunging cliffs form on coasts **III** to **IV** if the rocks are very strong and on coast **V** if the rocks are moderate to very strong.

## 9.7 SUMMARY

- Rocky coasts are eroding coasts. The main processes operating on rocky coasts are mechanical wave erosion, weathering, bio-erosion and mass movements.
- The characteristic landforms of rocky coasts are cliffs and shore platforms. Both landforms can be significantly affected by inheritance.
- The morphology of coastal cliffs depends primarily on the rate of downslope transport of material by mass movements and the ability of marine processes to remove the debris at the base of the cliff. Rock type (lithology and structure) and sea-level history are also important factors that control cliff morphology.
- Cliff erosion rates are mainly controlled by lithology.
- Shore platforms are sloping or sub-horizontal intertidal rock surfaces that form in front of an eroding coastal cliff. Platform lowering is accomplished by wave erosion and/or weathering.
- Limestone coasts found at higher latitudes experience a rather unique suite of physical/chemical/biological processes, resulting in the formation of characteristic limestone coastal geomorphology. These characteristic processes and landforms are brought about by the generally low energy wave conditions, higher temperatures and an enormously varied marine biota.
- A model for rocky coast evolution during a prolonged stationary sea level can be formulated on the basis of the type of initial coast and the relative strength of the rock.

## 9.8 FURTHER READING

Sunamura, T., 1992. *Geomorphology of Rocky Coasts*. Wiley, New York. [This is the most recent text on rocky coast geomorphology and is written from the point of view of an engineer as well as a geomorphologist. The text contains numerous equations linking process to form and quantifying wave forces and rock resistance, but is nevertheless very readable.]

Trenhaile, A.S., 1987. *The Geomorphology of Rock Coasts*. Oxford University Press, Oxford. [This text is a must for anyone interested in learning more about rocky coasts.]

Stephenson, W.J., 2000. Shore platforms: Remain a neglected coastal feature. *Progress in Physical Geography*, **24**, 311–327. [Review paper on shore platforms containing a critical analysis of the 'wave versus weathering' debate.]

Trenhaile, A.S., 1997. *Coastal Dynamics and Landforms*. Oxford University Press, Oxford. [Chapter 11 of this text gives a brief, but comprehensive overview of rocky coast morphology and processes.]

# CORAL REEFS AND ISLANDS

## 10.1 INTRODUCTION

**Coral reefs** represent many thousands, if not millions of years, of *in situ* production and deposition of calcium carbonate ($CaCO_3$). They extend over about $2 \times 10^6$ km$^2$ of the tropical oceans and are the largest biologically

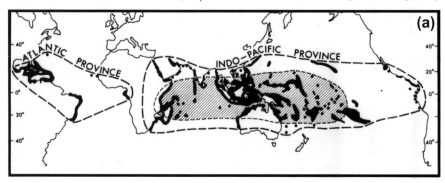

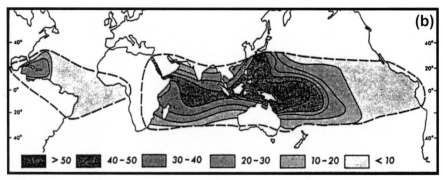

Figure 10.1 – Global distribution of: (a) coral reefs and (b) number of reef-building coral genera. The hatched region in (a) includes the area of the most prolific reef development and almost all coral atolls. [From Davies, 1980.]

constructed formations on Earth. There is a fundamental difference between coral reefs and the coastal environments discussed in the previous four chapters. Coral reefs represent a delicate balance between biological construction and physical processes that act to breakdown and control the morphology and shape of reef platforms. Without the presence of organisms, coral reefs would not exist.

Despite the fundamental importance of marine organisms in constructing coral reefs, the scope of this chapter is geomorphological rather than ecological/biological. It does not go into detail about the different organisms found associated with coral reefs, nor does it deal with the important role of coral reef ecosystems in terms of primary productivity and nutrient recycling. For a comprehensive and succinct treatment of these topics refer to Stoddart (1969) and Viles and Spencer (1995), respectively.

To the geomorphologist, coral reefs represent three-dimensional structures that consist of a living coral reef veneer overlying vast sequences of dead cemented calcium carbonate. These reefs are produced by the *in situ* constructive abilities of a range of organisms, most notably stony (or sclleractinian) corals assisted by other types of corals, coralline algae and other calcareous algae. Corals are commonplace throughout the world's oceans, however, **reef-building corals** (or hermatypic) corals are restricted to tropical or sub-tropical latitudes, particularly between latitudes 30°N and 30°S (Figure 10.1).

## 10.2 ECOLOGY

### 10.2.1 Corals: Building blocks of coral reefs

Contrary to common belief, corals are animals, not plants. They are carnivorous suspension feeders that use their tentacles to trap living zooplankton in the water. The basic unit of the coral is the **polyp**, which sits within an external skeleton of calcium carbonate secreted by the organism. The calcium carbonate is extracted by the organism from sea water. Certain corals live in solitude, in which case a single polyp produces a single skeleton. More frequently, however, a compound coral structure is associated with a colony of polyps. Such a colony grows by repeated division of its member polyps by asexual budding. As new polyps develop, old ones die, and as time proceeds enormous masses of calcium carbonate rock are formed consisting of the skeletal remains of millions of tiny animals. Corals also reproduce sexually during brief periods of the year. The resulting coral larvae have limited swimming ability and they are largely at the mercy of ocean currents. The dispersion and subsequent settling of the larvae are therefore determined by current circulation patterns and the duration of the planktonic phase. When coral larvae find a suitable substrate, they attach themselves to it and metamorphose into a juvenile polyp. Coral larvae establish most successfully on a firm, rocky substrate and although they may colonize loose rubble and even fine sediments, they can not initiate reef development where sediments are mobile.

An important aspect of the ecology of reef-building corals is their symbiotic relationship with small single-celled algae, known as **zooxanthellae**, which inhabit the endodermal tissue of the coral polyp. The zooxanthellae benefit from this relationship by being able to photosynthesize in a protected, nutrient-rich micro-environment. The zooxanthellae also give the coral colonies their distinctive colour. In return, the zooxanthellae remove metabolic waste, provide the coral with nutrients and most importantly, influence the rate of calcification, or skeletal deposition, as the colony grows.

## 10.2.2 Environmental controls on coral distribution and growth

The presence/absence and abundance of coral species is strongly controlled by global and local environmental conditions. A range of environmental variables directly affects zooxanthellae and, due to the strong symbiotic association with corals, influences coral survivorship. Two of the most critical environmental constraints to coral growth (and ultimately reef growth)

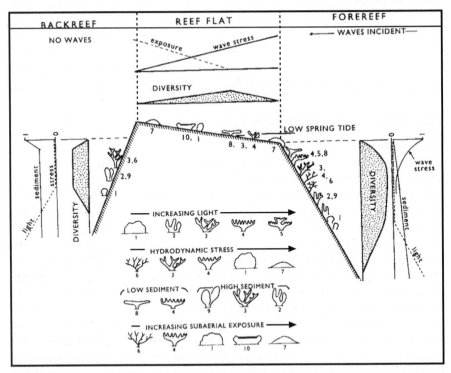

Figure 10.2 – Variations in coral community forms and diversity across a reef in response to changing environmental conditions. [From Chappell, 1980.] [Copyright © 1980 Macmillan Magazines, reproduced with permission.]

are water temperature and light. Locally, factors such as water turbidity, salinity, wave action and nutrient levels can also be important. These environmental factors are discussed below and summarized in Figure 10.2.

Sea temperature exerts a major control on the global distribution of coral reefs and the abundance of reef-building coral species (Figure 10.1). The lowest temperature that reef-building corals are able to withstand is 18°C and the limit of coral reef occurrence lies very close to the 20°C sea surface isotherm for the coldest month. Coral growth is also limited by high temperature and many reef builders will not survive very long in water with temperatures in excess of 34°C. Reef-building corals, and hence coral reefs, are therefore restricted to tropical and subtropical settings. Coral reefs occur in two main provinces (Figure 10.1):

- The **Indo-Pacific province** is centred on Indonesia and the Philippines and characterized by high biodiversity (80 genera, 500 species).
- The **Atlantic province** is mainly located in the Caribbean and characterized by a lower biodiversity (20 genera, 65 species).

Within each of these provinces, the environmental control of the sea water temperature is evident. Firstly, coral species-richness decreases with increase in latitude, a clear result of decreasing water temperatures away from the equator. Secondly, coral reefs are generally less developed along the eastern coasts of oceans (*i.e.*, west coast of continents), which is ascribed to frequent upwelling of cool water (colder than 18°C) along these coasts. The difference in species-richness between the Indo-Pacific province and the Atlantic province is unrelated to temperature and may perhaps be ascribed to different site-specific conditions (Holocene sea-level history, wave conditions and tidal range). Part of the explanation may also be that fewer genera and species simply reflect the much smaller area of shallow coastal water in the Caribbean.

Short-term elevation in water temperature can severely stress corals causing them to expel their zooxanthellae. Without zooxanthellae, corals lose their pigment and take on a white 'bleached' appearance. This phenomenon is known as **coral bleaching**. A moderate bleaching event will reduce the calcification rates and reproductive ability of the coral, but may not necessarily cause widespread coral mortality. Severe bleaching, however, may cause excessive coral mortality and damage to the coral reef system. Increased frequency and severity of bleaching events is one of the primary concerns of reef scientists regarding the effects of global climate change (and increased ocean temperatures) on reef systems.

Coral growth is also controlled by light because the symbiotic zooxanthellae depend on adequate light levels for photosynthesis. Light levels change dramatically with depth and this is why reef-building corals prefer shallow and clear water. For example, in Jamaica, light intensity at a depth of 25 m in clear water is only 1% of the surface illumination (Viles and Spencer, 1995). The decrease in light level with depth causes a reduction in coral diversity from depths as shallow as 10 meter. Below 20 meter, the number of coral species falls rapidly. Although reef-building coral may be found up to depths of 100 m, the maximum depth at which reefs are being built is rarely more than 50 m.

Coral growth is constrained by water level (approximately mean low water spring level). Above this limiting water level, zooxanthellae are susceptible to increased exposure as direct solar radiation can elevate temperature and wind can cause corals to be come desiccated. The control of water level is clearly seen on reef flats where **microatolls** are commonly found. These are small (less than 6 m in diameter) individual sub circular coral colonies with living sides and dead surfaces. The elevation of their dead surface is directly related to exposure to the atmosphere (low tide level) and fossil (raised) microatolls have been used extensively for determining past sea levels (Woodroffe and McLean, 1990).

Locally, water turbidity and salinity can be important factors in controlling reef growth (Hopley, 1982). Turbidity refers to the presence of fine sediment suspended in the water and affects corals in two ways. First, turbidity affects water clarity and light penetration, thereby reducing coral growth rates and the depth to which corals can grow. Second, high turbidity levels are also associated with high sedimentation levels (settling out of the suspended material) and this may 'smother' corals. Corals can survive where salinity ranges from 27 to 40 parts per thousand, but coral growth is restricted outside this range. The combination of high turbidity and low salinity is often the explanation for the absence of coral reefs in the vicinity of river mouths in tropical and subtropical regions.

Exposure to wave action is another important factor controlling coral growth and distribution across reef surfaces. One would perhaps expect energetic wave action to limit coral growth due to the physical forces associated with breaking waves. However, the opposite is the case. Corals have adapted their morphology to differences in wave energy so that massive and encrusting forms are able to survive in high wave energy environments on reefs. In these high energy zones the vigorous mixing of the water due to energetic waves prevents the deposition of suspended sediments, preserves a proper oxygen-carbon dioxide balance and brings in zooplankton within reach of the essentially sedentary reef community. As a result, coral reefs on the exposed, windward sides of tropical islands are generally better developed than on the more sheltered leeward sides. Major tropical storms can, however, significantly damage coral reefs.

Nutrient levels can also influence coral growth and survival. Coral reef communities thrive when nutrient levels are low. **Eutrophication** (nutrient enrichment) may result in the replacement of (hard) coral reef communities with (soft) macroalgae communities. Such a shift from hard to soft shallow water communities may result from sewage disposal in the lagoons of coral atolls. Of particular concern is the fact that re-establishment of the coral reef community may require far higher levels of water quality with respect to nutrients, salinity, turbidity and light than is required to maintain an established reef community.

## 10.2.3 Generalized cross-section through a coral reef

A large number of reef-building coral species exist and depending on environmental and genetic factors, corals are able to assume a variety of forms. Ten generalized coral growth forms are indicated in Figure 10.2, and these coral shapes exhibit a distinct zonation across a coral reef depending primarily on water depth (light level), wave action, sediments in the water and exposure to the atmosphere (Chappell, 1980). Three principal zones can be identified: (1) the forereef, which is the outer or seaward portion of the reef; (2) the reef flat, which is sub-horizontal and sometimes exposed during spring low tide; and (3) the backreef, which is the inner or landward portion of the reef.

The slopes of the **forereef** are generally very steep with gradients of 30–40°. The lengths of these slopes are highly variable and may range from several metres in the case of fringing reefs to several km for oceanic coral atolls. In any case, the forereef is exposed to the highest wave energy levels. Optimal coral growth conditions prevail here due to intense water mixing, and species diversity is high with a large variety of growth forms occurring. Coralline algae can build low ridges on the outer edge of the forereef on the windward side of coral reefs exposed to vigorous wave action. These algal ridges can stand up to 1 m above low tidal level and provide protection of the reef flat behind the ridge.

The **reef flat** lies to the lee of the algal ridge and generally forms the widest part of a coral reef. Its width can vary from less than 100 m to several km and its characteristic feature is its sub-horizontal morphology. On the reef flat, wave action is limited and exposure to light during low tide levels imposes a significant environmental stress to the coral. As a result, species diversity on the reef flat is limited. The reef flat generally consists of a pavement cemented by calcareous algae and incorporating various detrital elements derived from the forereef. Parts of the reef flat may also consist of thin, mobile sheets of sand, gravel or rubble.

The physiography of the **backreef** located behind the reef flat depends on the type of reef under consideration. On barrier reefs and coral atolls, a lagoon is found behind the reef flat, whereas in case of fringing reefs, the backreef merges with the land. Wave action on the backreef is insignificant, resulting in limited water mixing and deposition of sediments. In addition, freshwater and sediments associated with streams draining the hinterland on fringing and barrier reefs provides additional environmental stresses. Conditions for coral growth are thus not optimal on the backreef and as a result species diversity is limited.

## 10.3 REEF GROWTH

The growth rate of individual corals may range from 1–20 cm yr$^{-1}$ depending on the coral species in question. Consequently, the total rate of **coral**

**growth** varies spatially across a reef system according to the ecological zonation of a reef. However, coral growth does not equate to **reef platform growth**. The growth of a reef platform occurs at different space and timescales to growth of individual corals. The distinction between individual coral colony growth and reef platform growth is critical for understanding the geomorphic development of reefs.

It is a common perception that tropical reefs are composed primarily of coral preserved in their growth position. This perception is misleading as in many reefs corals comprise a relatively minor proportion of the reef matrix. Marine organisms, other than corals, have the ability of secreting calcium carbonate, and in many instances they also add their skeletal or other remains to the overall architecture of the reef. In fact, coral skeletons seldom form the major element in the material from which a reef is formed. The basic framework of the reef is thus a porous structure, provided by corals and coralline algae, cemented and sealed by a host of encrusting/cementing marine organisms (calcareous algae, sponges, bryozoans) and infilled by reef-derived sediments from other calcareous organisms (foraminifera, molluscs, echinoids, calcareous plants).

Constructional activity on the reef is always offset by the activities of organisms on the reef that destroy carbonate substrates. A number of destructional processes are involved and include physical breakdown (through mechanical wave stress) and biological breakdown. Bio-erosional processes (borers and grazers) effectively limit reef framework construction and produce vast quantities of sediments on reef fronts and reef tops. As a consequence of the reworking and cementation of carbonate material into the reef fabric, rates of reef platform growth are much lower than that of individual corals.

Coral reef communities are foci of high gross primary productivity, despite the fact that the surrounding waters are characterized by very low concentrations of key nutrients, specifically phosphate and nitrate (Viles

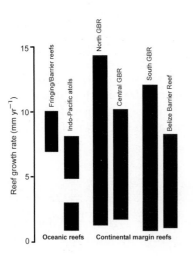

Figure 10.3 – Rates of Holocene reef growth based on radiometrically-dated core material. GBR refers to Great Barrier Reef. [Modified from Spencer, 1994.]

and Spencer, 1995). From a geological/geomorphological perspective, the most important consequence of the reef metabolic processes is **calcification**. Field studies have indicated typical rates of calcium carbonate production of 10 kg of $CaCO_3$ $m^{-2}$ $yr^{-1}$ for coral thickets, 4 kg of $CaCO_3$ $m^{-2}$ $yr^{-1}$ associated with typical Indo-Pacific reef flats, and 0.8 kg of $CaCO_3$ $m^{-2}$ $yr^{-1}$ on lagoon floor and rubble substrates (Kinsey, 1985). The production of calcium carbonate can be converted into a potential vertical growth rate. Since the density of calcium carbonate is 2.9 g $cm^{-3}$ and assuming that the average porosity of reef sediment is 50%, a calcification rate of 10 kg $m^{-2}$ $yr^{-1}$ corresponds to a potential upward growth of 7 mm $yr^{-1}$. A comparison of rates of reef growth from cores for different tectonic settings and reef types suggests a wide range of reef responses over the last 8000 years during the Holocene (Figure 10.3). Generally, rates of reef growth are in the order of 1–10 mm $yr^{-1}$.

## 10.4 TYPES OF CORAL REEF

Coral reefs occur in two major oceanographic settings:

- **Shelf reefs** are found on the continental shelf in water depths less than *c.* 200 m and usually have foundations of continental crust. They are relatively closely located to the mainland or islands and may be significantly affected by fresh water and sediments resulting from river outflow.
- **Oceanic reefs** may rise several km from the ocean floor. They have volcanic foundations and experience limited influence from terrestrial processes.

Historically, coral reef environments have been subdivided into fringing reefs, barrier reefs and coral atolls. Although some kinds of reefs can not be accommodated within this simple scheme, this division still forms the basis for many classifications and will be discussed below. The scheme is least suitable for reefs on continental shelves, which are much more variable than oceanic reefs in terms of shape, and biologic and geomorphic structure.

**Fringing reefs** extend out from the coast (mainland or island) (Figure 10.4a). They tend to be narrow where the submarine slope is steep, and wide where it is gentle. Fringing reefs are usually thin, and most fringing reefs comprise veneers of coral of Holocene age over platforms of non-reef origins. The forereef of a fringing reef is usually steep with rich coral growth and it is here that the reef is most actively developing. The reef flat is often strewn with sand and rubble, and is relatively inactive in terms of coral growth. A fringing reef is often breached by passages and these usually occur opposite the mouth of streams where the inflow of fresh water and sediment inhibit coral growth. Some fringing reefs, especially in East Africa are characterized by the presence of a small narrow depression, referred to as the **boat channel**, between the reef and the land (Guilcher, 1988). The depression, which may be 100–200 m wide and up to 3 m deep, develops

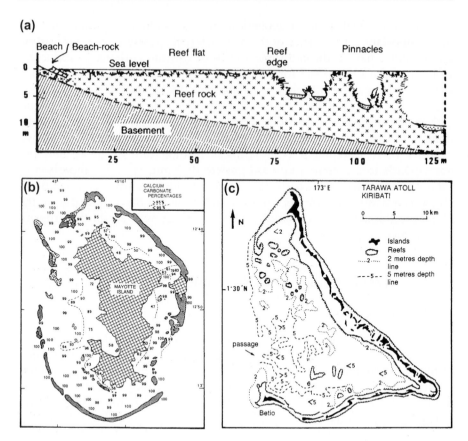

**Figure 10.4** – (a) Fringing reef with incipient boat channel, Aqaba harbour, Jordan, Red Sea. (b) Barrier reef of Mayotte Island, Indian Ocean. (c) Tarawa Atoll, Kiribati, Micronesia. [From Guilcher, 1988.] [Copyright © 1988 John Wiley & Sons, reproduced with permission.]

where deposition of sediment from the land inhibits coral growth near the coast.

**Barrier reefs** are fairly narrow, elongated structures, separated from the coast (mainland or island) by lagoons that are generally less than 30 m deep (Figure 10.4b). Barrier reefs may be continuous or fragmented into smaller units by passages of varying width and depth. These passages enable the influx of water and sediment from the offshore region into the lagoon. The nature of sediments in the lagoon behind the barrier reef can be used to indicate the relative contributions of sediment supply from the land (siliclastic material) and sediment supply from corals and associated marine organisms (carbonate material). Studies of the proportion of $CaCO_3$ in lagoonal sediments in the Great Barrier Reef have indicated that terrigenous sedimentation dominates the inner part of the lagoon, while marine sedimentation is prevalent in the outer parts.

Coral atolls consist of reefs that surround one, or a number of central lagoons (Figure 10.4c). There are 425 coral atolls in the world, the vast majority of which are in the Indian and Pacific Oceans, with only a few isolated atolls in the Caribbean (Stoddart, 1965). Atolls can have circular, elliptical or horse-shoe shapes and vary greatly in size. The large atolls of the Maldives, for example, have diameters in excess of 75 km, but most atolls have diameters of about 10 km. In some instances small atolls, referred to as **faros** or **atollons** are present on the margin of larger atolls. Coral atolls generally have an asymmetric planform shape. The reef platform on the windward side of atolls is often wider and better developed, while outlets to the ocean are generally on the leeward side. The lagoon floor is a smooth depositional surface from which pinnacles and ridges of live coral may protrude. Small platform-like reefs, known as **patch reefs**, may also be present in the lagoon. In contrast with sediments in lagoons associated with barrier reefs, sediment in atoll lagoons are entirely calcareous. Sand and gravel deposits and islands may be found on the top of the reef flats where reef-derived sediment is washed in through the entrances or over the bordering reef. The morphodynamics of these deposits and islands will be discussed later in this chapter.

## 10.5 FORMATION OF CORAL REEFS

During the famous voyage of H.M.S. *Beagle* from 1832–1836, Charles Darwin realized that reef-building corals only live in relatively shallow water (Darwin, 1842). He inferred that the vertical thickness of oceanic barrier and coral atolls is far greater than the maximum depth at which these animals flourish. As corals flourish in shallow water he argued that the great thickness of coral associated with reef structures could only be accounted for by the progressive subsidence of the platform on which the reefs had originated.

**Darwin's subsidence theory** provides a genetic sequence for the development of the three main types of coral reef (Figure 10.5). A fringing reef surrounding a slowly sinking island grows upwards, inwards and outwards as the sea level rises relative to the land. As the land behind the reef subsides, the environmental conditions in the lagoon, characterized by relatively high sedimentation rates, fresh water runoff and limited water mixing, are not favourable for coral reef growth and result in the death of

**Figure 10.5** – Darwin's theory of atoll formation.

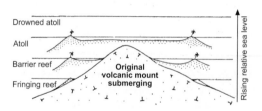

reef-building organisms. As a result, a lagoon forms. The width and depth of this lagoon increase as subsidence proceeds and in this way a fringing reef develops into a barrier reef. If the barrier reef encircles the island and the island itself sinks below sea level a coral atoll results. Should the rate of subsidence exceed the rate of upward growth of coral, a drowned atoll may result.

Darwin's theory makes intuitive sense – without a subsiding substrate it is difficult to envisage how otherwise an oceanic atoll can rise several km from the ocean floor. When coupled with **plate tectonics**, Darwin's subsidence theory also explains the occurrence of atoll island chains in the Pacific (Box 10.1). There are also some morphological arguments that can be put forward to support Darwin's theory. For example, if the reef had not developed under the influence of rising relative sea levels, one would expect the lagoon associated with barrier reefs and atolls to be completely filled with

## Box 10.1 – Formation of Pacific Island Chains

Many coral atolls in the Pacific are often found in linear chains, some of which (e.g., the Hawaiian and Society Islands) demonstrate stages in the Darwinian sequence of coral reef formation (fringing reefs – barrier reefs – coral atolls) along their length. Darwin's theory of coral reef formation combined with plate tectonics explains the formation of such atoll chains (Scott and Rotondo, 1983). Figure 10.6 shows a proposed model for coral atoll development in the Pacific. The Pacific plate migrates in a west-northwesterly direction, away from the mid-oceanic ridge, at a rate of 80–150 mm yr[-1] whilst at the same time subsiding at a rate of 0.02–0.03 mm yr[-1]. Irregular volcanic activity associated with a fixed hot spot results in the formation of a chain of volcanic islands. Coral reefs form on these slowly migrating and subsiding islands and during their travel, fringing reefs are gradually being transformed into barrier reefs, followed by coral atolls. As the coral atolls continue to migrate towards the west-northwest, the reduction in water temperature causes a decrease in coral growth rates. At some stage, vertical reef growth can no longer keep up with the subsiding substrate and the coral atoll drowns, resulting in a **guyot**.

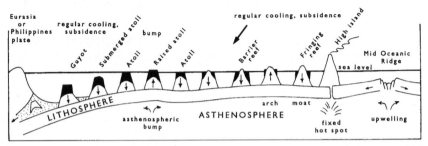

Figure 10.6 – Model of reef evolution on the Pacific Plate. [From Guilcher, 1988; modified from Scott and Rotondo, 1983.] [Copyright © 1988 John Wiley & Sons, reproduced with permission.]

reef-derived and terrestrial sediments. Additionally, the coast in the vicinity of barrier reefs and coral atolls generally bears the hallmarks of a subsiding coastline, including drowned river valleys (estuaries) and the absence of coastal cliffs. However, the most conclusive evidence in support of Darwin's theory has come from drilling cores through coral atoll reef platforms. As Darwin himself suggested, if his theory were valid, then a core through a coral atoll should demonstrate a thickness of the coral reef deposit in excess of the maximum depth of coral growth (c. 50 m).

The first attempt to drill a coral atoll, with the specific objective to confirm or refute Darwin's hypothesis, was undertaken in 1896–1898 on Funafuti atoll in the Pacific Ocean. After several attempts, 333 m of coral rock was drilled, with the upper 194 m in coral limestone, and the rest in dolomite. One would think that this provided conclusive evidence that the coral atoll had developed under the influence of a rising relative sea level. However, opponents to Darwin's theory argued that the drilling was made too close to the edge of the atoll and had passed through coral rock detritus and not *in situ* coral. Subsequent cores and seismic data, however, conclusively demonstrated the general validity of Darwin's subsidence theory. For example, drilling on Eniwetak atoll in the Marshall Islands (Pacific Ocean) demonstrated that this atoll consists of more than 1250 m of shallow-water coral limestone on top of a volcano that rises 3.2 km from the ocean floor (Ladd *et al.*, 1967).

## 10.6 REEF MORPHOLOGY AND QUATERNARY SEA-LEVEL CHANGES

Darwin's theory adequately explains the formation of oceanic atolls over very long time scales (millions of years). Of note, the typical rate of subsidence of atoll foundations ranges from 0.01–0.1 mm $yr^{-1}$. Superimposed upon this gradual subsidence of the coral reef basement have been fluctuations in sea level of around 120–150 m amplitude over the Quaternary (Chapter 2). These oscillations in water level have taken place at rates of 1–10 mm $yr^{-1}$, exerting a strong control on coral and reef growth. These alternate periods of subaerial exposure during glacial lowstands, followed by drowning and reef re-establishment during interglacial highstands are thought to be the overriding control on the morphological expression of contemporary coral reefs.

### 10.6.1 Effect of sub-aerial weathering during glacial lowstands

The geologist Daly was the first to be concerned with the effect of relative sea-level changes on coral reef development (Daly, 1915). According to his **glacial control theory**, the foundations of modern coral reefs were formed during periods of low Pleistocene sea level and hence the shape of contemporary coral reef systems is largely inherited. His main supporting argument was that the floors of lagoons on atolls and barrier reefs are sub-horizontal and occur in a narrow depth range of 50 to 90 m. Daly contended that these lagoon floors represented an erosional surface attributed to marine abrasion during glacial periods and pre-glacial weathering. Once sea level began to rise following the last glacial, coral re-colonization and vertical reef growth would produce a barrier reef or coral atoll depending on the extent of the planation surface. There are a number of problems associated with Daly's argument. For example, sea levels during glacial periods were considerably lower than supposed by Daly. In addition, lagoon depths are much more variable then envisaged by Daly and have concave, rather than sub-horizontal profiles.

However, Daly was correct to point out the importance of reef emergence during sea-level low-stands and the subsequent changes to reef morphology, as there is no doubt that Holocene reef morphology is to a large degree dependent on the Pleistocene surface it inherited. For example, it has been shown that many deep passages through barrier reefs and coral atolls are remnants of Pleistocene river valleys that drained from the emerged lagoon into a lowered sea (Guilcher, 1988). Coral reefs emerged above sea level during glacial periods have been subjected to **karst processes** (solution of calcium carbonate). The importance of karst processes in affecting coral reef morphology has been stressed by Purdy (1974) in what has become known

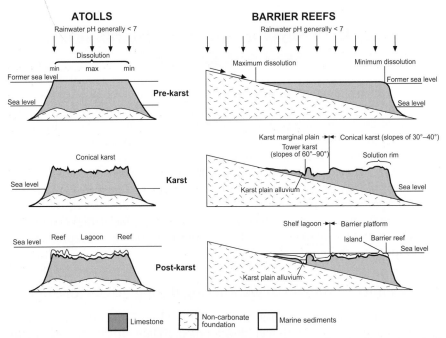

**Figure 10.7** – The evolution of atolls and barrier reefs according to the antecedent karst theory. [Modified from Purdy, 1974.]

as the **antecedent karst theory**. According to this theory, modern coral reefs are thin accretions over older reefs that have been strongly modified by subaerial karst processes. The role played by karstic processes in determining the shape of coral reefs is illustrated in Figure 10.7. Purdy proposed that the shape of many, if not most, reefs is karst-induced. Rainfall solution would be most rapid towards the interior of the surface of steep-sided limestone platforms (*i.e.*, emerged coral reefs), and would be inhibited at the edges due to rapid runoff. Such karst-eroded surfaces, with raised rims and central solution depressions, are transformed in atolls when drowned by rising sea level and re-colonized by coral. Barrier reefs may also be karst-induced, where the lagoon is the result of enhanced rainfall solution compared to the seaward edge of the reef.

It is difficult to determine the importance of karst processes in shaping modern coral reefs. There is no doubt that karst processes have operated on emerged coral reefs during glacial periods. However, the rate of subaerial weathering of limestone surfaces is considered less than 0.5 mm yr$^{-1}$ and there may not have been sufficient time for prominent karst landscapes to have developed during the last glacial. On the other hand, clear evidence of karstification can be identified in many reef environments (Guilcher, 1988), suggesting that at least locally, the effect of karst can be significant.

## 10.6.2 Coral reef growth strategies: Keep-up, catch-up and give-up

Changing relative and/or absolute sea levels are important factors in coral reef development. Of particular interest is the ability of vertical reef growth to keep pace with falling land levels or rising sea levels. Numerous studies on Caribbean reefs and the nearby tropical Atlantic have provided information on the various ways in which these reefs have responded to sea-level rise. A synthesis of this work by Neuman and McIntire (1985) has defined three types of reef strategies according to their response to the Holocene sea-level rise:

- **Keep-up reefs** grow upwards fast enough to keep pace with the rising sea level and are able to maintain shallow, frame-building communities throughout the sea-level rise.
- **Catch-up reefs** begin as shallow water reefs and become deeper as the rate of sea-level rise exceeds the vertical accretion rate. When the rate of sea-level rise decreases, they catch-up with the sea level. The stratigraphy of catch-up reefs is characterized by an upward-deepening sequence that develops when the rising sea level leaves the reef surface behind, followed by an upward-shallowing sequence which forms when the reef surface catches-up with the sea level.
- **Give-up reefs** fall behind rising sea level, and are eventually drowned. Their stratigraphy consists of an upward-deepening sequence followed by a deep-water facies.

While formulated in Caribbean reef systems, similar patterns of Holocene reef growth have been identified in the Indo-Pacific reef province.

Differences in growth strategy depend on a number of factors, including the rate of sea-level rise, the time lag for recolonization of the inundated reef surface and the potential vertical reef growth. Spencer (1994) suggests the give-up/catch-up threshold lies near a rate of sea-level rise of 20 mm yr$^{-1}$ and the catch-up/keep-up threshold at 8–10 mm yr$^{-1}$. The duration of sea-level rise and the depth below which reef growth ceases dictates whether a coral reef that can not keep-up has the opportunity to catch-up. The most common growth strategy of coral reefs in the Great Barrier Reef during the Holocene was the catch-up mode (Hopley, 1994). Initially, coral reef growth could not keep pace with rapidly rising sea level during the early stages of the Holocene, but after the rate of sea-level rise decreased around the mid-Holocene, most reefs caught up with sea level over a few thousands of years. Many Caribbean shelf reef systems, however, exhibited give-up behaviour during the Holocene sea-level rise (Macintyre, 1988). The demise of these reefs is generally attributed to a reduction in vertical growth rates (and subsequent drowning) due to a decrease in water quality after flooding of the continental shelf at the start of the Holocene (high turbidity and nutrient levels).

The reef growth strategy has important implications for the geomorphic development of reefs and the development of sedimentary deposits on the reef surface. While reefs grow vertically to achieve their maximum elevation with respect to sea level, the majority of calcified material is retained in the reef framework. However, once reefs attain their maximum vertical growth limit, excess calcification is shed from the reef as detrital sediments. The sediments become subsequently available for reef island formation and lagoon infilling.

## 10.6.3 A model for Quaternary development of oceanic atolls

McLean and Woodroffe (1994) proposed a five-stage pattern of oceanic atoll development over an interglacial-glacial-interglacial cycle (Figure 10.8). Such a pattern is presumed to have also occurred in a similar manner over previous sea-level cycles. The model is based on observation made on Cocos (Keeling) Islands (Box 10.2). Because the rates of vertical reef growth vary between (and within) atolls, considerable variation is expected in the timing of the different stages. Nevertheless, the model appears to be qualitatively applicable to many other atolls.

When sea level was high during the last interglacial, an atoll similar to that presently found would have existed (Stage 1). The extent to which the lagoon was infilled and to what extent the rim contained reef islands probably varied from atoll to atoll, as it does between atolls today. However, since the last interglacial appears to have contained a period of around 12,000 years of relatively stable sea level, it seems likely that many atolls were

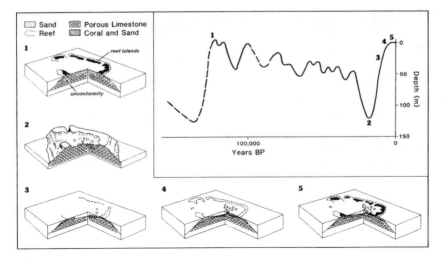

**Figure 10.8** – A summary of the late Quaternary evolution of a coral atoll. The sea-level curve for the last 140,000 years is from Chappell and Shackleton (1986), whereas the Holocene sea-level curve is from Woodroffe et al. (1994). Stage 1 is the Last Interglacial with an atoll like that at present. Stage 2 is the peak of the glacial. Stages 3–5 are at the end of the Holocene transgression. [From McLean and Woodroffe, 1994.] [Copyright © 1994 Cambridge University Press, reproduced with permission.]

filled in to a far greater extent than they have been during the last 6,000 years of relatively stable sea level. As the sea became progressively lower, the atoll would have been exposed as an emergent limestone island (Stage 2). During this emergence, it underwent solution, giving rise to a highly eroded karst surface. Solutional features, such as blue holes, were formed, or enlarged during this emergence, and internal drainage is likely to have concentrated solution into the lagoon.

During the Holocene sea-level rise, the last interglacial reef platform was inundated before sea level reached its present level because the atoll had subsided by c. 10–20 m. The first part of the Holocene (8,000–4,500 years BP) was a phase of rapid vertical reef growth as the reefs tried to catch-up with a rapidly rising sea level (Phase 3). Around 6,000 years BP, sea level in the Indo-Pacific had approximately reached its present level. The stable sea-level conditions allowed vertical reef growth to catch up with the sea level. Over a relatively short period (4,500–3,000 years BP) the reef flat consolidated through reworking infill and cementation of detrital material (Phase 4). After formation of the reef flat, islands started to develop on its surface (Phase 5). This phase of reef island formation, which started around 3,500 years ago, is continuing to the present.

# BOX 10.2 – COCOS (KEELING) ISLANDS

*I went on shore. The strip of dry land is only a few hundred yards wide; on the lagoon side we have the white beach, the radiation from which in such a climate is very oppressive; & on the outer coast a solid broad flat of coral rock, which serves to break the violence of the open ocean. Excepting near the lagoon where there is some sand, the land is entirely composed of rounded fragments of coral. […] These strips of land are raised only to the height to which during gales of wind the surf can throw loose fragments; their protection is due to the outward & lateral increase of the reef, which must break off the sea. The aspect and constitution of the Islets at once calls up the idea that the land & the ocean are here struggling for the mastery: although terra firm has obtained a footing, the denizens of the other elements think their claim at least equal.*

Charles Darwin
Diary of the Voyage of the H.M.S. *Beagle*
entry for 2 April 1836 when landing on Cocos (Keeling) Islands

The Cocos (Keeling) Islands are located in the Indian Ocean and comprise the South Keeling Islands and North Keeling, an isolated atollon, 27 km to the north. The islands rise as a single feature from the ocean floor from a depth of about 5 km. They are dominated by the southeast tradewinds, which blow strongly over most of the year, and maintain a large swell particularly on the southern and eastern margins of the atoll. The islands experience semi-diurnal, microtidal tides with a spring tidal range of 1.2 m. The Cocos (Keeling) Islands are perhaps the best studied atoll in the world, and in fact, it was the only atoll that Darwin visited on his voyage on the H.M.S. *Beagle*.

The South Keeling Islands have been most intensively studied and consist of 26 reef islands situated on a near-continuous reef flat surrounding a lagoon (Figure 10.9a, b). The reef front is relatively barren of living stony corals and the reef crest has a veneer of algae. The reef flat is covered by 1–2 m of water at high tide and much of dries at low tide. Twelve shallow passages, with depths up to 1–1.5 m below mean sea level, connect the ocean-side reef flat and lagoon to the south and the east of the atoll. Intertidal sand aprons extend laterally between 500 and 1500 m lagoonward of the passages. To the northeast and northwest the lagoon opens to the ocean via deep (15–20 m below MSL) and wide (200–500 m wide) passages. The northern part of the lagoon averages around 15 m deep and is mostly covered with sand. The southern part of the lagoon is shallow, but contains a network of 'blue holes', individual deep holes with coral rims and a muddy infill. Extensive sand flats and sand aprons occur around the margin of the lagoon.

Woodroffe and co-workers (for a most recent paper see Woodroffe et al., 1999) have conducted extensive research into the morphology, stratigraphy and Holocene development of the reef islands on South Keeling Islands (Figure 10.9c, d). Their results have indicated that the islands have accreted within the last 3,000–4,000 years, since sea level has reached a level close to present and after development of the reef flat. Island morphology was found to consist of an ocean-ward ridge, a less distinct lagoonward ridge and a low-lying depression. Of the models proposed in Figure 10.13, the data support either central core or oceanward accretion, with deceleration around 2,000 years BP. The dating also suggests that sediments on the reef flat have a short residence time, and are relatively swiftly incorporated in the reef island.

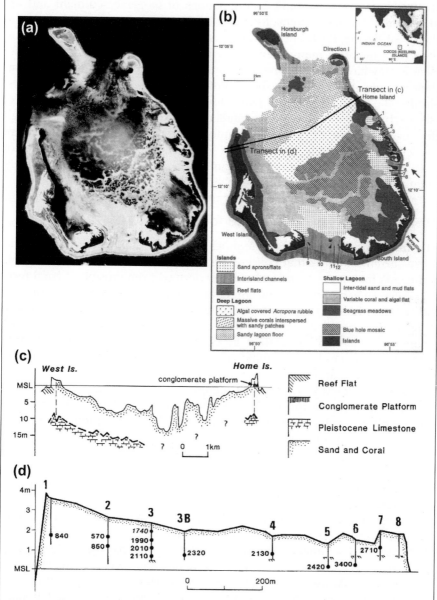

**Figure 10.9** – Cocos (Keeling) Islands: (a) aerial photograph; (b) division into physiographic zones [from Kench, 1997]; (c) transect across the lagoon between West and Home Island [from Woodroffe et al., 1999]; and (d) transect across West Island showing radio carbon dates (in years BP) on coral shingle and sand grains [from Woodroffe et al., 1999]. [Copyright © 1997, 1999 Elsevier Science, reproduced with permission.]

## 10.7 REEF ISLANDS

**Reef islands** are wave-built accumulations of biogenic sediment and are akin to mini-barrier systems. Reef islands are common on reef platforms in the Great Barrier Reef, but also occur on coral atolls, where depending on the size of the atoll, many islands may be present on the atoll rim. The reef islands may have complex shapes, often including spits extending out from their lee side. At some stage, the islands may become vegetated by grasses, shrubs and ultimately trees. Dune formation is not uncommon on reef islands, but the summits of the islands rarely exceed 5 m above spring high tide level. The low elevation and limited width of reef islands makes them particularly vulnerable to rising sea level. As outlined in the model of Quaternary atoll development (Figure 10.8), reef islands are relatively young features and have formed during the second half of the Holocene.

### 10.7.1 Controls on reef island formation

The formation of reef islands requires an ample supply of material and wave-current patterns that concentrate sediment on the reef. The biogenic sediments that make up reef islands are derived from erosion of coral reefs by waves and the activities of borers and grazers. Both wave erosion and bio-erosion produce large quantities of loose reefal material, ranging in size from mud to boulders. Part of this material will move down the forereef (coarse material) or be washed away (fine material). However, a large amount of this eroded reef material will be thrown onto the reef flat by waves, in particular during storms. Sediment transport processes on the reef flat may subsequently organize this material into reef islands.

The main physical processes that operate on reef flats, and which are responsible for sediment transport and reef island formation, are gravity waves and net currents (wind-, wave- and tide-driven). Similar to sediment transport processes in the nearshore zone of wave-dominated shorefaces, wave-induced bed shear stresses entrain and mobilize sediments on reef flats, while net currents move the sediment and determine transport pathways.

An important aspect is the amount of incident wave energy that is able to propagate across the reef surface, because it controls the size and quantity of the transported sediments. The height of the waves across the reef surface is governed by the interaction between the tidal water level and the elevation of the reef flat. At low tide, when a large proportion of the reef flat may be emergent, incident waves break on the reef crest and only a limited amount of wave energy is transmitted across the reef flat. In contrast, at high tide, particularly during storms and/or spring tides, water level on the reef surface may be 2–3 m deep, allowing a greater amount of wave energy to propagate across the reef surface. Nelson (1994) found that the maximum value of the ratio of wave height to water depth ($H/h$) never exceeds 0.55 on reef

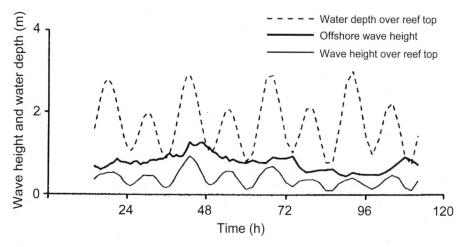

**Figure 10.10** – Time series of wave and tide data measured on the reef flat of John Brewer Reef, Australia. The data clearly demonstrate that the wave height on the reef flat is modulated by the tide, *i.e.*, waves on the reef flat are higher during high tide than low tide conditions. [Modified from Nelson, 1994.]

platforms. Thus, wave height on the reef flat is strongly controlled by the local water depth. As a result, tidal modulation of the wave height is apparent, with larger waves at high tide and low waves during low tide conditions (Figure 10.10).

Tidal currents, and to a lesser degree wind-driven currents, dominate the water circulation within lagoons and passages. Of more relevance to sediment transport, however, are the currents on the reef flat, and these are primarily wave-driven. Wave breaking at the reef crest results in wave set-up (Chapter 4), and this may force water to flow from the reef flat into the lagoon. At the same time, this wave-driven flow may move biogenic sediment towards the lagoon.

## 10.7.2 Models of reef island deposition

Islands formed of reef debris can be divided into two types:

- **Motus** are long and narrow islands found mainly on the windward side of coral atolls. Their location is controlled by the diminished capacity of currents to transport sediment across the reef and as a consequence they are found in linear chains that parallel the reef rim. The morphology of motus is characterized by ridges comprising gravel to cobble-size material at the oceanside and sand ridges at the lagoonside (Figure 10.11). The coarse material comprising the outer part of motus has been transported across the reef flat under storm conditions, while the sand-size material is deposited either through bypassing of sediment in passages between islands, or by locally generated wind waves in the lagoon.

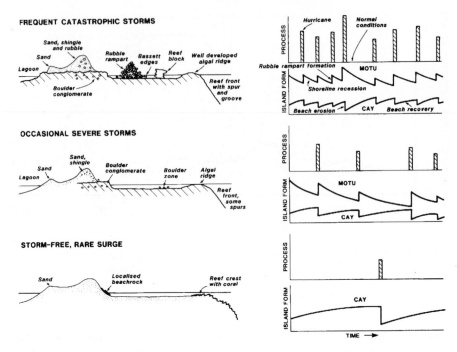

**Figure 10.11** – The morphology of reefs and reef islands in areas of different storm occurrence, and the response of form to process. Storm events are indicated by bars in the process diagram. Rapid change during storms is followed by redistribution of coarse material on motus, and slow recovery on cays between storms. [From McLean and Woodroffe, 1994; modified from Bayliss-Smith, 1988.] [Copyright © 1994 Cambridge University Press, reproduced with permission.]

- **Cays** are smaller than motus and consist of sand and gravel, rather than gravel and cobbles. They are generally found on the leeward side of coral atolls, where lower incident wave energy dictates that only smaller-sized sediment can be transported and deposited.

The occurrence of fine-grained cays and coarse-grained motus is strongly linked to the incident wave climate: cays are primarily found in low-energy wave settings and motus are characteristic of high-energy atolls (Bayliss-Smith, 1988). It should be pointed out, however, that the distinction between cays and motus is by no means sharp.

Islands that form on individual reef platforms, rather than coral atolls, commonly have similar shapes to the reef platform upon which they are deposited. Flood (1986) accounted for the formation of such islands through the process of wave refraction around reefs. For example, on circular reefs, waves refracting around the reef platform converge on a focal point on the reef top. As the energy of the waves and wave-driven currents decreases across the reef platform, sediment is deposited in these 'focal points'. Over time, continued deposition allows such islands to build radially towards the

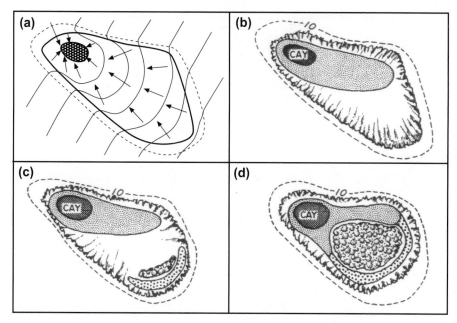

**Figure 10.12** – Evolution of a low wooded island on a reef platform. (a) Wave refraction on the reef platform causes sediments to converge at the back of the platform, resulting in the formation of a small island. [Modified from Flood, 1986.] (b) The island is colonized by vegetation. (c) A shingle ridge develops at the wind-ward side of the reef platform. (d) Mangroves start colonizing the area between the shingle ridge and the sandy cay. [From Bird, 2000.] [Copyright © 2000 John Wiley & Sons, reproduced with permission.]

reef edge. Figure 10.12 schematically shows such development of a reef island due to wave refraction, and also illustrates the further development of the island into a so-called **low wooded island**, a type of reef island typically found in the Great Barrier Reef, Australia.

It is evident that the formation of reef islands post-dates the development of the reef flat on which they form. Little is known, however, about the pattern and mode of sediment accretion on reef islands or their chronology of development following the formation of the reef flat. Woodroffe *et al.* (1999) has indicated a number of ways in which reef islands on atolls may have developed over the last 3,000–4,000 years (Figure 10.13). An important aspect of reef island evolution is whether they have developed gradually or episodically, and whether they are still accreting. In addition, the pattern of sediment deposition can vary greatly. Reef islands may have developed from a central core, but can also have developed by oceanward, lagoonward or vertical accretion. Research on Cocos (Keeling) Islands suggest that these islands developed from a central core and/or by oceanward accretion (Box 10.2).

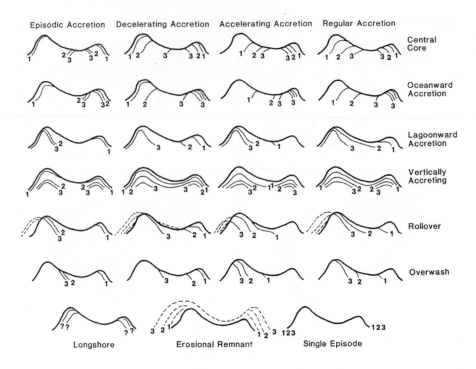

**Figure 10.13** – Schematic scenarios of reef island accumulation indicated by isochrons of deposition (in 1,000's years BP). Ocean is to the left and the lagoon is to the right. [From Woodroffe *et al.*, 1999.] [Copyright © 1999 Elsevier Science, reproduced with permission.]

## 10.7.3 Reef island morphodynamics

The morphodynamics of reef islands have received little attention. However, reef island size, shape and position on reef platforms are known to be sensitive to changes in physical processes (waves), storminess and sea level. Bayliss-Smith (1988) developed a simple framework of reef island behaviour based on the size of material that comprises the islands and whether islands are found in storm or non-storm environments (Figure 10.11). In this framework, the magnitude and frequency of storms play a vital role in controlling reef island morphology:

■ Extreme, catastrophic storms (cyclones or hurricanes) are necessary as a mechanism for coarse sediment production, but such events will also cause widespread shoreline erosion and will strip islands from fine sediment during overwash episodes. Cays are particularly vulnerable to storm erosion and may even completely disappear.

■ More frequent, lower-magnitude storms are also likely to erode cays. However, the more energetic wave conditions will enable the onshore transport of the coral rubble present on the reef flat, so as to reconstruct eroded motu beaches and shorelines. The accretion of storm debris onto old shorelines can lead to a net enlargement of islands, both in width and, through littoral drift and spit development, in length.

- The net effect of small storms and calm weather conditions will probably be erosive on motus. This is because wave action will remove sediment from the islands without there being a compensating replenishment from new coral rubble thrown up on to the reef flat. Motus will therefore be shrinking under relatively calm conditions, at least on their seaward coasts, but meanwhile coral growth on the forereef will be building up a reservoir of coarse sediment which eventually will become available when the next extreme storm event strikes. On cays, however, the effect of small storms and calm weather conditions will probably be growth of the islands through the influx of sand-size material.

Due to their low elevation, reef islands are generally considered very vulnerable to the effects of sea-level rise. Despite the acuteness of the problem, not much is known about the response of reef islands to rising sea levels. Kench and Cowell (2000) carried out numerical experiments and indicated that the morphological response of reef islands to sea-level rise may range from the standard Bruun response to barrier migration. The most unstable, and therefore most vulnerable islands have low-lying seaward margins and are narrow so that the entire island can undergo washover. The most stable, and therefore least vulnerable islands are wide with higher elevated seaward margins. In this case, washover processes are limited to the island margin and recession of the seaward ridge results in island narrowing. These islands are also expected to undergo vertical aggradation to keep pace with sea-level rise. This occurs because washover events carry sediments from the beach onto the island. In addition to reef island morphology, the production and supply of biogenic material is of critical importance in determining reef island response to sea-level rise. Because of this, secondary effects of sea-level rise are likely to be of first-order importance to island response (Cowell and Kench, 2000). These secondary effects involve changes in littoral drift patterns, caused in turn by changes to wave transformation due to increased water depths over the reefs. These effects may be compounded by directional shifts in the regional wave regime, also a potential by-product of global climate change.

## 10.8 SUMMARY

- Coral reefs represent many thousands, if not millions, of years of *in situ* production and deposition of calcium carbonate ($CaCO_3$). The most important environmental constraints to coral growth are water temperature and light. Locally, factors such as water turbidity, salinity, wave action and nutrient levels can also be important. Corals are commonplace throughout the world's oceans, however, reef-building corals are restricted to tropical or sub-tropical latitudes.

- Oceanic reef systems develop under the influence of a slowly subsiding land surface (rising relative sea level). They start their existence as fringing reefs along the margins of volcanic islands and slowly develop into barrier reefs, and then coral atolls as the land surface continues to fall.

- During the Quaternary, coral reef systems were alternately exposed to the atmosphere during glacial lowstands, followed by drowning and reef re-establishment during interglacial highstands. During reef emergence, karst processes may have played an important role in re-shaping the surface of the coral reefs. Consequently, reef morphology is strongly influenced by its antecedent morphology.

- The response of coral reefs to the post-glacial sea-level rise falls into three categories: (1) keep-up reefs grew upwards fast enough to keep pace with the rising sea level; (2) catch-up reefs were initially left behind by rising sea levels, but managed to catch-up when the rate of sea-level rise slowed down; and (3) give-up reefs fell behind rising sea level and eventually 'drowned'.

- Reef island development and lagoon infilling are the major constructional processes in reef environments once they reach their vertical growth limit with respect to sea level. In the Indo-Pacific region, these islands have developed during the last 3,500 years.

- Reef islands are the result of the transport of biogenic sediments by physical (wave and tide) processes on reef platforms. The energy level of the incident wave field controls the size of material that comprises islands, and the location of the islands. In high wave energy settings, gravel motus are deposited on the windward side of reef platforms. In low wave energy settings, sand cays tend to be constructed on the windward side of the reef platforms.

## 10.9 FURTHER READING

Guilcher, A., 1988. *Coral Reef Geomorphology*. Wiley, Chichester. [This is the only text available that deals specifically with coral reef geomorphology.]

Hopley, D., 1994. Continental shelf reef systems. In: R.W.G. Carter and C.D. Woodroffe (editors), *Coastal Evolution*, Cambridge University Press, Cambridge, 303–340. [Relatively recent review of coral reef systems on continental shelves.]

McLean, R.F. and Woodroffe, C.D., 1994. Coral atolls. In: R.W.G. Carter and C.D. Woodroffe (editors), *Coastal Evolution*, Cambridge University Press, Cambridge, 267–302. [Relatively recent review of coral atolls.]

Viles, H. and Spencer, T., 1995. *Coastal Problems*. Arnold, London. [Chapter 6 of this excellent text on coastal problems gives an succinct ecological perspective on coral reef systems.]

# 11

# MANAGEMENT OF COASTAL EROSION

## 11.1 INTRODUCTION

Coastlines are the most important and intensely used of all areas settled by humans. Many countries show above-average concentrations of population near the coast and two-thirds of the world's largest cities are located on coasts (Figure 11.1). Presently, almost 40% of the world's human population live within 100 km of the coast (Cohen *et al.*, 1997), but because many coastal populations are expanding faster than national populations, it is estimated that up to 75% of the world's population could be living within 60 km of the shoreline by 2020 (Edgren, 1993). Finally, Nichols and Mimura

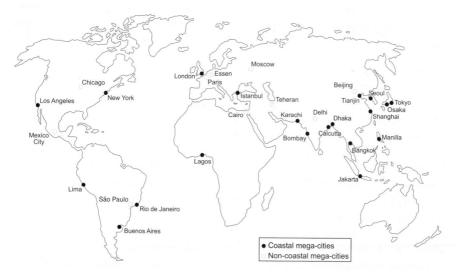

Figure 11.1 – Global distribution of 'world cities'.

(1998) predict that 600 million people will occupy coastal floodplain land below the 1,000-year flood level by 2100. There are many reasons for increased population densities near the coast, and these include historical settlement, trading or political linkages, climate, availability of fertile alluvial soils, proximity to fish stocks and, more recently, aesthetic and recreational reasons (Carter, 1988).

Despite the benefits of living near the coast, humans have an uneasy relationship with the coast and throughout history we have tried to ignore it, adjust to it, or control it – more often than not unsuccessfully (Carter, 1988). All these activities come under the broad heading of **coastal management**. Proper management of the coast is becoming increasingly important due to increased pressures on the coast noted above and the fact that about 70% of our sandy coastlines are eroding (Bird, 1985). To add to this, eustatic sea level is rising and is anticipated by many to do so at an increasing rate, due to anthropogenic emissions of Greenhouse gases (Chapter 2).

This chapter presents an overview of coastal management and the impact of sea-level rise on coastal environments. The field of coastal management is indeed vast and covers a wide range of issues. Here, we limit our discussion to the management of coastal morphology and sediment transport processes, and in particular coastal erosion. A broader treatment of the topic that covers additional material can be found in Kay and Alder (1999).

## 11.2 COASTAL MANAGEMENT FRAMEWORK

From an anthropogenic point of view, the coastal zone is essentially a resource to be used and exploited by humans, whereas from an environmental perspective, the coastal zone is an environment often adversely affected by human activities (French, 1997). The coastal zone is used for various activities, ranging from nature conservation to waste disposal (Table 11.1). Most coasts support multiple activities, and interactions inevitably occur between two or more coastal uses. These interactions are often conflicting, or even mutually exclusive, and management is required to plan and co-ordinate the different uses of the coastal zone to avoid conflicts. Most human activities adversely affect the coastal environment. Industrial activities and waste disposal result in water pollution, land claim destroys intertidal habitats and coastal defence structures interfere with natural sediment cycling. The damage incurred by the coastal environment is important in itself, but may also reduce the resource value of the coast. To minimize these adverse environmental impacts the coast needs to be properly managed. Management is also concerned with non-anthropogenic processes that affect coasts, such as coastal erosion due to sea-level rise.

Coastal management can be subdivided into three broad areas (Carter, 1988):

**Table 11.1** Typical uses of the coastal zone

| Coastal uses |
| --- |
| Tourism |
| Residential |
| Recreation |
| Industry and commercial |
| Resource exploitation |
| Agriculture |
| Infrastructure |
| Waste disposal |
| Aquaculture |
| Fishing |
| Nature reserves |
| Military and strategic |
| Coastal defence |

- **Policy** relates to the political and administrative framework through which coastal management is regulated, principally through legislation and education.
- **Planning** is the process of allocation of environmental, ecological, social or economic resources. Planning may be negative in that it discourages development, or positive in that it encourages it.
- **Practice** covers the techniques needed for implementation of planning decision, or for undertaking restorative or remedial works. Practices range from building coastal protection structures to planting beach grass.

The three areas fall within a major feedback loop, as further planning/policy decision depend on the performance of the management practices.

In the past, coastal management was mainly concerned with single issues that could be dealt with by a single authority. This is no longer the case. The increased complexity of coastal management issues and the varying spatial and temporal scales at which they operate, bring in many different organizations with an interest in the management of a coastline. These organizations typically include administrative authorities (councils, government agencies, environmental organizations), industry and other interest groups (residents, tourism). For effective management of the coast, an integrated approach should be adopted and the term 'integrated coastal management' is used to indicate this approach (Cicin-Sain and Knecht, 1998).

**Integrated coastal management** (ICM) can be defined broadly as a dynamic process in which a co-ordinated strategy is developed and implemented for the allocation of resources to achieve the conservation and sustainable use of the coastal zone. In a geographical sense, ICM embraces upland watersheds, the shoreline (beaches, dunes and wetlands), estuaries and nearshore waters. ICM requires (Bijlsma *et al.*, 1996):

- integration of programs for economic development, environmental quality development and land use;
- integration of programs for sectors such as food production (including agriculture and fishing), energy, transportation, water resources, waste disposal and tourism;
- integration of all the tasks of coastal management (planning, analysis, implementation, operation, maintenance, monitoring and evaluation) performed continuously over time;
- integration of responsibilities for various tasks of management among levels of government (international, national, regional, state/provincial and local) and between the public and private sectors;
- integration of available resources for management (*i.e.*, personnel, funds, materials, equipment); and
- integration among disciplines (physical science, social science, economics, engineering, political science and law).

The ICM concept is increasingly being applied to coastal management problems in both developed and developing countries. An example of an English ICM initiative is given in Box 11.1 (for more examples refer to Kay and Alder, 1999). ICM has been singled out by the Intergovernmental Panel on Climate Change (IPCC) as a key tool for dealing with the threat of accelerating sea-level rise in low-lying coastal areas.

**Sustainability** is an essential attribute of any ICM initiative. Sustainable development is development that meets the needs of the present without compromising the ability of future generations to meet their own needs (World Commission on Environment and Development, 1987). In other words, sustainable coastal management means that all human activities should be non-destructive and that the resources we exploit should be renewable. Against such a benchmark, it is clear that many practices are not sustainable and that our descendants will not inherit the opportunities that the present generation enjoys. However, as Kay and Alder (1999) point out, sustainability is not a set of prescriptive actions, rather it is a 'way of thinking' about our use of the coastal zone and the resulting impacts. In a general sense, application of the concept of sustainability to coastal management has resulted in a management approach with a longer-term view and more holistic perspective.

The output of ICM consists of a range of statutory and non-statutory policies, plans, strategies and initiatives. Statutory initiatives are legally binding and are a very powerful means to direct practices in the coastal zone. An example of a statutory initiative at a national level is the Dynamic Preservation Strategy adopted by the Dutch national government in 1991. This strategy included a legal provision that prescribed that the Dutch coastline be maintained at its 1990 position, irrespective of uncertain future developments (Koster and Hillen, 1995). In other words, land losses due to coastal erosion are considered unlawful and have to be compensated for by beach nourishment. On a local level, councils can use bylaws to control

# Box 11.1 – Coastal Management in the UK: Regional Coastal Management Groups

For historical reasons, coastal management in the UK was until very recently cumbersome, muddled and unlikely to be effective (Carter, 1988). To illustrate one of the difficulties, coastal protection from erosion is the responsibility of local authorities, whereas sea defences against flooding are the responsibility of the Environment Agency. Since the mid-1980s, major changes in attitude and procedures relating to coastal management have taken place in the UK (Hooke and

**Figure 11.2** – Coastal groups in England and Wales (in 1995). [From Hooke and Bray, 1995.] [Copyright © 1995 Blackwell Publishers, reproduced with permission.]

Bray, 1995). The historic lack of co-ordination and fragmentary nature of coastal management in the UK is rapidly being redressed and at present, shoreline management plans have now been drawn up for the entire coastline. An important role in improving coastal management in the UK continues to be played by regional coastal groups.

The Standing Conference on Problems Associated with the Coastline (SCOPAC) was established in 1986 and was the first regional coastal group (Bray et al., 1997). SCOPAC comprises 29 local authorities and agencies (including district and county councils, the Environment Agency, English Nature), and covers a 200 km segment of the south coast of England from Lyme Regis to Brighton (coasts of Dorset, Hampshire and West Sussex) (Figure 11.2). The group was formed when the coastal authorities realized the need for closer co-operation between neighbouring organizations/agencies. This realization came in light of increasing shoreline problems and the *ad hoc* nature of local management solutions, which often adversely affected an adjacent authority's coast.

SCOPAC is a non-statutory organization; it can advise and promote best practice, but individual member authorities make the important decisions. The integrative nature of the initiative is exemplified by its aims and objectives (Hooke and Bray, 1995):

- To ensure a fully co-ordinated approach to all coastal engineering works and related matters between neighbouring authorities on the central section of the south coast of England.
- To eliminate the risk of coastal engineering work carried out by one authority adversely affecting the coastline of a neighbouring authority.
- To exchange information on the success or failure of specific types of engineering projects.
- To establish close liason with Government and other bodies concerned with coastal engineering projects.
- To identify aspects of overall coastal management where further research work is required and to promote such research.

SCOPAC has been successful in fulfilling most of its aims, and this has prompted other authorities around the British coastline to create local groups (Figure 11.2). The setting up of these coastal management groups has been done, as far as possible according to the distribution of natural coastal cells. These cells are stretches of coastline which behave as essentially self-contained sub-systems, with identifiable inputs, transfers, stores and outputs of sediment (Hooke and Bray, 1995). By delineating group boundaries coincident with natural system boundaries, it is hoped that more holistic and fully integrated management may be possible. As noted by Haslett (2000), this approach to integrated coastal management has great potential for developing sustainable management plans.

activities in the coastal zone. Non-statutory initiatives, on the other hand, are not legally binding, but are advisory. They are generally in the form of **coastal management plans**, although not all management plans are non-statutory. Coastal management plans can chart out a course for the future development of a stretch of coast and/or assist in resolving current

| management issue | engineered solution(s) | types | description | problems | illustration |
|---|---|---|---|---|---|
| cliff erosion | seawalls | vertical wall | a wall constructed out of rock blocks, or bulkheads of wood or steel, or simply semi-vertical mounds of rubble in front of a cliff | rock walls are highly reflective, bulkheads less so. Loose rubble however, absorbs wave energy | 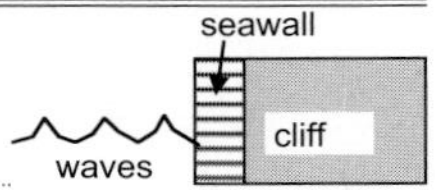 seawall / cliff / waves |
| | | curved wall | a concrete constructed concave wall | quite reflective, but the concave structure introduces a dissipative element | 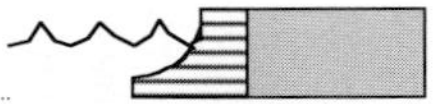 wave-return wall |
| | | stepped | a rectilinear stepped hard structure, as gently sloping as possible, often with a curved wave-return wall at the top | the scarps of the steps are reflective, but overall the structure is quite dissipative | 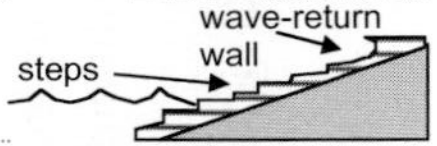 steps |
| | | revetment | a sloping rectilinear armoured structure constructed with less reflective material, such as interlocking blocks (tetrapods), rock-filled gabions, and asphalt | the slope and loose material ensure maximum dissipation of wave energy | 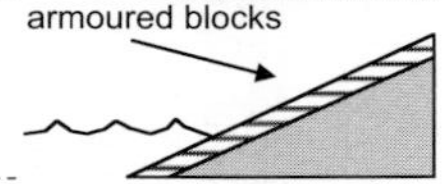 armoured blocks |
| coastal inundation | seawalls | earth banks | a free-standing bank of earth and loose material, often at the landward edge of coastal wetlands | may be susceptible to erosion, and overtopped during extreme high-water events | 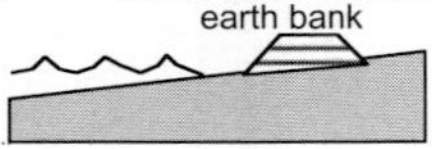 earth bank |
| | tidal barriers | | barriers built across estuaries with sluice gates that may be closed when threatened by storm surge | extremely costly, and relies on reliable storm surge warning system (e.g. Thames Barrier) | |
| beach stabilization | groynes | | shore-normal walls of mainly wood, built across beaches to trap drifting sediment | starve downdrift beaches of sediment | 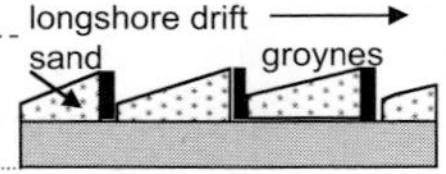 longshore drift / groynes / sand |
| | beach nourishment | | adding sediment to a beach to maintain beach levels and dimensions | sediment is often rapidly removed through erosion and needs regular replenishing; often sourced by dredging coastal waters | waves / sand |
| offshore protection | breakwaters | | structures situated offshore that intercept waves before they reach the shore. Constructed with concrete and/or rubble | very costly and often suffer damage during storms | 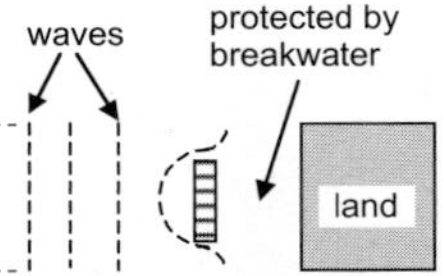 protected by breakwater / land |
| tidal inlet management | jetties | | walls built to line the banks of tidal inlets or river outlets in order to stabilise the waterway for navigation | the jetties protrude into the sea and promote sediment deposition on the updrift side, but also sediment starvation and erosion on the downdrift side | 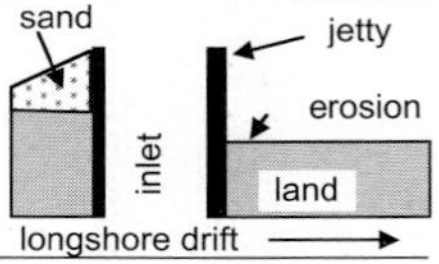 land / jetty / erosion / land / inlet / sand / longshore drift |

**Figure 11.3** – Summary of engineered coastal protection works. [From Haslett, 2000] [Copyright © 2000 Taylor & Francis Books, reproduced with permission.]

management problems. This dual benefit is the greatest strength of coastal management plans – they can have an eye to the future, but still be firmly based in the present (Kay and Alder, 1999).

## 11.3 COASTAL MANAGEMENT PRACTICE

The previous section discussed some of the policy and planning aspects of coastal management; this section will focus on management practice. Coastal engineers commonly carry out the practical element of coastal management, although coastal geomorphologists are increasingly involved in the process. **Coastal engineering** includes a large range of activities, such as the building of tidal barrages to prevent coastal flooding, harbour design and beach nourishment. We limit our discussion here to coastal engineering practices used to combat coastal erosion. A summary of coastal protection works is given in Figure 11.3.

Over most of the last century, the construction of large structures made of timber, concrete or rocks was considered the best way to combat coastal erosion. This approach to coastal protection is termed 'hard engineering' and has resulted in large sections of coastline now being 'cast in stone' (*e.g.*, Japan). During the last few decades there has been increased concern with regards to the environmental problems associated with hard stabilization structures. Consequently, there has been a shift towards more environment-friendly methods of coastal protection. These new methods are termed 'soft engineering'. A number of hard and soft engineering measures for coastal protection will now be discussed.

### 11.3.1 'Hard' shore-based structures

The most commonly used structures for shore and cliff protection are seawalls and revetments. These are built parallel to the shoreline and are designed to protect the land behind them. **Sea walls** are massive structures constructed of concrete, steel sheets or timber. They generally have a vertical face, but may also be curved or stepped. **Revetments** are generally more modest in size and are inclined. They may be built from natural stone, known as rip-rap, or concrete armour units. The vertical or concave faces of seawalls reflect the incident wave energy back to sea. While this seems desirable, wave reflection may lead to scouring of the beach in front of the seawall, followed by undermining and eventually collapse of the seawall. Rip-rap revetments are more permeable than solid sea walls and induce less wave reflection than vertical seawalls.

Seawalls and revetments can be used to protect coastal development in the face of erosion. However, a range of environmental problems may arise from them (Kraus and McDougal, 1996). They are static features and therefore conflict with the dynamic beach changes and impede the exchange of sediment between land and sea. In addition to their unsightly visual

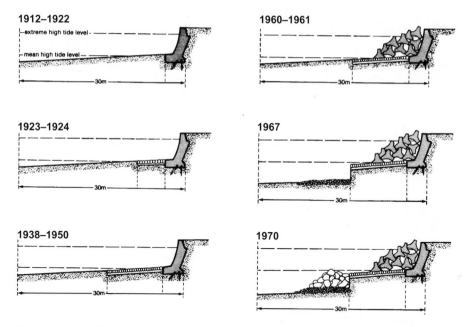

**1912–1922**

extreme high tide level

mean high tide level

30m

**1923–1924**

30m

**1938–1950**

30m

**1960–1961**

30m

**1967**

30m

**1970**

30m

**Figure 11.4** – Development of the beach and seawall of Westerland, Sylt Island, Germany. [From Kelletat, 1992.] [Copyright © 1992 Coastal Education & Research Foundation, reproduced with permission.]

appearance, there are two specific problems associated with seawalls and revetments. First, wave reflection induced by the structure may result in scour and the subsequent lowering of the sand level of the fronting beach (Figure 11.4). Second, they may accelerate erosion of adjacent, unprotected coastal properties because they affect the littoral drift process. The pragmatic solution to these problems is to extend the structure. However, in time, this may exacerbate the problem. The potential adverse impacts of seawalls and revetments need to be recognized and evaluated whenever installation of a structure is under consideration (Komar, 1998).

**Embankments** (or **dikes**) are used for flood defence, rather than erosion protection. They are commonly found along estuarine shores and salt marshes, and are constructed between mean spring high tide level and highest astronomical tide level. They are usually built of unconsolidated material (clay). The main problem associated with embankments is a process called **coastal squeeze** (French, 1997). If a salt marsh backed by steep slope experiences rising sea levels, the different vegetation zones are compressed against the slope (Figure 11.5). This will significantly reduce the width of the salt marsh habitat.

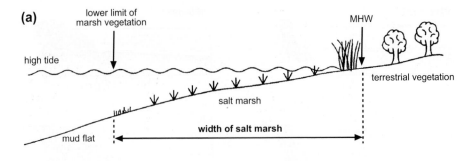

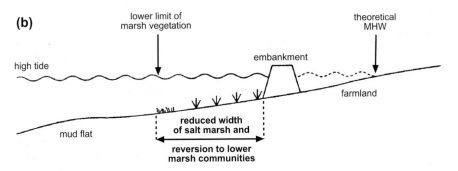

Figure 11.5 – Effect of embankment construction on intertidal habitats (a) before and (b) after construction. [From French, 1997.] [Copyright © 1997 Taylor & Francis Books, reproduced with permission.]

## 11.3.2 'Hard' shore-attached structures

**Groynes** can be constructed from concrete, rip-rap, steel or timber, and are designed to trap a portion of the longshore sediment transport to build out a buffering beach. The construction of groynes therefore aims to protect the coast against coastal erosion whilst maintaining a beach. Groynes are typically inserted perpendicular to the beach and often a series of groynes, referred to as a groyne field, is deployed to protect a large area from erosion (Figure 11.6). An important element in the design of groynes is the ratio between groyne length and groyne spacing. For sandy beaches a ratio of 1:4 is recommended, while for a gravel beach is smaller ratio of 1:2 is more appropriate. Another important factor is the length of the groynes relative to the width of the surf zone. The conventional practice is that the groyne length should be approximately 40–60% of the average surf zone width, allowing the structure to trap some, but not all, of the littoral drift (Komar, 1998). Once the area between two groynes is filled, the subsequent longshore transport will be able to pass around the end of the structure, which is still in the surf zone.

Figure 11.6 – (a) Two groynes on City Beach, Perth, Western Australia. Littoral drift is from top to bottom and has led to saturation of both groynes allowing sediment to spill around the tips of the groynes. [Copyright © 2000 Werner Barthel, reproduced with permission.] (b) Wooden groynes on a gravel beach, Snettisham, Norfolk, UK. [Photo G. Masselink.]

Groynes are not intended to completely block longshore sediment transport, but trap just that amount required for maintaining a 'healthy' beach between the structures. However, the trapping efficiency of a groyne field is often almost 100%, leading to down-drift sediment starvation and erosion. Often the response to this problem is to extend the groyne field in the down-drift direction, but in doing so the problem is not solved but merely transferred to down-coast neighbours. Another problem associated with groynes is that rip currents tend to form adjacent to the groynes. Not only do such currents present a hazard to swimmers, but during storms these rips may move sediment beyond the surf zone into water so deep that it cannot readily return to replenish the beach. The use of groynes in coastal protection has declined in recent years due to these problems and also their unsightly aspect. However, in some cases groynes may be the most suitable technique to combat local erosion, in particular if combined with beach nourishment (Section 11.3.4).

## 11.3.3 'Hard' offshore structures

**Breakwaters** are shore-parallel structures placed seaward of the shoreline and are designed to dissipate part of the incident energy, reducing the direct impact of storm waves. They are used for harbour design to reduce wave action in the harbour, but of more interest is their implementation in coastal protection. When used as a coastal-protection structure, breakwaters are placed just offshore from the surf zone and are referred to as **detached breakwaters**. They are usually built as a series, analogous to a groyne field, to protect a long stretch of coast. Important parameters in the design of a series of breakwaters are the gap distance between individual breakwaters, the length of the breakwaters and their offshore distance (Pope and Dean, 1986). To reduce construction costs and improve the aesthetic aspect of the structure, submerged breakwaters may be used with their tops located just below sea level. This method works well in microtidal environments but

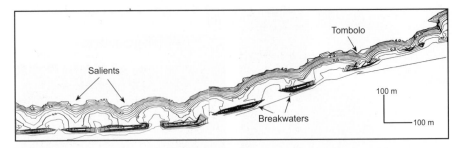

Figure 11.7 – Shoreline response to the placement of detached breakwaters on Elmer beach, Littlehampton, England. The diagrams shows the digital terrain model of the beach 6 months after construction of the breakwaters and indicates pronounced salients in the lee of the structures. Additional beach planform changes after initial 6 months were limited suggesting that the beach had assumed an equilibrium configuration. [Modified from King *et al.*, 2000.] [Copyright © 2000 Coastal Education & Research Foundation, reproduced with permission.]

when the tidal range is large, breakwaters submerged at high tide are left 'high and dry' during low tide.

Detached breakwaters provide a sheltered beach area behind them, with calmer water than found on the open coasts. Wave shadowing and wave refraction/diffraction patterns all contribute to the deposition of sediment in the lee of the structure (Figure 11.7). The effect of a detached breakwater on the coastline is similar to that of an offshore island and the shoreline may form a tombolo, a salient or present only limited modification of the shoreline configuration. The type of morphological response depends on the breakwater design parameters, with tombolo formation being promoted by relatively long breakwaters placed at a relatively short distance from the shore (Pope and Dean, 1986).

It seems that detached breakwaters provide a good alternative to groyne fields. They help to build out the beach and reduce coastal erosion, but they do not block the littoral drift (unless tombolos develop), nor induce offshore losses of sediment. The main problem associated with the construction of breakwaters is their large cost, because they are exposed to extreme wave action. Because they are placed in the most energetic part of the nearshore zone, they are also prone to failure, necessitating frequent maintenance.

## 11.3.4 'Soft' approach to coastal protection: Beach nourishment

**Beach nourishment** involves the artificial deposition of sediment (sand or gravel) on the beach or in the nearshore zone to advance the shoreline seaward (Box 11.2). As a rule of thumb, the size of the sand used for nourishment must be equal to or somewhat coarser than that of the local sediment to minimize rapid sediment loss offshore (Stive *et al.*, 1991). Beach

# Box 11.2 – Restoration of Miami Beach

Miami Beach is one of the most developed and, in financial terms, one of the most valuable coastlines in the world. The city is built on a narrow barrier island that has a history of dramatic shoreline changes, including opening and closing of tidal inlets. The development of Miami Beach from the 1920s onward occurred against a backdrop of continuing coastal erosion problems. Almost the entire beach frontage was privately owned and this resulted in the remedial works being largely haphazard and inconsistent. By 1965, after 40 years of largely uncontrolled development the Miami shoreline (both ocean and lagoon side) was almost destroyed. The loss of an important recreational beach, a steepening shoreface, pollution of nearshore reefs and backshore lagoons, combined to produce a major environmental disaster (Carter, 1988).

**Figure 11.8** – The restoration of Miami Beach by beach nourishment: (a) February 1978 just prior to the project; and (b) October 1979 after completion of the project. [From CERC, 1984.] [Reproduced with permission from U.S. Army Corps of Engineers.]

The solution to the coastal erosion problem was beach nourishment. This project took place from 1976 to 1981 and represents the largest beach nourishment scheme that has been undertaken in the United States (Figure 11.8). The nourishment project cost more than US$60 million and involved the placement of 13 million $m^3$ of sand dredged from offshore over a 16 km stretch of coastline (Komar, 1998). The nourishment produced a dry beach 55 m wide at an elevation of 3 m above mean low water. The objectives of the project were to provide an expanded recreational beach that additionally served to protect the highly-developed beachfront area from hurricane waves and storm surge.

The project is viewed as having been highly successful in that the nourished beach has functioned for more than 20 years and has survived two major hurricanes. The project has also been a huge success in economical terms. The cost for coastal protection and beach nourishment is US$3 million per year, while the income generated by the Miami Beach economy from foreign visits is estimated at US$2 billion per year.

nourishment treats the symptoms of coastal erosion (loss of beach) without addressing the real cause (deficit in the overall sediment budget and/or rising sea level). Therefore, sediment will continue to be lost from the beach and treatment will have to be repeated at regular intervals. To reduce sediment losses following beach nourishment, groynes may be placed at the boundaries of the nourished area.

One of the main problems with beach nourishment is the oft-perceived failure of nourishment schemes by the public. Immediately following placement of sand or shingle on the beach, waves will redistribute the sediment to form a more natural cross-shore profile. The first few post-nourishment storms will inevitably transport large quantities of sediment from the subaerial beach to the nearshore, resulting in a significant narrowing of the subaerial beach. This is often perceived by the general public as a failure of the nourishment scheme and a waste of money. However, the sediment is not 'lost', but often remains resident in the nearshore. Komar (1998) emphasizes the need for better monitoring programs to follow beach nourishment, and also the need for more effective public education as to what is expected from beach-nourishment projects.

Beach nourishment is often the preferred option for coastal protection, because the method is relatively cost-efficient and the results appear more natural. Beach nourishment is also more aligned with sustainable coastal management because there are few adverse effects. As a result, beach nourishment is now very widely used (cf., Bird, 1996).

## 11.3.5 'Soft' approach to coastal protection: Construction of artificial dunes

Coastal dunes provide a natural buffer to coastal erosion. This has been well recognized in the Netherlands, where dunes are an essential part of

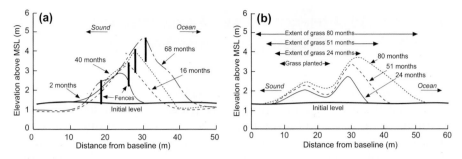

**Figure 11.9** – Construction of artificial dunes. (a) Sand accumulation by a series of four single-fence lifts, Outer Banks, North Carolina. (b) Sand accumulation following planting with American beachgrass, Ocracoke Island, North Carolina. [Modified from CERC, 1984.]

coastal protection, and where maintenance of a 'healthy' dune system is considered a priority for coastal management (Louise and van der Meulen, 1991). The natural protection of sandy coastlines can be improved by encouraging the growth of dunes in areas where dunes are either absent or discontinuous.

**Dune construction** is usually achieved through the use of fences (brush-wood or wooden palings) that are strategically placed at the back of the beach. The fences disrupt the airflow, thereby promoting sediment deposi-tion, generally on both sides of the fence. CERC (1984) recommends installing straight fences parallel to the coast with a 'porosity' of 50%, preferably coinciding with the natural dune vegetation or foredune line prevalent in the area. Once the fence is at 'full capacity' (*i.e.*, almost buried by sand), dune growth can be maintained by lifting the fences. Fence-built dunes must be stabilized with vegetation or the fence will deteriorate and release the sand. A well-designed and implemented fence deployment in an area with abundant aeolian sediment transport can achieve rates of vertical dune growth in excess of 1 m yr$^{-1}$ (Figure 11.9).

Dunes can also be constructed by planting vegetation. A large number of aspects need to be considered, including selection of the plant species, tim-ing of planting, planting density, planting width, use of fertilizer (Ranwell and Boar, 1968). Generally, dune development induced by planting is not as rapid as that due to fencing, but once established, most dunes and dune plants grow steadily.

Closely associated with dune construction is **dune stabilization**, which is directed towards securing bare sand surface in the dunes and repairing gaps in coastal dune ridges. Both fences and vegetation are frequently used to stabilize dunes. Gaps in the dunes may also be filled using earthmoving equipment followed by planting to stabilize the bare sand.

## 11.4 IMPACT OF SEA-LEVEL RISE ON COASTAL ENVIRONMENTS

In Chapter 2 we showed that during the last century, global temperatures have increased by 0.3–0.6°C and eustatic sea level has risen by *c*. 0.2 m. It is likely that these two trends are causally related and that the continuing warming of the Earth will result in sustained sea-level rise, due to a combination of thermal expansion of ocean water and melting of glaciers and ice caps. The rise in global temperature is generally attributed to increased atmospheric concentrations of so-called Greenhouse gases (mainly $CO_2$), due to the burning of fossil fuels. Since humankind is unlikely to stop burning fossil fuels in the near future, concentrations of Greenhouse gases in the atmosphere are expected to continue to rise over the next century. Climate models estimate that the global temperature may therefore rise by another 2°C by 2100, with sea levels rising 0.11–0.77 m over the same period (Church *et al.*, 2001).

The ongoing and accelerating rise in eustatic sea level is of major concern to coastal nations for obvious reasons. Sea-level rise may lead to coastal erosion, flooding of coastal wetland areas and salt contamination of coastal aquifers. As a result, human activities in the coastal zone will be disrupted, traditional transport routes will be severed and agricultural land will become unsuitable. To plan our future use of the coastal zone and mitigate sea-level induced problems, we need to be able to predict the response of coastal environments to rising sea level.

### 11.4.1 Effect of sea-level rise on coasts in general

One way of assessing the potential risk of rising sea levels is to produce **coastal inundation maps**, which indicate areas threatened by sea-level rise (Figure 11.10). Such maps are generated by projecting a predicted sea-level rise (for example 2 m rise over 200 years) onto a contour map and shading those areas that would be flooded by such a sea-level rise. Alternatively, such maps just indicate low-lying land and labels these as being at risk from being flooded. Although such analysis appeals to the general public and may provide a compelling argument for increased spending on coastal protection measures, it has a weak scientific basis. The response of the coasts to rising sea level can not be defined simply in terms of inundation of the sea upon the land, nor by just shifting the land-sea contour by an amount corresponding to the projected vertical increase in global sea level (Bijlsma *et al.*, 1996). Such an approach ignores the fact that coastal environments are morphodynamic systems that respond to, rather than passively undergo, changing environmental conditions.

In reality, coastal environments respond to rising sea levels by redistributing coastal sediment. Usually, a rising sea level will result in coastal erosion and retreat of the shoreline. However, it is also possible that the shoreline

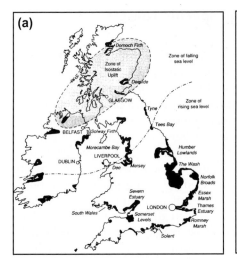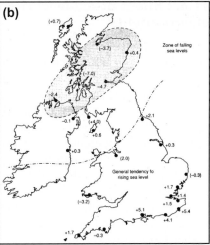

**Figure 11.10** – Map of Great Britain showing: (a) areas of isostatic uplift (grey shading) and those threatened by sea-level rise (black shading); and (b) rates of relative sea-level rise in mm per year. [From Briggs et al., 1997; modified from Boorman et al., 1989.] [Copyright © 1997 Taylor & Francis Books, reproduced with permission.]

position is maintained or even moved seaward. A crucial factor is played by the sediment budget. A coastal region with a positive sediment budget may accrete, rather than erode under rising sea-level conditions. Similarly, coastal erosion due to rising sea levels will be exacerbated in case of a negative sediment budget.

Hence, it should be acknowledged that the impact of sea-level rise on coastal environments is complex and by no means easy to assess. The coastal response is strongly site-dependent and predictions of coastal response to rising sea levels require careful analyses of the local coastal mor-phodynamics and sediment budget situation. Even if the coastal system is thought to be well understood, predictions may not be very reliable due to feedback mechanisms and thresholds within the systems, and the uncertain-ties associated with the future environmental conditions (rate of sea-level rise and climate change). In addition, geomorphological responses to a ris-ing sea level will also reflect other aspects of climate change, such as changes in storminess and wave direction. Of particular relevance to nearshore sediment budgets is the possibility that changes in the incident wave climate, due to climate change, may modify the rate and direction of the littoral drift.

We will now move on to briefly discuss the potential impact of rising sea levels on the five main types of coastal environments we have identified in this book (river-dominated, tide-dominated, wave-dominated, coral reefs and rocky coasts).

## 11.4.2 Effect of sea-level rise on specific coastal environments

River-dominated coastal environments (deltas) develop when sediment brought down to the coast by rivers accumulates more rapidly than can be removed by waves, tides and currents. Most deltas are subsiding under the weight of accumulating sediment (sediment isostasy), a process that often is enhanced by compaction and artificial groundwater extraction. Delta survival can therefore be considered a battle between fluvial sedimentation versus coastal emergence (Baumann *et al.*, 1984). Along a natural delta coast, most of the shoreline is prograding because the sediment discharge by the river tributaries exceeds land losses induced by rising sea levels. Eroding sections do occur along natural delta coasts, but are mainly confined to locations of former tributary mouths that have been abandoned following channel avulsion or channel switching. The vast majority of the world's delta systems are, however, no longer entirely natural, and have a much reduced sediment discharge due to damming of rivers and extraction of sediment from the riverbed. For example, sediment transport by the Nile, Indus and Ebro rivers has been reduced by 95% and in the Mississippi by half in the past 200 years, mostly since 1950 (Day *et al.*, 1997). Many deltas are therefore currently experiencing severe and widespread coastal erosion problems. The future does not bode well for these deltas. It is unlikely that the fluvial sediment supply will be turned on again. At the same time, the land surface will continue to fall and eustatic sea level will continue to rise. It is very likely that these deltas will continue to erode, at increasing rates, as sea level rises.

Tide-dominated coastal environments comprise estuaries, tidal flats, salt marshes and mangroves. According to Bird (1993), as sea level rises, estuaries will tend to widen and deepen. Tides will penetrate further upstream, tidal range may increase and tidal currents may increase in strength as well. However, Pethick (1993) demonstrated that along the southeast coast of Britain, where relative sea-level rise is 4–5 mm yr$^{-1}$, estuarine channels are becoming wider and shallower. Salt marsh and mangrove environments (and to a lesser extent tidal flats) grow through processes of vertical accretion of sediment on the wetland surface and progradation at the edges. The response of these environments to sea-level rise is affected by organic and inorganic sediment supply, and primarily depends on the rate of vertical accretion in relation to the rate of sea-level rise (French, 1993; Woodroffe, 1995). If the vertical accretion rate exceeds the rate of sea-level rise, salt marshes and mangroves can continue to accrete and keep up with the rising sea level. If the sediment supply is very large, progradation may even occur under transgressive conditions. On the other hand, if the accretion rate is less than the rate of sea-level rise, salt marshes and mangroves will be flooded by the rising sea level resulting in increased wave action (due to increased water depths). Subsequently, this may lead to erosion of the surface, assisted by widening and deepening of the tidal creeks.

Wave-dominated coastlines generally retreat as a result of sea-level rise, either through erosion, migration or overstepping of the barrier (Section 8.5.3). Barrier migration is promoted on low-gradient substrates, while an erosional response is more common on steep substrates (Roy *et al.*, 1994). Barrier migration also appears the dominant mode of response to rising sea levels for gravel barriers (Orford *et al.*, 1995). Barriers can also prograde under transgressive conditions if there is a large sediment supply to the coast. For example, the barrier system in the Netherlands underwent extensive coastal progradation during the mid-Holocene despite rising sea levels (Beets *et al.*, 1992). This occurred because sufficient sediment was available either through longshore transport from the main river systems or onshore transport from the bed of the North Sea. So again, we see that the rate of sediment supply to the coast plays a crucial role in determining the response of wave-dominated sandy and gravel coasts to sea-level rise. Bird (1993) argues that with future sea-level rise there will be tendencies for eroding shorelines to erode further, stable shorelines to begin to erode, and accreting shorelines to wane or stabilize.

The response of coral reefs to sea-level rise depends primarily on whether vertical reef growth can keep up with the rising sea level. Reef accretion rates range from less than 1 mm $yr^{-1}$ to a maximum slightly in excess of 10 mm per year (Spencer, 1994). A rate of 10 mm $yr^{-1}$ is commonly taken as the maximum sustained vertical reef accretion rate (Buddemeier and Smith, 1988). The present best estimates for global sea-level rise over the next century are well within the range of typical reef growth rates. However, coral reef ecosystems are vulnerable to a wide range of environmental stresses (Viles and Spencer, 1995). Coral reefs should generally be able to keep up with the predicted rate of sea-level rise, provided that other factors such as increased water temperature and damaging anthropogenic influences are not acting simultaneously (McLean *et al.*, 2001).

Coral atolls and reef islands are considered particularly vulnerable to sea-level rise (and climate change), due to their low-lying nature. The response of these environments to sea-level rise depends on the balance between reef growth, sediment supply from reef flat to island and the sea-level rise. Differences in response can be further expected between different localities, particularly between islands within and beyond cyclone regions. Recent reviews on coral islands have emphasized their variability and resilience (McLean and Woodroffe, 1993). Despite this, coral islands remain among the most sensitive environments to long-term sea-level rise and climate change. This is especially the case where these effects are superimposed on destructive short-term effects such as hurricanes, damaging human activities and declining environmental quality.

The effect of sea-level rise on rocky coasts should theoretically be the least difficult to predict. As sea level rises, deepening nearshore waters will submerge part of the existing intertidal zone, thereby increasing wave attack at the base of the cliff and hence cliff erosion (Bird, 1993). This will be particularly relevant if the rise in sea level is accompanied by an increase in rainfall (promoting cliff failure) and storminess. However, on resistant, vertical cliffs, the sea level may just rise vertically without any effect on cliff erosion. In any area where cliff erosion is related to a 'conveyor belt' system, as in southeast England (Box 9.1), the situation is likely to be much more complex because of the linkage between cliff and beach dynamics. Relatively few studies have attempted to predict the response of coastal cliffs to sea-level rise, and those that do highlight the complexity of factors involved with such predictions (Bray and Hooke, 1997).

## 11.4.3 Managing sea-level rise

It is now well established that sea level is rising and, due to global warming, is expected to rise at an accelerated rate over the next century. The majority of the world's coastlines are already suffering from erosion and coastal retreat is likely to become even more widespread in the future. An important concept in the management of the effects of sea-level rise is **coastal vulnerability**, which refers to the ability to cope with the consequences of sea-level rise (Box 11.3). Vulnerable coasts, for example densely populated barrier islands, obviously require a different management approach to less vulnerable coasts.

## Box 11.3 – Coastal Vulnerability

Recently, Klein and Nichols (1999) proposed an innovative framework for determining the vulnerability of coastal systems to climate change and sea-level rise (Figure 11.11). The framework makes an important distinction between **natural vulnerability** and **socio-economic vulnerability**.

Analysis of coastal vulnerability starts with a notion of the natural system's susceptibility to the biophysical effects of sea-level rise by looking at the coastal system's potential to be affected by sea-level rise. For example, a subsiding delta is expected to be more susceptible to the adverse effects of sea-level rise than an emerging rock coast. The natural capacity of the coastal system to deal with the impacts of sea-level rise is also important in determining the natural vulnerability. It consists of resistance, which describes the ability of a coastal system to avoid disturbance, and resilience, which is a measure of the system's capacity to respond to the consequences of disturbance. Many natural features contribute to coastal resilience by providing ecological buffers (coral reefs, salt marshes and mangrove forests) and morphological protection (sand and gravel beaches, barriers and coastal dunes) (McLean et al., 2001).

The natural response of a coastal system to a disturbance is termed autonomous adaptation. An example of such adaptation is the onshore migration of barrier systems, which continues to provide protection to the hinterland from flooding and erosion, in the face of a rising sea level. Human activities often reduce resistance and resilience, thereby increasing natural vulnerability. However, planned adaptation can reduce natural vulnerability by enhancing the system's resistance and resilience. An example of planned adaptation is managed retreat. This involves

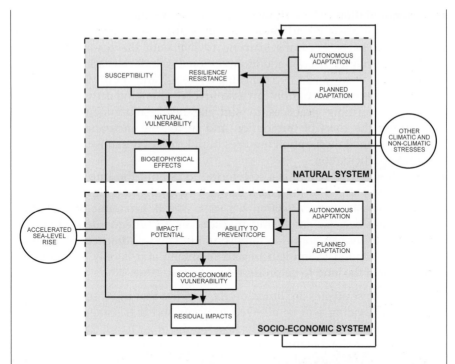

**Figure 11.11** – A conceptual framework for coastal vulnerability assessment. [Modified from Klein and Nicholls, 1999.]

allowing the coastline to recede to a new line of defence to restore natural coastal processes and thereby natural coastal systems, such as mud flats, salt marsh areas and dunes. Once established, these natural systems can help to meet protection standard further inland so that the human contribution to safety provision can be reduced. Thus managed retreat helps to reduce coastal vulnerability by increasing coastal resilience.

There are four principal management options available to cope with coastal erosion due to sea-level rise:

- **Do nothing** – This option is only viable if the coastline under question is undeveloped and nothing is at stake by giving up the land to coastal erosion.
- **Managed retreat** – This option, also referred to as realignment, generally involves the relocation of coastal communities and/or industry, with a prohibition on further development through the use of set-back limits. In this strategy, risks are minimized and costs of protection are avoided. However, social and economic costs associated with relocation and compensation are potentially high. The retreat option requires a strong governmental role with supportive legislation.

- **Accommodation** – This allows continued occupancy and use of vulnerable coastal areas by adapting to, rather than protecting fully against, adverse impacts. It means learning to live with the sea-level rise and coastal flooding. Accommodation options include elevating buildings, enhancing storm and flood warning systems, and modifying drainage. The accommodation option can also involve changing activities, such as changing farming practices to suit the new environment, or simply accepting the risks of inundation and increasing insurance premiums. The accommodation option requires high levels of organization and community participation.
- **Protection** – This option is also referred to as **hold-the-line** and involves physically protecting the coast through hard engineering structures or beach nourishment. Protection has clear social, economic and political advantages because assets and investments are safeguarded while economic activity can continue largely unhindered. Protection is the most expensive option to implement and maintain, and is only economically justifiable if the land to be protected is of great value.

The first three strategies are based on the premise that increased land losses and coastal flooding will be allowed to occur and that some coastal functions and values will be changed or lost. On the other hand, these strategies help to maintain the dynamic nature of the coast and thus allows it to adjust to rising sea levels naturally. It is beneficial to allow as many coastal regions to retreat naturally as possible. This is because erosion of these natural areas will liberate sediments, which may lessen the impact of sea-level rise on those areas that are not allowed to retreat naturally. Hence, the first three options are most sustainable from a geomorphological point of view (although not necessarily from a socio-economic perspective).

Which management option is most suitable on any given coast depends more on socio-economic and political factors, than on physical factors. Economic evaluation principles, especially **risk assessment** and **benefit-cost analysis**, can provide useful tools in deciding whether to retreat, accommodate or protect. Drawing up a balance sheet of benefits and costs for shoreline protection is, however, not trivial. The principal problem is not doing the sums, but allocating price tags to intangible quantities like scenic beauty, pollution adsorption capacity, wild life, and even human life. The best that can be expected is some kind of economic appraisal, assessing the costs and benefits of various protection strategies.

## 11.5 SUMMARY

- The aims of coastal management are to facilitate the human use of the coastal zone, whilst minimizing the environmental impacts of such use, and to protect human interests at the coast from natural and human related processes. Key attributes of coastal zone management are integration and sustainability.

- The practical element of coastal zone management is concerned with the design, implementation and maintenance of coastal protection works. The two main approaches to coastal protection are 'hard' and 'soft' engineering.
- The ongoing and accelerating rise in eustatic sea level is of major concern to coastal nations for obvious reasons. To plan future use of the coastal zone and mitigate sea-level induced problems, we need to predict the response of coastal environments to rising sea levels. Unfortunately, the impact of sea-level rise on coastal environments is complex and by no means easy to assess. Coastal responses are strongly site-dependent and predictions require careful analyses of the local coastal morphodynamics and sediment budget situation.
- There are four principal management options available to deal with coastal erosion: (1) do nothing; (2) managed retreat; (3) accommodation; and (4) protection. Which management option is most suitable on any given coast depends more on socio-economic and political factors, than on physical factors.

## 11.6 FURTHER READING

Bird, E.C.F., 1993. *Submerging coasts: The Effects of a Rising Sea Level on Coastal Environments*. John Wiley, Chichester, UK. [A useful review of the response of coastlines to rising sea level.]

Kay, R. and Alder, J., 1999. Coastal Planning and Management. E & FN Spon, Routledge, London. [Up-to-date and comprehensive book on coastal management and planning with a large number of case studies from around the world.]

Komar, P.D., 1998. *Beach Processes and Sedimentation*. Prentice Hall, New Jersey. [Chapter 12 of this text provides extensive detail on coastal protection measures.]

Nordstrom, K.F., 2000. Beaches and Dunes of Developed Coasts. Cambridge University Press, Cambridge. [Topical and authoritative book discussing the effect of human activity on sandy coastlines.]

Viles, H. and Spencer, T., 1995. *Coastal Problems*. Arnold, London. [Excellent text dealing specifically with environmental problems in the coastal environment.]

# REFERENCES

Aagaard, T. and Masselink, G., 1999. Chapter 4: The surf zone. In: A.D. Short (editor), *Beach Morphodynamics*, Wiley and Sons, London, 72–118.

Allen, G.P., Salomon, J.C., Bassoulet, P., Du Penhoat, Y., and De Grandpre, C., 1980. Effects of tides on mixing and suspended sediment transport in macrotidal estuaries. *Sedimentary Geology*, **26**, 69–90.

Allen, J.R.L., 1984. *Sedimentary Structures: Their Character and Physical Basis.* Elsevier, Amsterdam.

Allen, J.R.L., 1985. *Principles of Physical Sedimentology*. George Allen and Unwin, London.

Allen, J.R.L., 1994. Fundamental properties of fluids and their relation to sediment transport. In: K. Pye (editor), *Sediment Transport and Depositional Processes*, Blackwell Scientific Publications, Oxford, 25–60.

Allen, J.R.L., 1996. Shoreline movement and vertical textural patterns in salt marsh deposits: Implications of a simple model for flow and sedimentation over tidal marshes. *Proceedings of the Geologist's Association*, **107**, 15–23.

Allen, P.A., 1997. *Earth Surface Processes*. Blackwell Science, Oxford.

Allison, R.J. (editor), 1990. *Landslides of the Dorset Coast*. British Geomorphological Research Group, Field Guide, 128.

Anthony, E.J., 1995. Beach-ridge development and sediment supply: examples from West Africa. *Marine Geology*, **129**, 175–186.

Arens, S.M., 1994. *Aeolian Processes in the Dutch Foredunes*. PhD Thesis, University of Amsterdam.

Augustinus, P.G.E.F., 1989. Cheniers and chenier plains: A general introduction. *Marine Geology*, **90**, 219–229.

Avoine, J., 1981. *L'Estuare de la Seine: Sediments et Dynamique Sedimentaire.* These, Docteur de Specialite. Université de Caen, France.

Bagnold, R.A., 1941. *The Physics of Blown Sand and Desert Dunes.* Methuen, London.

Bagnold, R.A., 1963. Mechanics of marine sedimentation. In: M.N. Hill (editor), *The Sea*, Vol. 3, Wiley-Interscience, New York, 507–528.

Bagnold, R.A., 1966. *An Approach to the Sediment Transport Problem from General Physics*. Professional Paper no. 422-I, United States Geological Survey.

Bailard, J.A., 1982. Modeling on-offshore sediment transport in the surf zone. *Proceedings 18th Coastal Engineering Conference*, ASCE, 1419–1438.

Bartrum, J.A., 1916. High water rock platforms: A phase of shoreline erosion. *Transactions New Zealand Institution*, **48**, 132–134.

Bascom, W.H., 1953. Characteristics of natural beaches. *Proceedings 4th International Conference on Coastal Engineering*, ASCE, 163–180.

Bascom, W.N., 1951. The relationship between sand size and beach-face slope. *Transactions, American Geophysical Union*, **32**, 866–874.

Bascom, W.H., 1980. *Waves and Beaches*. Anchor Books, Garden City.

Bates, C.C., 1953. Rational theory of delta formation. *Bulletin of the American Association of Petroleum Geologists*, **37**, 2119–2161.

Battjes, J.A., 1974. Surf similarity. *Proceedings 14th International Conference on Coastal Engineering*, ASCE, 466–480.

Bauer, B.O. and Allen, J.R., 1995. Beach steps: An evolutionary perspective. *Marine Geology*, **123**, 143–166.

Bauer, B.O. and Sherman, D.J., 1999. Coastal dune dynamics: Problems and prospects. In: A.S. Goudie, I. Livingstone and S. Stokes (editors), *Aeolian Environments, Sediment and Landforms*, Wiley and Sons, Chichester, 71–103.

Bauer, B.O., Yi, J., Namikas, S.L. and Sherman, D.J., 1998. Event detection and conditional averaging in unsteady aeolian systems. *Journal of Arid Environments*, **39**, 345–375.

Baumann, R.H., Day, J.W. and Miller, C., 1984. Mississippi delta wetland survival: Sedimentation versus coastal submergence. *Science*, **224**, 1093–1095.

Bayliss-Smith, T.P., 1988. The role of hurricanes in the development of reef islands, Ontong Java Atoll, Solomon Islands. *Geographical Journal*, **154**, 377–391.

Been, K., and Sills, G.C., 1981. Self weight consolidation of soft soils: An experimental and theoretical study. *Geotechnique*, **31**, 519–535.

Beets, D.J., van der Valk, L. and Stive, M.J.F., 1992. Holocene evolution of the coast of Holland. *Marine Geology*, **103**, 423–443.

Berger, A.L., 1992. Astronomical theory of paleoclimates and the last glacial-interglacial cycle. *Quaternary Science Reviews*, **11**, 571–582.

Bijlsma, L., Ehler, C.N., Klein, R.T.J., Kulshrestha, S.M., McLean, R.F., Mimura, N., Nicholls, R.J., Nurse, L.A., Perez Nieto, H., Stakhiv, E.Z., Turner, R.K. and Warrick, R.A., 1996. Chapter 9: Coastal zones and small islands. In: Intergovernmental Panel on Climate Change, *Climate Change 1995: Impacts, Adaptations, and Mitigation of Climate Change: Scientific-Technical Analyses*, Contribution of Working Group II to the Second Assessment Report of the Intergovernmental Panel on Climate Change, Cambridge University Press, Cambridge, 289–324.

Bird, E.C.F., 1984. *Coasts: An Introduction to Coastal Geomorphology* (3rd Edition). Blackwell, Oxford.

Bird, E.C.F., 1985. *Coastline Changes: A Global Review*. Wiley-Interscience, Chichester, UK.

Bird, E.C.F., 1993. *Submerging coasts: The Effects of a Rising Sea Level on Coastal Environments*. Wiley and Sons, Chichester, UK.

Bird, E.C.F., 1996. *Beach Management*. Wiley and Sons, Chichester.

Bird, E.C.F., 2000. *Coastal Geomorphology: An Introduction*. Wiley and Sons, Chichester.

Bishop, P. and Cowell, P., 1997. Lithological and drainage network determinants of the character of drowned, embayed coastlines. *Journal of Geology*, **105**, 685–699.

Boggs, S., 1995. *Principles of Sedimentology and Stratigraphy* (2nd Edition). Prentice Hall, New Jersey.

Boorman, L.A., Goss-Custard, J.D. and McGrorty, S., 1989. *Climate Change, Rising Sea Level and the British Coast*, London, HMSO.

Bowen, A.J., Inman, D.L. and Simmons, V.P., 1968. Wave 'setdown' and setup. *Journal of Geophysical Research*, **73**, 2569–2577.

Brander, R.W., 1999. Field observations on the morphodynamic evolution of a low-energy rip current system. *Marine Geology*, **157**, 199–217.

Brander, R.W. and Short, A.D., 2000. Morphodynamics of a large-scale rip current system, Muriwai Beach, New Zealand. *Marine Geology*, **165**, 27–39.

Bray, M.J., 1997. Episodic shingle supply and the modified development of Chesil beach. *Journal of Coastal Research*, **13**, 1035–1049.

Bray, M.J. and Hooke, J.M., 1997. Prediction of soft-cliff retreat with accelerating sea-level rise. *Journal of Coastal Research*, **13**, 453–467.

Bray, M., Hooke, J. and Carter, D., 1997. Planning for sea-level rise on the south coast of England: Advising the decision-makers. *Transactions Institute of British Geographers*, **22**, 13–30

Bretschneider, C.L., 1952. The generation and decay of wind waves in deep water. *Transactions American Geophysical Union*, **33**, 381–389.

Briggs, D., Smithson, P., Addison, K. and Atkinson, K., 1997. *Fundamentals of the Physical Environment* (2nd Edition). Routledge, London.

Brunsden, D. and Goudie, A., 1997. *Classic Landforms of the West Dorset coast*. Geographical Association, Sheffield.

Bruun, P., 1954. *Coastal Erosion and the Development of Beach Profiles*. U.S. Army Beach Erosion Board, Technical Memorandum, 44.

Bruun, P., 1962. Sea level rise as a cause of shore erosion. *Journal of Waterway, Port, Coastal and Ocean Engineering*, ASCE, **88**, 117–130.

Bruun, P., 1988. The Bruun rule of erosion: A discussion on large-scale two and three dimensional usage. *Journal of Coastal Research*, **4**, 626–648.

Bryant, E., 2001. *Tsunami: The underrated hazard*. Cambridge University Press, Cambridge.

Buddemeier, R.W. and Smith, S.V., 1988. Coral reef growth in an era of rapidly rising sea level: Predictions and suggestions for long-term research. *Coral Reefs*, **7**, 51–56.

Carr, A.P. and Graff, J., 1982. The tidal immersion factor and shore platform development. *Transactions of the Institute of British Geographers*, **7**, 240–245.

Carter, R.W.G., 1988. *Coastal Environments*. Academic Press, London.

Carter, R.W.G. and Orford, J.D., 1993. The morphodynamics of coarse clastic beaches and barriers: A short- and long-term perspective. *Journal of Coastal Research*, Special Issue 15, 158–179.

Carter, R.W.G. and Woodroffe C.D., 1994. Coastal evolution: An introduction. In: R.W.G. Carter and C.D. Woodroffe (editors), *Coastal Evolution*, Cambridge University Press, Cambridge, 1–31.

Cartwright, D.E., 1969. Deep sea tides. *Science Journal*, **5**, 60–67.

Cartwright, D.E., 1999. *Tides: A Scientific History*. Cambridge University Press, Cambridge.

CERC, 1984. *Shore Protection Manual* (4th Edition). Coastal Engineering Research Center, Waterway Experiment Station, Corps of Engineers, Vicksburg.

Chandler, J.H. and Brunsden, D., 1995. Steady state behaviour of the Black Ven mudslide: The application of archival analytical photogrammetry to studies of landform change. *Earth Surface Processes and Landforms*, **20**, 255–275.

Chappell, J., 1980. Coral morphology, diversity and reef growth. *Nature*, **286**, 249–252.

Chappell, J. and Shackleton, N.J., 1986. Oxygen isotopes and sea level. *Nature*, **324**, 137–140.

Chappell. J., Rhodes, E.G., Thom, B.G. and Wallenski, E., 1982. Hydro-isostasy and the sea-level isobase of 5500 BP in north Queensland, Australia. *Marine Geology*, **49**, 81–90.

Church, J.A., Gregory, J.M., Huybrechts, P., Kuhn, M., Lambeck, K., Nhuan, M.T., Qin, D. and Woodworth, P.L., 2001. Chapter 11: Changes in sea level. In: Intergovernmental Panel on Climate Change, *Climate Change 2001: the Scientific Basis*, Contribution of Working Group I to the Third Assessment Report of the Intergovernmental Panel on Climate Change, Cambridge University Press, Cambridge, 639–693.

Cicin-Sain, B. and Knecht, R.W., 1998. *Integrated Coastal and Ocean Management.* Island Press, Washington, D.C.

Clarke, J.A., Farrell, W.E. and Peltier, W.R., 1978. Global changes in postglacial sea level: A numerical calculation. *Quaternary Research*, **9**, 265–278.

Coco, G., O'Hare, T.J. and Huntley, D.A., 2000. Investigation of a self-organisation model for beach cusp formation and development. *Journal of Geophysical Research*, **105**, 21991–22002.

Cohen, J.E., Small, C., Mellinger, A., Gallup, J. and Sachs, J., 1997. Estimates of coastal populations, *Science*, **278**, 1211–1212.

Colella, A. and Prior, D.B., 1990. *Coarse-grained Deltas.* Special Publication Number 10 of the International Association of Sedimentologists, Blackwell Science, Oxford.

Coleman, J.M., 1982. *Deltas: Processes of Deposition and Models for Exploration.* International Human Resources Development Corporation, Boston.

Coleman, J.M., Walker, H.J., and Grabau, W.E., 1998. Sediment instability in the Mississippi River delta. *Journal of Coastal Research*, **14**, 872–881.

Cooper, J.A.G., 1994. Lagoons and microtidal coasts. In: R.W.G. Carter and C.D. Woodroffe (editors), *Coastal Evolution*, Cambridge University Press, Cambridge, 219–266.

Corey, A.T., 1949. *Influence of shape on the fall velocity of sand grains.* Unpublished MS thesis, A&M College, Colorado.

Cowell, P.J. and Kench, P.S., 2000. The morphological response of atoll islands to sea-level rise. Part 1: Modifications to the Shoreface Translation Model. *Journal of Coastal Research*, Special Issue 34, 633–644.

Cowell, P.J. and Thom, B.G., 1994. Morphodynamics of coastal evolution. In: R.W.G. Carter and C.D. Woodroffe (editors), *Coastal Evolution*, Cambridge University Press, Cambridge, 33–86.

Cowell, P.J., Hanslow, D.J. and Meleo, J.F., 1999. The shoreface. In: A.D. Short (editor), *Handbook of Beach and Shoreface Morphodynamics*, Wiley and Sons, Chichester, 39–71.

Dalrymple, R.W., Zaitlin, B.A., and Boyd, R., 1992. Estuarine facies models: Conceptual basis and stratigraphic implications. *Journal of Sedimentary Petrology*, **62**, 1130–1146.

Daly, R.A., 1915. The glacial-control theory of coral reefs. *American Academy of the Arts and Science*, **51**, 155–251.

Dana, J.D., 1849. *Geology.* Putnam, New York, Report US Exploration expedition 1838–1842, 10, 35–38.

Darwin, C., 1842. *The Structure and Distribution of Coral Reefs.* Smith, Elder and Co., London.

Davies, J.L., 1980, *Geographical Variation in Coastal Development* (2nd Edition). Longman, New York.

Davis, R.A., 1994. Barrier island systems – a geologic overview. In: R.A. Davis (editor), *Geology of Holocene Barrier Island Systems*, Springer-Verlag, New York, 1–46.

Davis, R.A. and Hayes, M.O., 1984. What is a wave-dominated coast? *Marine Geology*, **60**, 313–329.

Davis, R.A., Fox, W.T., Hayes, M.O. and Boothroyd, J.C., 1972. Comparison of ridge and runnel systems in tidal and non-tidal environments. *Journal of Sedimentary Petrology*, **42**, 413–421.

Day, J.W., Martin, J.F., Cordoch, L. and Templet, P.H., 1997. System functioning as a basis for sustainable management of deltaic ecosystems. *Coastal Management*, **25**, 115–153.

Dean, R.G., 1973. Heuristic models of sand transport in the surf zone. *Proceedings 1st Australian Coastal Engineering Conference*, IEA, 208–214.

Dean, R.G., 1977. *Equilibrium Beach Profiles: US Atlantic and Gulf Coasts*. Ocean Engineering Technical Report, 12.

Dean, R.G., 1987. Coastal sediment processes: Toward engineering solutions. *Coastal Sediments'87*, ASCE, 1–24.

Dean, R.G., 1991. Equilibrium beach profiles: Characteristics and applications. *Journal of Coastal Research*, **7**, 53–84.

Defant, A., 1958. *Ebb and Flow*. The University of Michigan Press.

Defant, A., 1961. *Physical Oceanography*. Volume 2, Pergamon Press.

Dyer, K.R., 1986. *Coastal and Estuarine Sediment Dynamics*. Wiley and Sons, Chichester.

Dyer, K.R., 1994. Estuarine sediment transport and deposition. In: K. Pye (editor), *Sediment Transport and Depositional Processes*. Blackwell Scientific Publications, Oxford, 25–60.

Dyer, K.R., 1998. *Estuaries: A Physical Introduction*. Wiley and Sons, Chichester.

Edgren, G., 1993. Expected economic and demographic developments in coastal zones world-wide. *World Coast '93*, National Institute for Coastal and Marine Management, Coastal Zone Management Centre, Noordwijk, The Netherlands, 367–370.

Eisma, D., 1993. *Suspended Matter in the Aquatic Environment*. Springer-Verlag, Berlin.

Elliott, T., 1986. Deltas. In: H.G. Reading (editor), *Sedimentary Environments and Facies*, Blackwell Science, Oxford.

Emery, K.O. and Kuhn, G.G., 1982. Sea cliffs: Their processes, profiles and classification. *Geological Society of America Bulletin*, **93**, 644–654.

Eronen, M., 1983. Late Weichselian and Holocene shore displacement in Finland. In: D.E. Smith and A.G. Dawson (editors), *Shorelines and Isostasy*, Academic Press, London, 183–207.

Eysink, W.D., 1990. Morphological response of tidal basins to changes. *Proceedings 22nd International Conference on Coastal Engineering*, ASCE, 1948–1961.

Fairbanks, R.G., 1989. A 17,000-year glacio-eustatic sea level record: Influence of glacial melting rates on the Younger Dryas event and deep-ocean circulation. *Nature*, **342**, 637–642.

Fairbridge, R.W., 1961. Eustatic changes in sea-level. *Physics and Chemistry of the Earth*, **4**, 99–185.

Fairbridge, R.W., 1983. Isostasy and eustasy. In: D.E. Smith and A.G. Dawson (editors), *Shorelines and Isostasy*, Academic Press, London, 3–25.

Flood, P.G., 1986. Sensitivity of coral cays to climatic variations, southern Great Barrier Reef. *Coral Reefs*, **5**, 13-18.

Folk, R.L. and Ward, W., 1957. Brazos River bar: A study in the significance of grain size parameters. *Journal of Sedimentary Petrology*, **27**, 3–26.

Francis, J.R.D., 1973. Experiments on the motion of solitary grains along the bed of a water stream. *Philosophical Transactions of the Royal Society, London (A)*, **332**, 443–471.

French, J.R., 1993. Numerical simulation of vertical marsh growth and adjustment to accelerated sea-level rise, North Norfolk, UK. *Earth Surface Processes and Landforms*, **18**, 63–81.

French, J.R., Clifford, N.J., and Spencer, T., 1993. High frequency flow and suspended sediment measurements in a tidal wetland channel. In: N.J. Clifford, J.R. French, and J. Hardisty (editors), *Turbulence: perspectives on flow and sediment transport*, Wiley and Sons, Chichester, 249–278.

French, J.R., Spencer, T., Murray, A.L. and Arnold, A.S., 1995. Geostatistical analysis of sediment deposition in two small tidal wetlands, Norfolk, U.K. *Journal of Coastal Research*, **11**, 295–570.

French, P.W., 1997. *Coastal and Estuarine Management*. Routledge, London.

Friedrichs, C.T., and Aubrey, D.G., 1988. Non-linear tidal distortion in shallow well-mixed estuaries: A synthesis. *Estuarine, Coastal and Shelf Science*, **27**, 521–545.

Galloway, W.E., 1975. Process framework for describing the morphologic and stratigraphic evolution of deltaic depositional systems. In: M.L. Broussard (editor), *Deltas, Models for Exploration*. Houston Geological Society, Houston, 87–98.

Galvin, C.J., 1968. Breaker type classification on three laboratory beaches. *Journal of Geophysical Research*, **73**, 3651–3659.

Gornitz, V., 1993. Mean sea level changes in the recent past. In: R.A. Warrick, E.M. Barrow and T.M.L. Wigley (editors), *Climate and Sea Level Change: Observations, Projections and Implications*, Cambridge University Press, Cambridge, 25–44.

Gornitz, V., 1995. Sea level rise: A review of recent past and near-future trends. *Earth Surface Processes and Landforms*, **20**, 7–20.

Goudie, A., 1990. *Geomorphological Techniques* (2nd Edition). Unwin Hyman, London.

Goudie, A., 1995. *The Changing Earth*. Blackwell, Oxford.

Gourlay, M.R., 1968. *Beach and Dune Erosion Tests*. Delft Hydraulics Laboratory, Report no.M935/M936.

Gribble, C.D., and Hall, A.J., 1999. *Optical Mineralogy: Principles and Practice*. University College London, London.

Griggs, G.B. and Trenhaile, A.S., 1994. Coastal cliffs and platforms. In: R.W.G. Carter and C.D. Woodroffe (editors), *Coastal Evolution*, Cambridge University Press, Cambridge, 425–450.

Guilcher, A., 1953. Essai sur la zonation et la distribution des formes littorales de dissolution du calcaire. *Annales Géographique*, **62**, 161–179.

Guilcher, A., 1988. *Coral Reef Geomorphology*. Wiley and Sons, Chichester.

Guza, R.T. and Inman, D., 1975. Edge waves and beach cusps. *Journal of Geophysical Research*, **80**, 2997–3012.

Guza, R.T. and Thornton, E.B., 1985. Observations of surf beat. *Journal of Geophysical Research*, **90**, 3161–3172.

Hallermeier, R.J., 1981. A profile zonation for seasonal sand beaches from wave climate. *Coastal Engineering*, **4**, 253–277.

Hansom, J.D., 1988. *Coasts*. Cambridge University Press, Cambridge.

Harris, P.T., 1988. Large-scale bedforms as indicators of mutually evasive sand transport and the sequential infilling of wide-mouthed estuaries. *Sedimentary Geology*, **57**, 273–298.

Harris, P.T., Hughes, M.G., Baker, E.K., Dalrymple, R.W., and Keene, J.B., 2003. Sediment export from distributary channels to the pro-deltaic environment in a tidally-dominated delta: Fly River, Papua New Guinea. *Continental Shelf Research*, in press.

Haslett, S.K., 2000. *Coastal Systems*. Routledge, London.

Hasselmann, K., Ross, D.B., Muller, P. and Sell, W., 1976. A parametric wave prediction model. *Journal of Physical Oceanography*, **6**, 200–228.

Hayes, M.O. and Boothroyd, J.C., 1969. Storms as modifying agents in the coastal environment. In: M.O. Hayes (editor), *Coastal Environments: NE Massachusetts*, Department of Geology, University of Massachusetts, Amherst, 290–315.

Hays, J.D., Imbrie, J. and Shackleton, N.J., 1976. Variations in the Earth's orbit: Pace maker of the ice age. *Science*, **235**, 1156–1167.

Hesp, P., 1999. The beach, backshore and beyond. In: A.D. Short (editor), *Handbook of Beach and Shoreface Dynamics*, Wiley and Sons, Chichester, 145–169.

Hesp, P.A. and Short, A.D., 1999. Barrier morphodynamics. In: A.D. Short (editor), *Handbook of Beach and Shoreface Morphodynamics*, Wiley and Sons, Chichester, 307–333.

Holman, R.A. and Bowen, A.J., 1982. Bars, bumps, and holes: Models for the generation of complex beach topography. *Journal of Geophysical Research*, **87**, 457–468.

Holman, R.A., 1983. Edge wave sand the configuration of the shoreline. In: P.D. Komar (editor), *CRC Handbook of Coastal Processes and Erosion*, CRC Press, Boca Raton, 21–33.

Holman, R.A., 1986. Extreme value statistics for wave run-up on a natural beach. *Coastal Engineering*, **9**, 527–544.

Hooke, J.M. and Bray, M.J., 1995. Coastal groups, littoral cells, policies and plans in the UK. *Area*, **27**, 358–368.

Hopley, D., 1982. *Geomorphology of the Great Barrier Reef: Quaternary Development of Coral Reefs*. Wiley-Interscience, New York.

Hopley, D., 1994. Continental shelf reef systems. In: R.W.G. Carter and C.D. Woodroffe (editors), *Coastal Evolution*, Cambridge University Press, Cambridge, 303–340.

Hoyt, J.H., 1967. Barrier island formation. *Geological Society of America Bulletin*, **78**, 1123–1136.

Hsu, J.R.C., Silvester, R. and Xia, Y.M., 1987. New characteristics of equilibrium shaped bays. *Proceedings 8th International Conference on Coastal Engineering*, ASCE, 140–144.

Hughes, M.G. and Cowell, P.J., 1987. Adjustment of reflective beaches to waves. *Journal of Coastal Research*, **3**, 153–167.

Hughes, M.G. and Turner, I., 1999. The beachface. In: A.D. Short (editor), *Handbook of Beach and Shoreface Morphodynamics*, Wiley and Sons, Chichester, 119–144.

Hughes, M.G., Harris, P.T., and Hubble, T.C.T., 1998. Dynamics of the turbidity maximum zone in a micro-tidal estuary: Hawkesbury River, Australia. *Sedimentology*, **45**, 397–410.

Hughes, M.G., Masselink, G. and Brander, R.W., 1997. Flow velocity and sediment transport in the swash zone of a steep beach. *Marine Geology*, **138**, 91–103.

Hunt, J.N., 1979. Direct solution of wave dispersion equation. *Journal of Waterways, Ports, Coastal Oceans Division*, **105**, 457–459.

Huntley, D.A., Guza, R.T., and Thornton, E.B., 1981. Field observations of surf beat: 1) Progressive edge waves. *Journal of Geophysical Research*, **86**, 6451–6466.

Hutchinson, J.N., 1973. The response of London Clay cliffs to differing rates of toe erosion. *Geologia Applicata e Idrogeologia*, **8**, 221–239.

Inman, D.L. and Brush, B.M., 1973. The coastal challenge. *Science*, **181**, 20–32.

Inman, D.L. and Nordstrom, K.F., 1971. On the tectonic and morphologic classification of coasts. *Journal of Geology*, **79**, 1–21.

Jelgersma, S., 1961. Holocene Sea-Level Changes in the Netherlands. *Mededelingen Geologische Stichting*, vol. C-IV (7).

Johnson, H.D. and Baldwin, C.T., 1986. Shallow siliclastic seas. In: H.G. Reading (editor), *Sedimentary Environments and Facies* (2nd Edition). Blackwell, London, 229–282.

Johnson, J.W., 1919. *Shore Processes and Shoreline Development*. Wiley, New York. [facsimile edition: Hafner, New York (1965).]

Kay, R. and Alder, J., 1999. Coastal Planning and Management. E & FN Spon, Routledge, London.

Kelletat, D., 1992. Coastal erosion and protection measures at the German North Sea coast. *Journal of Coastal Research*, **8**, 699–711.

Kench, P.S. and Cowell, P., 2000. The morphological response of atoll islands to sea-level rise. Part 2: Application of the modified Shoreface Translation Model. *Journal of Coastal Research*, Special Issue 34, 645–656.

Kench, P.S., 1997. Contemporary sedimentation in the Cocos (Keeling) Islands, Indian Ocean: Interpretation using settling velocity analysis. *Sedimentary Geology*, **114**, 109–130.

King, C.A.M., 1972. *Beaches and Coasts*. Edward Arnold, London.

King, D.M., Cooper, N.J., Morfett, and Pope, D.J., 2000. Application of offshore breakwaters to the UK: A case study at Elmer Beach. *Journal of Coastal Research*, **16**, 172–187.

Kinsey, D.W., 1985. Metabolism, calcification and carbon production: 1. Systems level studies. *Proceedings 5th International Coral Reef Symposium*, 505–526.

Kinsman, B., 1984. *Wind Waves* (2nd Edition). Dover Publications, New York.

Kirby, R., 1988. High concentration suspension (fluid mud) layers in estuaries. In: J. Dronkers and W. van Leussen (editors), *Physical Processes in Estuaries*. Springer-Verlag, 463–487.

Klein, R.J.T. and Nicholls, R.J., 1999. Assessment of coastal vulnerability to climate change. *Ambio*, **28**, 182–187.

Komar, P.D. and Inman, D.I., 1970. Longshore sand transport on beaches. *Journal of Geophysical Research*, **75**, 5514–5527.

Komar, P.D., 1976. *Beach Processes and Sedimentation*. Prentice-Hall, New Jersey.

Komar, P.D., 1998. *Beach Processes and Sedimentation* (2nd Edition). Prentice Hall, New Jersey.

Koster, M.J. and Hillen, R., 1995. Combat erosion by law: Coastal defence policy for The Netherlands. *Journal of Coastal Research*, **11**, 1221–1228.

Kraus, N.C. and McDougal, W.G., 1996. The effects of seawalls on the beach: Part 1, an updated literature review. *Journal of Coastal Research*, **12**, 691–701.

Kuenen, P.H., 1948. The formation of beach cusps, *Journal of Geology*, **56**, 34–40.

Ladd, H.S., Tracey, J.I. and Gross, M.G., 1967. Drilling on Midway Atoll, Hawaii. *Science*, **156**, 1088–1094.

Lambeck, K., 1993. Glacial rebound and sea-level change: An example of a relationship between mantle and surface processes. *Tectonophysics*, **223**, 15–37.

Larson, M. and Sunamura, T., 1993. Laboratory experiment on flow characteristics at a beach step. *Journal of Sedimentary Petrology*, **63**, 495–500.

Lawson, N.V., and Abernathy, C.L., 1975. Long term wave statistics off Botany Bay. *Proceedings 2nd Australian Conference on Coastal and Ocean Engineering*, 167–176.

Lear, R., and Turner, T., 1977. *Mangroves of Australia*. University of Queensland Press, Brisbane.

Leatherman, S.P., 1983. Barrier dynamics and landward migration with Holocene sea-level rise. *Nature*, **301**, 415–418.

Leeder, M., 1999. *Sedimentology and Sedimentary Basins: From Turbulence to Tectonics*. Blackwell Scientific Publications, Oxford.

Lettau, K. and Lettau, H., 1977. Experimental and micrometeorological field studies of dune migration. In: K. Lettau and H. Lettau (editors), *Exploring the World's Driest Climate*, University of Wisconsin-Madison, IES Report 101, 110–147.

Lewis, D.W., and McConchie, D., 1994. *Analytical Sedimentology*. Chapman & Hall, New York.

Livingstone, I. and Warren, A., 1996. *Aeolian Geomorphology*. Longman, Harlow.

Longuet-Higgins, M.S. and Stewart, R.W., 1962. Radiation stresses and mass transport in gravity waves with applications to surf beat. *Journal of Fluid Mechanics*, **13**, 481–504.

Longuet-Higgins, M.S. and Stewart, R.W., 1964. Radiation stresses in water waves: A physical discussion with applications. *Deep-Sea Research*, **11**, 529–562.

Louise, C.J. and van der Meulen, F., 1991. Future coastal defence in the Netherlands: Strategies for protection and sustainable development. *Journal of Coastal Research*, **7**, 1027–1041.

Macintyre, I.G., 1988. Modern coral reefs of the western Atlantic: New geological perspective. *American Association of Petroleum Geologists Bulletin*, **72**, 1360–1369.

Masselink, G. and Anthony, E., 2001. Location and size of intertidal bars on macrotidal ridge and runnel beaches. *Earth Surface Processes and Landforms*, **26**, 759–774.

Masselink, G., 1999. Alongshore variation in beach cusp morphology in a coastal embayment. *Earth Surface Processes and Landforms*, **24**, 335–347.

Masselink, G. and Pattiaratchi, C.B., 1998. Morphological evolution of beach cusp morphology and associated swash circulation patterns. *Marine Geology*, **146**, 93–113.

Masselink, G. and Short, A.D., 1993. The effect of tide range on beach morphodynamics and morphology: A conceptual beach model. *Journal of Coastal Research*, **9**, 785–800.

Masselink, G. and Turner, I., 1999. The effect of tides on beach morphodynamics. In: A.D. Short (editor), *Handbook of Beach and Shoreface Morphodynamics*, Wiley and Sons, Chichester, 204–229.

Matsukura, Y. and Matsuoka, M., 1991. Rates of tafoni weathering on uplifted shore platforms in Nojima-Zaki, Boso Peninsula, Japan. *Earth Surface Processes and Landforms*, **16**, 51–56.

May, J.P. and Tanner, W.F., 1973. The littoral power gradient and shoreline changes. In: D.R. Coates (editor), *Coastal Geomorphology*, New York State University Press, New York, 43–61.

McLean, R.F. and Woodroffe, C.D., 1993. Vulnerability assessment of coral atolls: The case of Australia's Cocos (Keeling) Islands. In: R.F. McLean and N. Mimura (editors), *Vulnerability Assessment to Sea Level Rise and Coastal Zone Management*, Proceedings of the IPCC/WCC'93 Eastern Hemisphere workshop. Department of Environment, Sports and Territories, Canberra, Australia, 99–108.

McLean, R.F. and Woodroffe, C.D., 1994. Coral atolls. In: R.W.G. Carter and C.D. Woodroffe (editors), *Coastal Evolution*, Cambridge University Press, Cambridge, 267–302.

McLean, R.F., Tsyban, A., Burkett, V., Codignotto, J.O., Forbes, D.L., Mimura, N., Beamish, R.J. and Ittekkot, V., 2001. Chapter 6: Coastal zones and marine ecosystems. In: Intergovernmental Panel on Climate Change, *Climate Change 2001: Impacts, Adaptation, and Vulnerability*, Contribution of Working Group II to the Third Assessment Report of the Intergovernmental Panel on Climate Change, Cambridge University Press, Cambridge, 343–379.

Mehta, A.J., 1989. On estuarine cohesive sediment suspension behaviour. *Journal of Geophysical Research*, **94**, 14,303–14,314.

Meyer-Peter, E., and Muller, R., 1948. Formulas for bedload transport. *Proceedings 2nd Congress of International Association of Hydraulic Research,* Stockholm, 39–64.

Milliman, J.D., and Meade, R.H., 1983. World-wide delivery of river sediments to the oceans. *Journal of Geology,* **91,** 1–21.

Mörner, N-A., 1987. Models of global sea level changes. In: M.J. Tooley and I. Shennan (editors*), Sea Level Changes, Blackwell,* Oxford, 332–355.

Munk, W.H. and Traylor, M.A., 1947. Refraction of ocean waves: A process linking underwater topography to beach erosion. *Journal of Geology,* **55,** 1–26.

Nelson, R.C., 1994. Depth limited design wave heights in very flat regions. *Coastal Engineering,* **23,** 43–59.

Nemec, W., 1990. Aspects of sediment movement on steep delta slopes. In: A. Colella and D.B. Prior (editors), *Coarse-grained Deltas.* Special Publication Number 10 of the International Association of Sedimentologists, Blackwell Science, Oxford, 29–73.

Neuman, A.C. and MacIntire, I., 1985. Reef response to sea level rise: Keep-up, catch-up or give-up. *Proceedings 5th International Coral Reef Symposium,* 105-110.

New South Wales Government, 1992. *Estuary Management Manual.* Government printer.

Nicholls, R.J., Birkemeier, W.A. and Lee, G., 1998. Evaluation of depth of closure using data from Duck, NC, USA. *Marine Geology,* **148,** 179–201.

Nicholls, R.J. and Mimura, N., 1998. Regional issues raised by sea-level rise and their policy implications. *Climate Research,* **11,** 5–18.

Nichols, M., and Poor, G., 1967. Sediment transport in a coastal plain estuary. *Journal of Waterways, Harbours and Coastal Engineering Division,* ASCE, **WW4,** 83–95.

Niedoroda, A.W. and Swift, D.J.P., 1991. Shoreface processes. In: J.B. Herbich (editor), *Handbook of Coastal and Ocean Engineering,* Volume 2, Gulf Publishing, Houston, 736–770.

Niedoroda, A.W., Swift, D.J.P. and Hopkins, T.S., 1985. The shoreface. In: R.A. Davis (editor), *Coastal Sedimentary Environments* (2nd Edition). Springer-Verlag, New York, 533–624.

Nielsen, P., 1992. *Coastal Bottom Boundary Layers and Sediment Transport.* World Scientific, Singapore.

Nielsen, P. and Hanslow, D.J., 1991. Wave runup distributions on natural beaches. *Journal of Coastal Research,* **7,** 1139–1152.

Nordstrom, K.F., 1980. Cyclic and seasonal beach response: A comparison of oceanside and bayside beaches. *Physical Geography,* **1,** 177–196.

Nordstrom, K.F., 2000. *Beaches and Dunes of Developed Coasts.* Cambridge University Press, Cambridge.

Nordstrom, K.F. and Jackson, N.L., 1993. The role of wind direction in eolian transport on a narrow sandy beach. *Earth Surface Processes and Landforms,* **18,** 675–685.

O'Brien, M.P., 1969. Equilibrium flow areas and inlets on sandy coasts. *Journal of Waterways, Harbours, and Coastal Engineering,* **15,** 43–52.

Open University, 1993. *Ocean Circulation.* Pergamon Press, Oxford.

Open University, 1994. *Waves, Tides and Shallow-Water Processes.* Pergamon Press, Oxford.

Orford, J.D., Carter, R.W.G. and Jennings, S.C., 1991. Coarse clastic barrier environments: Evolution and implications for Quaternary sea level interpretation. *Quaternary International ,* **9,** 87–104.

Orford, J.D., Carter, R.W.G. and Jennings, S.C., 1996. Control domains and morphological phases in gravel-dominated coastal barriers of Nova Scotia. *Journal of Coastal Research*, **12**, 589–604.

Orford, J.D., Carter, R.W.G., Jennings, S.C. and Hinton, A.C., 1995. Processes and timescales by which a coastal gravel-dominated barrier responds geomorphologically to sea-level rise: Story Head barrier, Nova Scotia. *Earth Surface Processes and Landforms*, **20**, 21–37.

Orme, A.R., 1985. The behaviour and migration of longshore bars. *Physical Geography*, **5**, 142–164.

Otvos, E.G. and Price, W.A., 1979. Problems of chenier genesis and terminology – an overview. *Marine Geology*, **31**, 251–263.

Pethick, J., 1984. *An Introduction to Coastal Geomorphology*. Edward Arnold, London.

Pethick, J., 1993. Shoreline adjustment and coastal management: physical and biological processes under accelerated sea-level rise. *Geographical Journal*, **149**, 162–168.

Petit, J.R., Jouzel, J., Raynaud, D., Barkov, N.I., Barnola, J-M., Basile, I., Benders, M., Chappellaz, J., Davis, M., Delaygue, G., Delmotte, M., Kotlyakov, V.M., Legrand, M., Lipenkov, V.Y., Lorius, C., Pépin, L., Ritz, C., Saltzman, E. and Stievenard, M., 1999. Climate and atmospheric history of the past 420,000 years from the Vostok ice core, Antarctica. *Nature*, **399**, 429–436.

Pettijohn, F.J., Potter, P.E., and Siever, R., 1987. *Sand and Sandstone*. Springer-Verlag, New York.

Pilkey, O.H., Young, R.S., Riggs, S., Smith, A.W.S., Wu, H. and Pilkey, W.D., 1993. The concept of shoreface profile of equilibrium: A critical review. *Journal of Coastal Research*, **9**, 255–278.

Pinet, P.R., 2000. *Invitation to Oceanography* (2nd Edition). Jones and Bartlett Publishers, Sudbury.

Pirazzoli, P.A., 1989. Recent sea-level changes in the North Atlantic. In: D.B. Scott, P.A. Pirazzoli and C.A. Honig (editors), *Late Quaternary Sea-level correlation and Applications*. Kluwer, Dordrecht, NATO ASI Series C, vol. 256, 153–167.

Pirazzoli, P.A., 1996. *Sea-Level Changes: The Last 20,000 Years*. Wiley and Sons, Chichester.

Pond, S. and Pickard, G.L., 1983. *Introductory Dynamical Oceanography* (2nd Edition). Pergamon, Oxford.

Pope, J. and Dean, J.L., 1986. Development of design criteria for segmented breakwaters. *Proceedings 20th International Conference on Coastal Engineering*, ASCE, 2144–2158.

Postma, G., 1990. Depositional architecture and facies of river and fan deltas: A synthesis. In: A. Colella and D.B. Prior (editors), *Coarse-grained Deltas*. Special Publication Number 10 of the International Association of Sedimentologists, Blackwell Science, Oxford, 13–28.

Powers, M.C., 1953. A new roundness scale for sedimentary particles. *Journal of Sedimentary Petrology*, **23**, 117–119.

Psuty, N.P., 1992. Spatial variation in coastal foredune development. In: R.W.G. Carter, T.G.F. Curtis and M.J. Sheehy-Skeffington (editors), *Coastal Dunes*, Proceedings 3rd European Dune Congress, Balkema, Rotterdam, 3–13.

Pugh, D.T, 1987. *Tides, Surges and Mean Sea-Level*. Wiley and Sons, Chichester.

Purdy, E.G., 1974. Reef configurations: Cause and effect. In: L.F. Laporte (editor), *Reefs in Time and Space*, Society of Economic Paleontologists and Mineralogists, Special Publication 18.

Pye, K., 1994. Properties of sediment particles. In : K. Pye (editor), *Sediment Transport and Depositional Processes*. Blackwell Scientific Publications, Oxford, 1–24.

Ranwell, D.S. and Boar, R., 1986. *Coastal Dune Management Guide*. Institute of Terrestrial Ecology, University of East Anglia, Norwich.

Reading, H.G., and Collinson, J.D., 1996. Clastic coasts. In: H.G. Reading (editor), *Sedimentary Environments: Processes, Facies and Stratigraphy*, Blackwell Science, Oxford, 154–231.

Reineck, H.-E., and Singh, I.B., 1980. *Depositional Sedimentary Environments: With Reference to Terrigenous Clastics*. Springer-Verlag, Berlin.

Reinson, G.E., 1984. Barrier island and associated strand-plain systems. In: R.G. Walker (editor), *Facies Models* (2nd Edition), Geoscience Canada Reprint Series 1, Geological Association of Canada, 119–140.

Roberts, N., 1998. *The Holocene* (2nd Edition). Blackwell Publishers, Oxford.

Robinson, L.A., 1977. Marine erosive processes at the cliff foot. *Marine Geology*, **23**, 257–271.

Roelvink, J.A. and Stive, M.J.F., 1989. Bar-generating cross-shore flow mechanisms on a beach. *Journal of Geophysical Research*, **94**, 4785–4800.

Roy, P.S., 1984. New South Wales estuaries: Their origin and evolution. In: B.G. Thom (editor), *Coastal Geomorphology in Australia*. Academic Press, Sydney, 99–122.

Roy, P.S., Cowell, P.J., Ferland, M.A. and Thom, B.G., 1994. Wave-dominated coasts. In: R.W.G. Carter and C.D. Woodroffe (editors), *Coastal Evolution: Late Quaternary Shoreline Morphodynamics*, Cambridge University Press, Cambridge, 121–186.

Roy, P.S., Thom, B.G., and Wright, L.D., 1980. Holocene sequences on an embayed high-energy coast: An evolutionary model. *Sedimentary Geology*, **26**, 1–19.

Sallenger, A.H. and Holman, R.T., 1985. Wave energy saturation on a natural beach with variable slope. *Journal of Geophysical Research*, **90**, 11939–11944.

Sanders, J.E. and Kumar, N., 1975. Evidence of shoreface retreat and in-place 'drowning' during Holocene submergence of barriers, shelf off Fire Island, New York. *Geological Society of America Bulletin*, **86**, 65–76.

Sanderson, P.G. and Eliot, I., 1996. Shoreline salients, cuspate forelands and tombolos on the coast of Western Australia. *Journal of Coastal Research*, **12**, 761–773.

Scott, G.A.J. and Rotondo, G.M., 1983. A model to explain the differences between Pacific plate island atoll types. *Coral Reefs*, **1**, 139–150.

Shackleton, N.J. and Opdyke, N.D., 1976. Oxygen-isotope and paleo-magnetic stratigraphy of Pacific core V28-239, Late Pliocene to latest Pleistocene. *Geological Society American Memoirs*, **145**, 449–464.

Shackleton, N.J., 1987. Oxygen isotopes, ice volumes and sea level. *Quaternary Science Reviews*, **6**, 183–190.

Shackleton, N.J., Berger, A. and Peltier, W.R., 1990. An alternative astronomical calibration on the Lower Pleistocene time scale based on ODP site 677. *Transactions of the Royal Society London*, **B318**, 679–688.

Shand, R.D., Bailey, D.G. and Shepard, M.J., 1999. An inter-site comparison of net offshore bar migration characteristics and environmental conditions. *Journal of Coastal Research*, **15**, 750–765.

Shepard, F.P., 1963. *Submarine Geology* (2nd Edition). Harper & Row, New York.

Shepard, F.P. and Inman, D.L., 1950. Nearshore circulation. *Proceedings 1st International Conference on Coastal Engineering*, ASCE, 50–59.

Sherman, D.J. and Bauer, B.O., 1993. Dynamics of beach-dune systems. *Progress in Physical Geography*, **17**, 413–447.

Sherman, D.J., Jackson, D.W.T., Namikas, S.L. and Wang, J., 1998. Wind-blown sand on beaches: An evaluation of models. *Geomorphology*, **22**, 113–133.

Shields, A., 1936. *Anwendung der Ähnlichkeits-Mechanik und der Turbulenz-forschung auf die Geschiebebewegung*. Preussische Versuchsanstalt für Wasserbau und Schiffbau, 26.

Short, A.D. (editor), 1999. *Handbook of Beach and Shoreface Morphodynamics*, Wiley and Sons, Chichester.

Short, A.D., 1979. Three dimensional beach-stage model. *Journal of Geology*, **87**, 553–571.

Short, A.D., 1985. Rip current type, spacing and persistence, Narrabeen Beach, Australia. *Marine Geology*, **65**, 47–71.

Short, A.D., 1992. Beach systems of the central Netherlands coast: Processes, morphology and structural impacts in a storm driven, multi-bar system. *Marine Geology*, **107**, 103–137.

Short, A.D., 1999. Chapter 2: Global variation in beach systems. In: A.D. Short (editor), *Beach Morphodynamics*, Wiley and Sons, London, 72–118.

Short, A.D. and Hesp, P.A., 1982. Wave, beach and dune interactions in south-eastern Australia. *Marine Geology*, **48**, 259–284.

Sleath, J.F.A., 1984. *Sea Bed Mechanics*. Wiley and Sons, New York.

Sonu, C.J., 1972. Field observation of nearshore circulation and meandering currents. *Journal of Geophysical Research*, **77**, 3232–3247.

Soulsby, R.L., 1983. The bottom boundary layer of shelf seas. In: B. Johns (editor), *Physical Oceanography of Coastal and Shelf Seas*, Elsevier, Amsterdam, 189–266.

Soulsby, R.L., 1997. *Dynamics of Marine Sands: A Manual for Practical Applications*. Thomas Telford, London.

Southard, J.B., and Boguchwal, L.A., 1990. Bed configurations in steady unidirectional water flows, Part 2: Synthesis of flume data. *Journal of Sedimentary Petrology*, **60**, 658–679.

Spencer, T., 1988. Coastal biogeography. In: H.A. Viles (editor), *Biogeography*, Blackwell, Oxford, 255–318.

Spencer, T., 1994. Tropical coral islands – an uncertain future? In: N. Roberts (editor), *The Changing Global Environment*, Blackwell, Oxford, 190–209.

Stanley, D.J. and Chen, Z., 1993. Yangtze delta, eastern China: I. Geometry and subsidence of Holocene depocenter. *Marine Geology*, **112**, 1–11.

Stephenson, W.J., 2000. Shore platforms: Remain a neglected coastal feature. *Progress in Physical Geography*, **24**, 311–327.

Stephenson, W.J. and Kirk, R.M., 1996. Measuring erosion rates using the micro-erosion meter: 20 years of data from shore platforms, Kaikoura Peninsula, South Island, New Zealand. *Marine Geology*, **131**, 209–218.

Stephenson, W.J. and Kirk, R.M., 1998. Rates and patterns of erosion on inter-tidal shore platforms, Kaikoura Peninsula, South Island, New Zealand. *Earth Surface Processes and Landforms*, **23**, 1071–1085.

Stephenson, W.J. and Kirk, R.M., 2000a. Development of shore platforms on Kaikoura Peninsula, South Island, New Zealand, Part One: The role of waves. *Geomorphology*, **32**, 21–41

Stephenson, W.J. and Kirk, R.M., 2000b. Development of shore platforms on Kaikoura Peninsula, South Island, New Zealand, Part II: The role of subaerial weathering. *Geomorphology*, **32**, 43–56.

Stive, M.J.F., Nicholls, R.J. and De Vriend, H.J., 1991. Sea-level rise and shore nourishment: A discussion. *Coastal Engineering*, **16**, 147–163.

Stoddart, D.R., 1965. The shape of atolls. *Marine Geology*, **3**, 369–383.

Stoddart, D.R., 1969. Ecology and morphology of recent coral reefs. *Biological Reviews*, **44**, 433–498.

Stone, G.W. and Donley, J.C. (editors), 1998. The world deltas symposium: A tribute to James Plummer Morgan (1919–1995). Special Thematic Section: *Journal of Coastal Research*, **14**, 695–916.

Sunamura, T. and Mizuno, O., 1987. A study on depositional shoreline forms behind an island. *Annual Report Institute Geosciences*, University of Tsukuba, Tsukuba, No. 13, 71–73.

Sunamura, T. and Takeda, I., 1984. Landward migration of inner bars. *Marine Geology*, **60**, 63–78.

Sunamura, T., 1976. Feedback relationship in wave erosion of laboratory rocky coast. *Journal of Geology*, **84**, 427–437.

Sunamura, T., 1992. *Geomorphology of Rocky Coasts*. Wiley and Sons, New York.

Svendsen, I.A., 1984. Mass flux and undertow in the surf zone. *Coastal Engineering*, **8**, 347–365.

Sverdrup, H.U. and Munk, W.H., 1946. Theoretical and empirical relations in forecasting breakers and surf. *Transactions American Geophysical Union*, **27**, 828–836

Takeda, I. and Sunamura, T., 1982. Formation and height of berms. *Transactions, Japanese Geomorphological Union*, **3**, 145–157.

Taylor, M. and Stone, G.W., 1996. Beach-ridges: A review. *Journal of Coastal Research*, **12**, 612–621.

Thom, B.G. and Hall, W., 1991. Behaviour of beach profiles during accretion and erosion dominated periods. *Earth Surface Processes and Landforms*, **16**, 113–127.

Thornton, E.B. and Guza, R.T., 1982. Energy saturation and phase speeds measured on a natural beach. *Journal of Geophysical Research*, **87**, 9499–9508.

Thornton, E.B. and Kim, C.H., 1993. Longshore current and wave height modulation at tidal frequency inside the surf zone. *Journal of Geophysical Research*, **98**, 16509–16519.

Thornton, E.B., Humiston, R.T. and Birkemeier, W., 1996. Bar/trough generation on a natural beach. *Journal of Geophysical Research*, **101**, 12097–12110.

Thurman, H.V., and Burton, E.A., 2001. *Introductory Oceanography* (9th Edition). Prentice Hall, New Jersey.

Tooley, M.J., 1993. Long term changes in eustatic sea level. In: R.A. Warrick, E.M. Barrow and T.M.L. Wigley (editors), *Climate and Sea Level Change: Observations, Projections and Implications*, Cambridge University Press, Cambridge, 81–107.

Trenhaile, A.S. and Mercan, D.W., 1984. Frost weathering and the saturation of coastal rocks. *Earth Surface Processes and Landforms*, **9**, 321–331.

Trenhaile, A.S., 1987. *The Geomorphology of Rock Coasts*. Oxford University Press, Oxford.

Trenhaile, A.S., 1997. *Coastal Dynamics and Landforms*. Oxford University Press, Oxford.

Trenhaile, A.S., 1999. The width of shore platforms in Britain, Canada and Japan. *Journal of Coastal Research*, **15**, 355–364.

Trenhaile, A.S., 2000. Modelling the development of wave-cut shore platforms. *Marine Geology*, **166**, 163–178.

Trenhaile, A.S., 2001. Modelling the effect of weathering on the evolution and morphology of shore platforms. *Journal of Coastal Research*, **17**, 398–406.

Tucker, M.E., 1995. *Sedimentary Petrology: An Introduction to the Origin of Sedimentary Rocks*. Blackwell Science, Oxford.

Turner, I., 1993. Water table outcropping on macro-tidal beaches: A simulation model. *Marine Geology*, **115**, 227–238.

Turner, I.L., 1995. Simulating the influence of groundwater seepage on sediment transported by the sweep of the swash zone across macro-tidal beaches. *Marine Geology*, **125**, 153–174.

Uncles, R.J., Barton, M.L., and Stephens, J.A., 1994. Seasonal variability of fine-sediment concentrations in the turbidity maximum region of the Tamar estuary. *Estuarine, Coastal and Shelf Science*, **38**, 19–39.

van Dyke, M., 1982. *An Album of Fluid Motion*. The Parabolic Press, Stanford.

Viles, H. and Spencer, T., 1995. *Coastal Problems*. Edward Arnold, London.

Villard, P.V., and Osborne, P.D., 2002. Visualisation of wave-induced suspension patterns over two-dimensional bedforms. *Sedimentology*, **49**, 363–378.

Warrick, E.M., 1993. Climate and sea level change: a synthesis. In: R.A. Warrick, E.M. Barrow and T.M.L. Wigley (editors), *Climate and Sea Level Change: Observations, Projections and Implications*, Cambridge University Press, Cambridge, 3–21.

Wemelsfelder, P.J., 1953. The disaster in the Netherlands caused by the storm flood of February 1, 1953. *Proceedings of the 4th Coastal Engineering Conference*, ASCE, 256–271.

Werner, B.T., 1999. Complexity in natural landform patterns. *Science*, **284**, 102–104.

Werner, B.T. and Fink, T.M., 1993. Beach cusps as self-organized patterns. *Science*, **260**, 968–971.

Whitford, D.J. and Thornton, E.B., 1993. Comparison of wind and wave forcing of longshore currents. *Continental Shelf Research*, **13**, 1205–1218.

Wiegel, R.L., 1964. *Oceanographical Engineering*. Prentice-Hall, Englewood Cliffs.

Wijnberg, K.M., 1996. On the systematic offshore decay of breaker bars. *Proceedings 25th International Conference on Coastal Engineering*, ASCE, 3600–3613.

Woodroffe, 1995. Response of tide-dominated mangrove shorelines in Northern Australia to anticipated sea-level rise. *Earth Surface Processes and Landforms*, **20**, 65–86.

Woodroffe, C.D. and McLean, R., 1990. Microatolls and recent sea level change on coral atolls. *Nature*, **344**, 531–534.

Woodroffe, C.D., McLean, R.F. and Wallensky, E., 1994. Geomorphology of the Cocos (Keeling) Islands. *Atoll Research Bulletin*, **402**, 1–33.

Woodroffe, C.D., McLean, R.F., Smithers, S.G. and Lawson, E.M., 1999. Atoll reef-island formation and response to sea-level change: West Island, Cocos (Keeling) Islands. *Marine Geology*, **160**, 85–104.

World Commission on Environment and Development, 1987. *Our Common Future*. Oxford University Press, Oxford.

Wright, L.D., 1976. Morphodynamics of a wave-dominated river mouth. *Proceedings 15th Coastal Engineering Conference*, ASCE, 1721–1737.

Wright, L.D., 1977. Sediment transport and deposition at river mouths: A synthesis. *Geological Society of America Bulletin*, **88**, 857–868.

Wright, L.D., 1978. River deltas. In: R.A. Davis (editor), *Coastal Sedimentary Environments*, Springer-Verlag, New York, 5–68.

Wright, L.D., 1989. Benthic boundary layers of estuarine and coastal environments. *Reviews in Aquatic Sciences*, **1**, 75.

Wright, L.D. and Short, A.D., 1984. Morphodynamic variability of surf zones and beaches: A synthesis. *Marine Geology*, **56**, 93–118.

Wright, L.D. and Thom, B.G., 1977. Coastal depositional landforms: A morphodynamic approach. *Progress in Physical Geography*, **1**, 412–459.

Wright, L.D., Coleman, J.M., and Thom, B.G., 1973. Processes of channel development in a high-tide-range environment: Cambridge Gulf-Ord River delta, Western Australia. *Journal of Geology*, **81**, 15–41.

Wright, L.D., Yang, Z.S., Bornhold, B.D., Keller, G.H., Prior, D.B. and Wiseman, W.J., 1986. Hyperpycnal plumes and plume fronts over the Huanghe (Yellow River) delta front. *Geo-Marine Letters*, **6**, 97–105.

Yasso, W.E., 1965. Plan geometry of headland-bay beaches. *Marine Geology*, **73**, 702–714.

Young, I.R. and Holland, G.J., 1996. *Atlas of Oceans, Wind and Wave Climate*. Elsevier, Oxford.

Zenkovich, V.P., 1967. *Processes of Coastal Development*. Oliver and Boyd, London.

Zingg, T., 1935. Beitrag zur Schotteranalyse. *Schweiz. Mineral. Petrog. Mitt.*, **15**, 39–140.

# LOCATION INDEX

# SUBJECT INDEX